T0391024

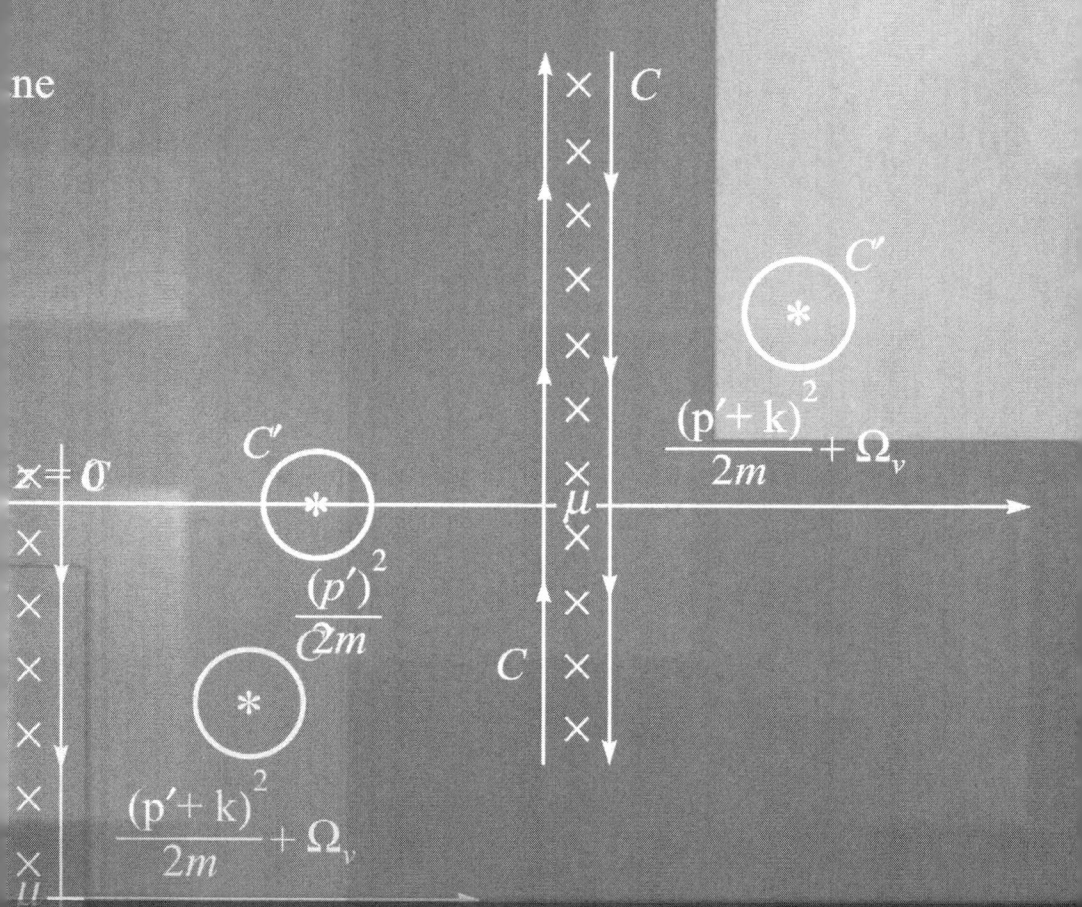

Annotations to Quantum Statistical Mechanics

Annotations to Quantum Statistical Mechanics

In-Gee Kim

PAN STANFORD ∏ PUBLISHING

Published by

Pan Stanford Publishing Pte. Ltd.
Penthouse Level, Suntec Tower 3
8 Temasek Boulevard
Singapore 038988

Email: editorial@panstanford.com
Web: www.panstanford.com

British Library Cataloguing-in-Publication Data
A catalogue record for this book is available from the British Library.

Annotations to Quantum Statistical Mechanics

Copyright © 2018 Pan Stanford Publishing Pte. Ltd.

ISBN 978-981-4774-15-4 (Hardback)
ISBN 978-1-315-19659-6 (eBook)

To my wife Gajean
and
my daughter Eugenie

Contents

Preface

Physicists around the world received the sad news of the demise of Professor Leo P. Kadanoff in October 2015. I had no personal interaction with him. I heard about him during my first statistical physics class when I was a college junior while I studied Kadanoff's block spin procedure that provides an insight into the renormalization group theory. Since my professional career has long been dedicated to investigating the electronic structure problems in solids, I usually studied the density functional theory, which is conventionally on top of the zero-temperature or the finite-temperature quantum many-body theory in equilibrium.

My first professional touch on the nonequilibrium statistical physics was during my post-doctoral experience at Northwestern University, Illinois, USA, in 2005, when I was struggling to develop a computer code, under the guidance of Miyoung Kim and Art Freeman, for calculating the Seebeck coefficients from the electronic structures of solids. The transport coefficients, such as electric conductivity, thermal conductivity, and thermoelectric power, are defined by the assumption that a system is in a near-equilibrium state, i.e., essentially in a nonequilibrium state close to equilibrium; this leads to a completely different physical formalism from the equilibrium physics with what I usually had dealt. At that time, I adapted a branch of Boltzmann equation, the so-called Bhatnagar–Gross–Krook (BGK) model in which the collision term is replaced by a simple parametric function of the distribution function. The BGK model has been known, erroneously in many textbooks for solid state physicists, as the Boltzmann equation. The code for the Seebeck coefficients based on the BGK model was written incompletely, so the remaining numerical problems were fixed by my friend Jung-Hwan Song, who unexpectedly passed away on June

15, 2011, and its first realistic application was done by Min Sik Park and Julia Medvedeva, who wrote the first draft of the manuscript for publishing in *Physical Review B* in 2010.

During my POSTECH period, I faced a bunch of problems on nonequilibrium statistical physics, but they are full of phenomenological and empirical treatments dedicated for metallurgical applications. I have had spent most of my efforts to build a research framework, the so-called multiscale simulation method, by organizing a research team consisting of Korean experts from the vast disciplines of electronic structure modeling, molecular dynamics modeling, phase field modeling, phase thermodynamics with databases, and dislocation dynamics modeling. Struggling to understand those theories, I realized that we need a rather smoothly unified theoretical framework derived from first principles. To this end, it is necessary to eliminate structural complications by arranging atoms to form crystals and solids. Such a system is nothing more than a very cold and dense plasma. In 2014, I decided to move to the New Mexico Consortium, Los Alamos, New Mexico, where I studied the two-component equilibrium quantum plasma physics, the classical and quantum kinetic theories for multicomponent systems, and the two-temperature molecular dynamics for calculating transport coefficients.

In the meantime, I carefully read Leo P. Kadanoff and Gordon Baym's book *Quantum Statistical Mechanics: Green's Function Methods in Equilibrium and Nonequilibrium Problems* (Benjamin, New York, 1962). Like many other classic books, especially *Frontiers in Physics: A Lecture Note and Reprint Series*, this book also explains nonequilibrium statistical physics in a systematic way. It contains essential concepts on statistical physics in terms of Green's functions with sufficient and rigorous details. However, as my friends agree with me, a lack of effort at the publisher's end reduced the readability of this book. The book was printed as a photocopy of the original manuscript, which was prepared with the help of a typewriter. In my humble opinion, a book prepared with careful typesetting helps a reader's brain to work smoothly because it does not have to work hard to interpolate text from bad printing. I have rewritten the text in the LaTeX2e format, fixed some typographical errors, corrected mistakes in equation numbers, drawn figures with modern computer programs, added my own footnotes to the text, and saved in my laptop. This rather tedious work was extremely helpful for me to understand the formalism of nonequilibrium quantum statistical mechanics.

During this rewriting and annotating, I felt the necessity to provide a short note on the second quantization chapter in front of the original text. Although there are no substantial paradigm shifts after the publication of the original text, the curricula of graduate schools have evolved since the 1960s. Graduate students of modern physics now learn relativistic quantum field theory and quantum many-body physics and have to work on their own research topics. It, therefore, becomes necessary for them to spend time on consistent study to make the knowledge of a topic concrete in their minds in addition to passing relevant examinations. My experience tells me that a systematically prepared summary is extremely useful for settling down the key knowledge of a subject.

I would like to appreciate Mr. Stanford Chong, Pan Stanford Publishing, for encouraging me to publish this rewritten text, which was prepared purely for personal purposes, in the form of a book, so that graduate students as well as senior researchers may benefit from these annotations on the classical text.

In-Gee Kim
Winter 2017

Preface of
Quantum Statistical Mechanics: Green's Function Methods in Equilibrium and Nonequilibrium Problems

These lectures are devoted to a discussion of the use of thermodynamic Green's functions in describing the properties of many-particle systems. These functions provide a method for discussing finite-temperature problems with no more conceptual difficulty than ground-state (e.g., zero-temperature) problems; the method is equally applicable to boson and fermion systems, equilibrium and nonequilibrium problems.

The first four chapters develop the equilibrium Green's function theory along the lines of the work of Martin and Schwinger. We use the grand-canonical ensemble of statistical mechanics to define thermodynamic Green's functions. These functions have a direct physical interpretation as particle propagates. The one-particle Green's function describes the motion of one particle added to the many-particle system; the two-particle Green's function describes the correlation motion of two added particles. Because they are propagators they contain much detailed dynamic information, and because they are expectation values in the grand-canonical ensemble they contain all statistical mechanical information. Several methods of obtaining the partition function from the Green's functions are discussed. We determine the one-particle Green's function from its equation of motion, supplemented by the boundary conditions appropriate to the grand-canonical ensemble. This equation of motion, which is essentially a matrix element of the second-quantized Schrödinger equation, gives the time derivative of the one-particle Green's function G in terms of the two-particle Green's function G_2. We physically motivate simple approximations, which express G_2 in terms of G, by making use of the propagator interpretation of the Green's functions.

Chapter 6 presents a formal method for generating Green's function approximations. This method is based on a consideration of the system in the presence of an external scalar potential. We also discuss here the relation between our equation of motion method and the more standard perturbative expansions.

Chapters 7, 8, and 9 outline a theory of nonequilibrium phenomena. We consider the deviations from equilibrium arising from the application of an external time- and space-dependent force field to the system. By making use of the results of Chapter 6 we show that every Green's function approximation for an equilibrium system can be generalized to describe nonequilibrium phenomena. In this way the Green's function equations of motion can be transformed into approximate quantum mechanical equations of transport. These are used, in Chapter 10, to derive generalizations of the Boltzmann equation. As examples of the nonequilibrium theory, we then discuss ordinary sound propagation and also the Landau theory of the low-temperature Fermi liquid.

Chapters 13 and 14 describe two approximations that have been extensively applied in the recent literature. A dynamically shielded potential is employed to discuss the properties of a Coulomb gas; the two-body scattering matrix approximation is developed for application to systems with short-range interactions.

An appendix and a list of references and supplementary reading are included at the end.

We should like to express our gratitude for the hospitality offered us at the Institutes for Theoretical Physics in Warsaw and Krakow, Poland, and Uppsala, Sweden, where these lectures were given in part. Special thanks are due Professor Niels Bohr of the Institute for Theoretical Physics in Copenhagen, where lectures were first delivered and finally written.

Leo P. Kadanoff
Gordon Baym
March 1962

Chapter 1

Physical Prerequisites

1.1 Basic Quantum Mechanics

The quantum revolution in the beginning of the 20th century changed our concepts of dynamics. The dynamic variables such as position $\mathbf{r}$ and momentum $\mathbf{p}$ are replaced by the corresponding position and momentum operators $\hat{\mathbf{r}}$ and $\hat{\mathbf{p}}$, respectively. When one would like to *observe* a dynamical variable, say ω, of a particle , one has to introduce a wavefunction ψ, which contains all the dynamical information of particles, and to apply the corresponding dynamical operator $\hat{\Omega}$ to the wavefunction ψ. Then one may obtain the desired dynamical value of the particle as the eigenvalue of the operator,

$$\hat{\Omega}\psi = \omega\psi. \tag{1.1}$$

This simple eigenvalue equation raises difficult philosophical problems.

Although it possesses a simple mathematical structure, Eq. (1.1) tells us that a dynamical property of particle is not a measurement independent of particle. In order to observe a dynamical variable of the particle, we have to perform an observation represented by the operator $\hat{\Omega}$ and then we have to apply the operation to the

Annotations to Quantum Statistical Mechanics
In-Gee Kim
Copyright © 2018 Pan Stanford Publishing Pte. Ltd.
ISBN 978-981-4774-15-4 (Hardcover), 978-1-315-19659-6 (eBook)
www.panstanford.com

corresponding spuriously defined wavefunction ψ, which looks like a *metaphysical object*. It seems like that our mother *nature* responds us based on our observational acts. We would like to understand the nature of measurement and the wavefunctions.

In the Hamiltonian dynamics,[a] all the dynamical properties are described in terms of the canonical coordinates x_r and the corresponding conjugate momenta p_s with $r, s = 1, 2, 3, \ldots, n$ where n is the degree of freedom. The corresponding quantum operators $\hat{x}_r$ and $\hat{p}_r$ follow the conditions

$$\hat{x}_r \hat{x}_s - \hat{x}_s \hat{x}_r = 0, \quad \hat{p}_r \hat{p}_s - \hat{p}_s \hat{p}_r = 0$$
$$\hat{x}_r \hat{p}_s - \hat{p}_s \hat{x}_r = i \hbar \delta_{rs}, \tag{1.2}$$

where $i = \sqrt{-1}$, $\hbar = h/2\pi$ is the rationalized Planck's constant, and δ_{rs} is the Kronecker delta.[b] This is known as the *fundamental quantum conditions*.[c] The fundamental quantum conditions state that the measurement order of two conjugate dynamical variables is important. The noncommutative operations of conjugate dynamical variables restrict the precision of measurements. The implication of the fundamental quantum conditions to the classical dynamical variables is so-called the *first quantization*.

Now one can prepare a quantum mechanical Hamiltonian operator $\hat{H}$, which is written in the form of operators described earlier and is essentially the same as the classical Hamiltonian H, with the care of the fundamental quantum conditions Eq. (1.2). When we operate a Hamiltonian operator $\hat{H}$ to the wavefunction ψ of a particle, we obtain the energy of the particle:

$$\hat{H} \psi = E \psi,$$

---

[a]The annotator presumably assumes that the readers have studied the classical dynamics at the level of L. D. Landau and E. M. Lifshitz, *Mechanics*, 3rd Ed. (Elsevier, Amsterdam, 2005) and/or H. Goldstein, C. P. Poole, and J. L. Safko, *Classical Mechanics*, 2nd Ed. (Addison-Wesley, Reading, Massachusetts, 1980).

[b]The annotator assumes that the readers have studied the mathematical physics at the level of George B. Arfken, Hans J. Weber, and Frank E. Harris, *Mathematical Methods for Physicists: A Comprehensive Guide*, 7th Ed. (Elsevier, Amsterdam, 2013).

[c]P. A. M. Dirac, *The Principles of Quantum Mechanics*, 4th Ed. (Clarendon Press, Oxford, 1998) p. 87.

where E is the total energy of the particle. It will be helpful if we investigate the simplest physical situation: the motion of a free particle.[d]

A free particle with mass m has the energy

$$E = \frac{\mathbf{p}^2}{2m}. \tag{1.3}$$

Let us align our coordinate system by putting the x-direction parallel to the particle motion. In quantum mechanics, we have a corresponding wavefunction ψ, which is believed to contain all the information for describing the dynamics of the free particle. The motion of transverse wave in time t is described by the equation

$$\frac{\partial^2 \psi}{\partial t^2} = \gamma \frac{\partial^2 \psi}{\partial x^2}, \tag{1.4}$$

where γ is the square of wave velocity. We may assume the wavefunction is one of the linear combinations of plane waves,

$$\cos(kx - \omega t), \ \sin(kx - \omega t), \ e^{i(kx - \omega t)}, \ e^{-i(kx - \omega t)},$$

as usual. Then the differential equation Eq. (1.4) satisfies if and only if

$$\gamma = \frac{\omega^2}{k^2} = \frac{E^2}{p^2} = \frac{p^2}{4m^2}.$$

The Planck–Einstein relations,

$$p = \hbar k, \ E = \hbar \omega, \tag{1.5}$$

enable us to write the wave equation Eq. (1.4) as

$$\frac{\partial \psi}{\partial t} = \gamma \frac{\partial^2 \psi}{\partial x^2},$$

where

$$\gamma = \frac{i\omega}{k} = \frac{i\hbar E}{p^2} = \frac{i\hbar}{2m}.$$

We arrive at the one-dimensional Schrödinger equation of a free particle,

$$i\hbar \frac{\partial \psi}{\partial t} = -\frac{\hbar^2}{2m} \frac{\partial^2 \psi}{\partial x^2}.$$

---

[d]The annotator follows the discussion of Leonard I. Schiff, *Quantum Mechanics*, 3rd Ed. (McGraw-Hill, New York, 1968) Chapters 1, 2, 3, and 6.

The extension to the three-dimensional case is straightforward to yield that

$$i\hbar\frac{\partial\psi}{\partial t} = -\frac{\hbar^2}{2m}\nabla^2\psi. \tag{1.6}$$

Considering the three-dimensional Einstein relation is $\mathbf{p} = \hbar\mathbf{k}$ and comparing with Eq. (1.3), one obtains the quantum operators of energy and momentum,

$$E \to i\hbar\frac{\partial}{\partial t}, \quad \mathbf{p} \to -i\hbar\nabla, \tag{1.7}$$

respectively.

When an external force $\mathbf{F}$ defined by the external potential V, such as

$$\mathbf{F}(\mathbf{r}, t) = -\nabla V(\mathbf{r}, t),$$

exerts on the particle, the total energy of the particle becomes

$$E = \frac{\mathbf{p}^2}{2m} + V(\mathbf{r}, t) \tag{1.8}$$

and one may have the Schrödinger equation as

$$i\hbar\frac{\partial\psi}{\partial t} = -\frac{\hbar^2}{2m}\nabla^2\psi + \hat{V}(\mathbf{r}, t)\psi, \tag{1.9}$$

where $\hat{V}$ is the potential operator corresponding to the classical potential V. We have a good machinery to solve Eq. (1.9), so we can obtain, in principle, the wavefunction of particle under the given boundary conditions. However, we have a big problem: *What is the wavefunction?*

There are many interpretations on the wavefunction, but we are going to accept the standard assumption, due to Born,[e] that the numerical value of a measurement is described by a probability function, which is related to the wavefunction ψ, which is a complex function. Since a probability must be real and nonnegative, one may think of a multiplication of its complex conjugate ψ^* to ψ as the *probability density*. As an example, one may obtain a probability density $P(\mathbf{r}, t)dxdydz$ to find the particle in the neighborhood of volume $dxdydz$ around the position $\mathbf{r}$ at time t as

$$P(\mathbf{r}, t) = \psi^*(\mathbf{r}, t)\psi(\mathbf{r}, t) = |\psi(\mathbf{r}, t)|^2. \tag{1.10}$$

---

[e]M. Born, *Z. Physik* **37**, 863 (1926); *Nature* **119**, 354 (1927).

This interpretation is termed the *Copenhagen interpretation.* Because the particle should be found one and only one in the space, the wavefunction *normalization* have to be

$$\int |\psi(\mathbf{r}, t)|^2 \, d\mathbf{r} = 1. \tag{1.11}$$

Copenhagen interpretation can be understood as follows: A dynamical variable ω can be measured as any value after an observation. Unfortunately, we have no prior knowledge which value will be measured before the observations. Instead we have an expectation value of the corresponding operator $\hat{\Omega}$,

$$\langle \omega \rangle = \int \psi^*(\mathbf{r}, t) \hat{\Omega} \psi(\mathbf{r}, t) d\mathbf{r}. \tag{1.12}$$

Let us investigate how the expectation values of the position operator $\hat{x}$ and the momentum operator $\hat{p}_x$ evolve in time. The time evolution of $\langle x \rangle$ is

$$\frac{d}{dt} \langle x \rangle = \frac{d}{dt} \int \psi^* \hat{x} \psi \, d\mathbf{r} = \int \psi^* \hat{x} \frac{\partial \psi}{\partial t} d\mathbf{r} + \int \frac{\partial \psi^*}{\partial t} \hat{x} \psi \, d\mathbf{r}$$

$$= -\frac{i}{\hbar} \int \psi^* \hat{x} \left(-\frac{\hbar^2}{2m} \nabla^2 \psi + V \psi \right) d\mathbf{r}$$

$$+ \frac{i}{\hbar} \int \left(-\frac{\hbar^2}{2m} \nabla^2 \psi^* + V \psi^* \right) \hat{x} \psi \, d\mathbf{r}$$

$$= \frac{i\hbar}{2m} \int \left[\psi^* \hat{x} \left(\nabla^2 \psi \right) - \left(\nabla^2 \psi^* \right) \hat{x} \psi \right] d\mathbf{r}$$

$$= \frac{i\hbar}{2m} \int \left[\psi^* \hat{x} \left(\nabla^2 \psi \right) + \left(\nabla \psi^* \right) \cdot \nabla \left(\hat{x} \psi \right) \right] d\mathbf{r}$$

$$- \frac{i\hbar}{2m} \oint_S (\hat{x} \psi \nabla \psi^*) \cdot \hat{n} \, dS.$$

Because the wave packet vanishes at infinity, the last surface integral term vanishes. Then we have

$$\frac{d}{dt} \langle x \rangle = \frac{i\hbar}{2m} \int \left[\psi^* \hat{x} \left(\nabla^2 \psi \right) + \left(\nabla \psi^* \right) \cdot \nabla \left(\hat{x} \psi \right) \right] d\mathbf{r}$$

$$= \frac{i\hbar}{2m} \int \psi^* \left[\hat{x} \nabla^2 \psi - \nabla^2 \left(\hat{x} \psi \right) \right] d\mathbf{r} + \frac{i\hbar}{2m} \oint_S \nabla \cdot \left(\psi^* \nabla \hat{x} \psi \right) \hat{n} dS$$

$$= -\frac{i\hbar}{m} \int \psi^* \frac{\partial \psi}{\partial x} d\mathbf{r}.$$

Therefore, we arrived at a relation

$$\frac{d}{dt} \langle x \rangle = \frac{1}{m} \langle p_x \rangle. \tag{1.13}$$

It is straightforward, in the same fashion, to have the time evolution of the expectation value of the momentum operator as

$$\frac{d}{dt}\langle p_x \rangle = \left\langle -\frac{\partial V}{\partial x} \right\rangle. \tag{1.14}$$

Equations (1.13) and (1.14) constitute the *Ehrenfest's theorem*[f] for the x-component. The Ehrenfest's theorem shows the analogy of the expectation values of $\hat{x}$ and $\hat{p}$ to the classical equations of motion:

$$\frac{d\mathbf{r}}{dt} = \frac{\mathbf{p}}{m}, \quad \frac{d\mathbf{p}}{dt} = -\nabla V.$$

Let us imagine a function $u_E(\mathbf{r})$, which satisfies an eigenvalue equation

$$\left[-\frac{\hbar^2}{2m}\nabla^2 + \hat{V}(\mathbf{r}) \right] u_E(\mathbf{r}) = E u_E(\mathbf{r}). \tag{1.15}$$

It also defines the Hamiltonian operator

$$\hat{H} = -\frac{\hbar^2}{2m}\nabla^2 + \hat{V}(\mathbf{r}), \tag{1.16}$$

and u_E is the eigenfunction of the Hamiltonian operator. Using this eigenfunction, we may write the wavefunction as

$$\psi(\mathbf{r}, t) = u(\mathbf{r})e^{-iEt/\hbar}. \tag{1.17}$$

Applying the energy operator $i\hbar\frac{\partial}{\partial t}$ to the wavefunction Eq. (1.17), we obtain

$$i\hbar\frac{\partial\psi}{\partial t} = E\psi. \tag{1.18}$$

So the constant E is an energy *eigenvalue* and the function ψ is an energy *eigenfunction* ψ, corresponding to the energy operator $i\hbar\frac{\partial}{\partial t}$. Since $|\psi|^2$ is constant in time, the energy eigenfunction ψ represents a *stationary state* of the particle of energy E.

The eigenfunction $u_E(\mathbf{r})$ satisfies the normalization condition $\int |u_E(\mathbf{r})|^2 d\mathbf{r} = 1$ for any discrete set of eigenfunctions labeled by E. For two different normalized eigenfunctions of the respective eigenvalues E and E' are orthogonal each other,

$$\int u_{E'}^*(\mathbf{r})u_E(\mathbf{r})d\mathbf{r} = \delta_{EE'} \tag{1.19}$$

---

[f]P. Ehrenfest, *Z. Physik* **45**, 455 (1927).

for non-degenerate energy eigenfunctions. When there is degeneracy identified by s and s', the orthonormality condition has to be

$$\int u_{E's'}^*(\mathbf{r})u_{Es}(\mathbf{r})d\mathbf{r} = \delta_{EE'}\delta_{ss'}. \tag{1.20}$$

We are able to expand any wavefunction $\psi(\mathbf{r})$ in terms of the energy eigenfunctions,

$$\psi(\mathbf{r}) = \sum_E A_E u_E(\mathbf{r}). \tag{1.21}$$

The coefficients in the expansion Eq. (1.21) can be obtained by the procedure

$$\int u_{E'}^*(\mathbf{r})\psi(\mathbf{r})d\mathbf{r} = \sum_E A_E \int u_{E'}^*(\mathbf{r})u_E(\mathbf{r})d\mathbf{r} = \sum_E A_E \delta_{EE'} = A_{E'}.$$

It is also important to note that the energy eigenfunctions u_E satisfy the closure property:

$$\sum_E u_E^*(\mathbf{r}')u_E(\mathbf{r}) = \delta(x-x')\delta(y-y')\delta(z-z') = \delta(\mathbf{r}-\mathbf{r}'). \tag{1.22}$$

In Copenhagen interpretation, we consider $P(E) = |A_E|^2$ as the probability of finding a particle described by the wavefunction $\psi(\mathbf{r})$ at the energy E, because

$$\sum_E P(E) = \sum_E \int u_E^*(\mathbf{r})\psi(\mathbf{r})d\mathbf{r} \int u_E(\mathbf{r}')\psi^*(\mathbf{r}')d\mathbf{r}'$$

$$= \iint \psi^*(\mathbf{r}')\psi(\mathbf{r}) \left[\sum_E u_E^*(\mathbf{r})u_E(\mathbf{r}') \right] d\mathbf{r}d\mathbf{r}'$$

$$= \iint \psi^*(\mathbf{r}')\psi(\mathbf{r})\delta(\mathbf{r}-\mathbf{r}')d\mathbf{r}d\mathbf{r}'$$

$$= \int |\psi(\mathbf{r})|^2 \, d\mathbf{r} = 1.$$

The energy eigenfunction expansion of the wavefunction enables us to separate the time dependence of the Schrödinger equation if the potential energy operator $\hat{V}$ is independent of time t. The wavefunction $\psi(\mathbf{r}, t)$ is expanded in energy eigenfunctions at the time t with the time-dependent expansion coefficients:

$$\psi(\mathbf{r}, t) = \sum_E A_E(t)u_E(\mathbf{r}), \quad A_E(t) = \int u_E^*(\mathbf{r})\psi(\mathbf{r}, t)d\mathbf{r}. \tag{1.23}$$

The expansion Eq. (1.23) is substituted into the Schrödinger equation (1.9) to yield

$$i\hbar \sum_E u_E(\mathbf{r}) \frac{d}{dt} A_E(t) = \sum_E A_E(t) E u_E(\mathbf{r}),$$

or, using the orthonormality of the u_E,

$$i\hbar \frac{d}{dt} A_E(t) = E A_E(t), \tag{1.24}$$

with the probability $P(E) = |A_E(t)|^2$ being constant in time. It is a simple procedure that the time integration to Eq. (1.24) is performed once with the initial condition at time t_0. The general initial value wavefunction is, therefore, written in the form:

$$\psi(\mathbf{r}, t) = \sum_E A_E(t_0) e^{-\frac{i}{\hbar} E(t-t_0)} u_E(\mathbf{r})$$

$$A_E(t_0) = \int u_E^*(\mathbf{r}') \psi(\mathbf{r}', t_0) d\mathbf{r}'. \tag{1.25}$$

We have another important eigenfunction expansion method by using the momentum eigenfunctions defined by the momentum eigenvalue equation

$$-i\hbar \nabla u_{\mathbf{p}}(\mathbf{r}) = \mathbf{p} u_{\mathbf{p}}(\mathbf{r}). \tag{1.26}$$

The generic solutions to the momentum eigenvalue Eq. (1.26), with the relation $\mathbf{p} = \hbar\mathbf{k}$, are written in the form of

$$u_{\mathbf{k}}(\mathbf{r}) \propto \exp(i\mathbf{k} \cdot \mathbf{r}).$$

These are eigenfunctions of the momentum operator with the eigenvalues $\hbar\mathbf{k}$. The proportionality constants are determined by the choice of normalization method.

The simple and commonly chosen normalization method is the box normalization, in which the probability to find a particle in a cubic box of volume L^3 with the length of each edge to be L is unity. The box normalization restricts the possible values of $\mathbf{k}$ to be a set of discrete values:

$$k_x = \frac{2\pi n_x}{L}, \quad k_y = \frac{2\pi n_y}{L}, \quad k_z = \frac{2\pi n_z}{L}, \quad n_{x,y,z} = 0, \pm 1, \pm 2, \cdots \tag{1.27}$$

and the proportionality constant becomes $L^{-3/2}$. The orthonormalization condition becomes

$$\int_{L^3} u_{\mathbf{q}}^*(\mathbf{r}) u_{\mathbf{k}}(\mathbf{r}) d\mathbf{r} = \delta_{k_x q_x} \delta_{k_y q_y} \delta_{k_z q_z} = \delta_{\mathbf{k}\mathbf{q}} \tag{1.28}$$

with the properly normalized momentum eigenfunctions

$$u_{\mathbf{k}}(\mathbf{r}) = L^{-3/2} \exp(i\mathbf{k} \cdot \mathbf{r}). \tag{1.29}$$

The continuity of $\mathbf{k}$ is assumed by taking the limit of the size of the box to be sufficiently large, $L \to \infty$. This limit is commonly taken at the end of calculations.

There is another normalization method by using the properties of delta function. To see this, we may consider the integral $\int u_{\mathbf{q}}^{*}(\mathbf{r})u_{\mathbf{k}}(\mathbf{r})d\mathbf{r}$ is the product of three integrals of each component:

$$
\begin{aligned}
\int_{-\infty}^{\infty} e^{i(k_x - q_x)x} dx &= \lim_{g \to \infty} \int_{-g}^{g} e^{i(k_x - q_x)x} dx \\
&= \lim_{g \to \infty} \frac{2 \sin g(k_x - q_x)}{k_z - q_x} \\
&= 2\pi \delta(k_x - q_x),
\end{aligned}
$$

where we employed the sinc function representation of delta function, $\delta(x) = \lim_{g \to \infty} \frac{\sin gx}{\pi x}$. We now give the proportionality to the momentum eigenfunctions defined in the infinite space to be

$$u_{\mathbf{k}}(\mathbf{r}) = (2\pi)^{-3/2} \exp(i\mathbf{k} \cdot \mathbf{r}), \tag{1.30}$$

which satisfy the orthonormality condition

$$\int u_{\mathbf{q}}^{*}(\mathbf{r})u_{\mathbf{k}}(\mathbf{r})d\mathbf{r} = \delta(k_x - q_x)\delta(k_y - q_y)\delta(k_z - q_z) = \delta(\mathbf{k} - \mathbf{q}). \tag{1.31}$$

When we choose the delta-function normalization scheme, one may expand the wavefunction in terms of the momentum eigenfunctions with the introduction of k-dependent energy E_k,

$$\psi(x, t) = \int dk A_k e^{-\frac{i}{\hbar} E_k t} u_k(x),$$

where we consider a one-dimensional motion for simplicity. The free particle Schrödinger equation in one dimension,

$$i\hbar \frac{\partial \psi}{\partial t} = -\frac{\hbar^2}{2m} \frac{\partial^2 \psi}{\partial x^2},$$

yields the energy–momentum relation

$$E_k = \frac{\hbar^2 k^2}{2m}. \tag{1.32}$$

The Planck–Einstein relation $E_k = \hbar\omega(k)$ transforms Eq. (1.32) to a dispersion relation,

$$\omega(k) = \frac{\hbar}{2m}k^2. \tag{1.33}$$

The efforts to obtain the dispersion relations for the interacting systems constitute a central pillar of modern physics, since a dispersion relation contains every physically relevant information about the stationary quantum system.

Let us consider how we can observe the dynamical variables precisely in quantum mechanics. This can be analyzed by considering the mean-square deviation of observations. The word "mean" implies the expectation value discussed in the Ehrenfest's theorem. Let us restrict our discussions to the one-dimensional free particle motion. The mean-square deviations of the position $(\Delta x)^2$ and the momentum $(\Delta x)^2$

$$(\Delta x)^2 = \langle (\hat{x} - \langle x \rangle)^2 \rangle = \langle x^2 \rangle - \langle 2x \langle x \rangle \rangle + \langle \langle x \rangle^2 \rangle = \langle x^2 \rangle - \langle x \rangle^2$$

$$(\Delta p)^2 = \langle (\hat{p} - \langle p \rangle)^2 \rangle = \langle p^2 \rangle - \langle p \rangle^2.$$

Introducing the measurement error operators

$$\hat{\alpha} \equiv \hat{x} - \langle x \rangle, \quad \hat{\beta} \equiv \hat{p} - \langle p \rangle = -i\hbar \left(\frac{d}{dx} - \left\langle \frac{d}{dx} \right\rangle \right),$$

one may obtain

$$(\Delta x)^2 (\Delta p)^2 = \int_{-\infty}^{\infty} \psi^* \hat{\alpha}^2 \psi \, dx \int_{-\infty}^{\infty} \psi^* \hat{\beta}^2 \psi \, dx$$

$$= \int_{-\infty}^{\infty} (\hat{\alpha}^* \psi^*) (\hat{\alpha} \psi) \, dx \int_{-\infty}^{\infty} (\hat{\beta}^* \psi^*) (\hat{\beta} \psi) \, dx \tag{1.34}$$

The right-hand side of the product of the mean-square deviation Eq. (1.34) is in the form of $\int_{-\infty}^{\infty} f^* f \, dx \int_{-\infty}^{\infty} g^* g \, dx$, with $f = \hat{\alpha}\psi$ and

$g = \hat{\beta}\psi$. Since the inequality

$$\int \left| f - g \frac{\int fg^* dx}{\int |g|^2 \, dx} \right|^2 dx \geq 0$$

holds for all infinite range integrals, but the equality holds only for $f = cg$ with constant c, the inequality

$$\int |f|^2 \, dx \int |g|^2 \, dx \geq \left| \int f^* g \, dx \right|^2$$

also holds. Then the product of the mean-square deviation Eq. (1.34) satisfies the inequality

$$(\Delta x)^2 \, (\Delta p)^2 \geq \left| \int (\hat{\alpha}^* \psi^*) \, (\hat{\beta}\psi) \, dx \right|^2 = \left| \int \psi^* \hat{\alpha}\hat{\beta}\psi \, dx \right|^2. \quad (1.35)$$

We consider the symmetric description of the operator product $\hat{\alpha}\hat{\beta}$ as in the right-hand side of Eq. (1.35)

$$\hat{\alpha}\hat{\beta} = \frac{1}{2} \left(\hat{\alpha}\hat{\beta} - \hat{\beta}\hat{\alpha} \right) + \frac{1}{2} \left(\hat{\alpha}\hat{\beta} + \hat{\beta}\hat{\alpha} \right)$$

and take care about the fact that the operator $\hat{\beta}$ is a differential operator to enable us in performing the integration by part with discarding the surface integrals and cross terms. By definition and the fundamental quantum conditions Eq. (1.2), it is easy to show that

$$\left(\hat{\alpha}\hat{\beta} - \hat{\beta}\hat{\alpha} \right) \psi = -i\hbar \left[\hat{x} \frac{d\psi}{dx} - \frac{d}{dx} \left(\hat{x}\psi \right) \right] = i\hbar\psi.$$

It is, then, straightforward to show the relation

$$(\Delta x)^2 \, (\Delta p)^2 \geq \frac{1}{4}\hbar^2 \quad \text{or} \quad \Delta x \cdot \Delta p \geq \frac{1}{2}\hbar. \quad (1.36)$$

This inequality is known as the *Heisenberg's principles of uncertainty.*[g]

One has to take care about the interpretation of the uncertainty principles. During the derivation of Eq. (1.36), we have not involved any interaction related with any observation experiments. The principle of uncertainty is the very nature of dynamics, once the fundamental quantization conditions hold. The commonly known explanation about the large disturbance of the motion due to the lightness of particle, invented by Heisenberg himself (!), is invalid to explain the principles of uncertainty.

---

[g]W. Heisenberg, *Z. Physik* **43**, 172 (1927).

1.2 Representations and Equations of Motion

The quantum mechanical operator relation Eq. (1.1) has been represented in the form

$$\hat{\Omega}v_\mu(\mathbf{r}) = \omega_\mu v_\mu(\mathbf{r}), \tag{1.37}$$

where $\hat{\Omega}$ can be the momentum operator Eq. (1.7) with eigenfunctions Eq. (1.29) or Eq. (1.30), or the Hamiltonian operator Eq. (1.16) with eigenfunctions defined by Eq. (1.15). It can also expand v_μ in terms of the elements of an orthonormal complete set of eigenfunctions w_κ with expansion coefficients $u_{\kappa\mu}$:

$$v_\mu(\mathbf{r}) = \int^\Sigma d\kappa \; u_{\kappa\mu} w_\kappa(\mathbf{r}), \tag{1.38}$$

where the symbol $\int^\Sigma d\kappa$ denotes both a summation $\sum_\kappa$ over discrete values of the subscript κ and an integration $\int d\kappa$ over the continuous part of its range. The orthonormality of $w_\kappa(\mathbf{r})$ yields the coefficients of transformation as

$$u_{\kappa\mu} = \int w_\kappa^*(\mathbf{r})v_\mu(\mathbf{r})d\mathbf{r}. \tag{1.39}$$

We can also expand w_κ in terms of v_μ:

$$w_\kappa(\mathbf{r}) = \int^\Sigma d\mu \; u_{\kappa\mu}^* v_\mu(\mathbf{r}). \tag{1.40}$$

We may regard $u_{\kappa\mu}$ as the elements of a typical transformation matrix $U U^\dagger$:

$$
\begin{aligned}
\left(U U^\dagger\right)_{\kappa\lambda} &= \int^\Sigma d\mu \; u_{\kappa\mu} u_{\lambda\mu} \\
&= \int^\Sigma d\mu \int w_\kappa^*(\mathbf{r})v_\mu(\mathbf{r})d\mathbf{r} \int v_\mu^*(\mathbf{r}')w_\lambda(\mathbf{r}')d\mathbf{r}' \\
&= \iint w_\kappa^*(\mathbf{r})\delta(\mathbf{r} - \mathbf{r}')w_\lambda(\mathbf{r}')d\mathbf{r}d\mathbf{r}' \\
&= \int w_\kappa^*(\mathbf{r})w_\lambda(\mathbf{r})d\mathbf{r} = (\mathbb{1})_{\kappa\lambda},
\end{aligned}
\tag{1.41}
$$

where $\mathbb{1}$ is the identity matrix. This is the definition of the unitary matrix U. So the expansion coefficients $w_{\kappa\lambda}$ are the elements of the unitary matrix. In general, the unitary matrix U transforms an operator $\hat{\Omega}$ from one *representation*, in which a mathematical object in *Hilbert space* is expanded in terms of a set of orthonormal eigenfunctions, to another:

$$U \hat{\Omega} U^\dagger = \hat{\Omega}'. \tag{1.42}$$

Let $\psi_\alpha(\mathbf{r})$ represent a particular state α of a system. We regard ψ_α as a matrix with one column, in which the rows are labeled by the coordinate $\mathbf{r}$. This column matrix ψ_α can be expanded in terms of the orthonormal complete set of $u_k(\mathbf{r})$ with the expansion coefficients $a_{\alpha k}$:

$$\psi_\alpha(\mathbf{r}) = \int^\Sigma dk\, a_{\alpha k} u_k(\mathbf{r}), \quad a_{\alpha k} = \int u_k^*(\mathbf{r}) \psi_\alpha(\mathbf{r}) d\mathbf{r}, \tag{1.43}$$

which can be written in the matrix form as

$$\psi_\alpha = U^\dagger a_\alpha, \quad a_\alpha = U \psi_\alpha, \tag{1.44}$$

respectively, where a_α is a one-column matrix with rows labeled by k. Just as the unitary matrix U transforms an operator $\hat{\Omega}$ into another representation Eq. (1.42), it also transforms a state function ψ_α to the corresponding representation through Eq. (1.44).

It is left as an exercise to show that the unitary transformation does not change the *norm* of the state function:

$$\int \psi_\alpha^*(\mathbf{r}) \psi_\alpha(\mathbf{r}) d\mathbf{r} = \int^\Sigma d\mathbf{r} \psi_\alpha^*(\mathbf{r}) \psi_\alpha(\mathbf{r}) = \psi_\alpha^\dagger \psi_\alpha, \tag{1.45}$$

where $\psi_\alpha^\dagger$ is the Hermitian adjoint of the one-column matrix ψ_α with the column labeled by $\mathbf{r}$. The norm of a state function is a special case of the *inner product* of two state vectors ψ_α and ψ_β, which is defined as

$$\left(\psi_\alpha, \psi_\beta\right) = \psi_\alpha^\dagger \psi_\beta = \int \psi_\alpha^*(\mathbf{r}) \psi_\beta(\mathbf{r}) d\mathbf{r} \tag{1.46}$$

and is also a number. We can consider the two state vectors ψ_α and ψ_β to be *orthogonal* if the inner product vanishes. The matrix element $\psi_\alpha^\dagger \hat{\Omega} \psi_\beta = \left(\psi_\alpha, \hat{\Omega}\psi_\beta\right)$ is then the inner product of the state vector ψ_α and $\hat{\Omega}\psi_\beta$.

An extremely convenient notation system to represent the state vectors and operators was invented by Dirac.[h] Since any state

---

[h] P. A. M. Dirac, *op. cit.* Section 6 and J. J. Sakurai and San Fu Tuan, *Modern Quantum Mechanics*, Revised Ed. (Addison-Wesley, Reading, Massachusetts, 1994) Chapter 1.

function or state vector can be transformed from one representation to another, i.e., no matter how the state function is written as ψ_α or a_α, we know those representations indicate a definite quantum state α. So we can write a quantum state α as a *ket vector* $|\alpha\rangle$ and its Hermitian conjugate *bra vector* $\langle\alpha|$. The inner product of the two state vectors is written as

$$\psi_\alpha^\dagger \psi_\beta = \langle\alpha|\beta\rangle \qquad (1.47)$$

and is called a bracket expression. Operations on a ket vector from the left with $\hat{\Omega}$ produce another ket vector

$$\hat{\Omega}\,|\beta\rangle = |\beta'\rangle \qquad (1.48a)$$

and operation on a bra from the right with $\hat{\Omega}$ produces another bra vector

$$\langle\alpha|\,\hat{\Omega} = \langle\alpha'|\,. \qquad (1.48b)$$

The matrix element of $\hat{\Omega}$ between states α and β is written as

$$\begin{aligned} \Omega_{\alpha\beta} &= \int \psi_\alpha^* \left(\hat{\Omega}\psi_\beta\right) d\mathbf{r} \\ &= \left(\psi_\alpha, \hat{\Omega}\psi_\beta\right) \\ &= \langle\alpha|\beta'\rangle = \langle\alpha\,|\hat{\Omega}|\,\beta\rangle \end{aligned} \qquad (1.49a)$$

or equivalently

$$\begin{aligned} \Omega_{\alpha\beta} &= \int \left[\hat{\Omega}^\dagger \psi_\alpha(\mathbf{r})\right]^* \psi_\beta(\mathbf{r}) d\mathbf{r} \\ &= \left(\hat{\Omega}^\dagger \psi_\alpha, \psi_\beta\right) \\ &= \langle\alpha\,|\hat{\Omega}|\,\beta\rangle = \langle\alpha|\hat{\Omega}|\beta\rangle\,. \end{aligned} \qquad (1.49b)$$

The matrix element of the Hermitian adjoint operator $\hat{\Omega}^\dagger$ is then given by

$$\left(\hat{\Omega}\right)_{\beta\alpha} = \Omega_{\alpha\beta}^* = \langle\beta\,|\hat{\Omega}^\dagger|\alpha\rangle = \langle\alpha\,|\hat{\Omega}|\,\beta\rangle^*\,. \qquad (1.50)$$

We may represent a quantum state by Dirac notation: The ket $|\mu\rangle$ to denote an eigenstate of $\hat{\Omega}$ with eigenvalue ω_μ. A specific example is that $|k\rangle$ denotes an energy eigenstate of Hamiltonian $\hat{H}$ with eigenvalue E_k. In the same way, we assign $|\mathbf{r}\rangle$ as an eigenstate of position operator $\hat{\mathbf{r}}$ with eigenvalue $\mathbf{r}$. We may write the energy eigenfunction of energy E_k as

$$u_k(\mathbf{r}) = \langle\mathbf{r}|k\rangle\,, \qquad u_k^*(\mathbf{r}) = \langle k|\mathbf{r}\rangle\,.$$

The Dirac notation representation of Eq. (1.43) becomes

$$\psi_\alpha(\mathbf{r}) = \langle \mathbf{r}|\alpha \rangle = \int^\Sigma dk\, a_{\alpha k} \langle \mathbf{r}|k \rangle = \int^\Sigma dk\, \langle k|\alpha \rangle \langle \mathbf{r}|k \rangle.$$

Since both factors $\langle k|\alpha \rangle$ and $\langle \mathbf{r}|k \rangle$ in the right-hand side are scalar, their positions are interchangeable to yield

$$\psi_\alpha(\mathbf{r}) = \int^\Sigma dk\, \langle \mathbf{r}|k \rangle \langle k|\alpha \rangle = \langle \mathbf{r}|\alpha \rangle.$$

Here we employed the completeness relation

$$\int^\Sigma dk\, |k \rangle \langle k| = 1, \tag{1.51}$$

which have a short-hand notation of summation convention

$$|k \rangle \langle k| = 1.$$

Now we may rewrite the Schrödinger equation with the Hamiltonian operator $\hat{H}$ in Dirac notation:

$$i\hbar \frac{d}{dt}|\alpha_S(t)\rangle = \hat{H}\,|\alpha_S(t)\rangle, \tag{1.52}$$

where the total time derivative is used due to the fact that there is no explicit coordinate dependence of the ket and the subscript S refers to the ket as viewed in the *Schrödinger picture*. It means the Schrödinger picture ket varies in time as a function in the ordinary differential equation. The fact that the Hamiltonian operator $\hat{H}$ is a Hermitian leads the Hermitian adjoint equation,

$$-i\hbar \frac{d}{dt}\langle\alpha_S(t)| = \langle\alpha_S(t)|\,\hat{H}. \tag{1.53}$$

The solutions to Eqs. (1.52) and (1.53) are obvious if $\hat{H}$ is independent of time:

$$|\alpha_S(t)\rangle = e^{-i\hat{H}t/\hbar}\,|\alpha_S(0)\rangle\,\langle\alpha_S(t)|\,\langle\alpha_S(0)|\,e^{i\hat{H}t/\hbar}. \tag{1.54}$$

One should take care about the order of operator products appearing in the infinite series representation of $e^{\pm i\hat{H}t/\hbar}$ as an infinite sum of powers of $\hat{H}$, which is composed of noncommutative operators $\hat{\mathbf{r}}$ and $\hat{\mathbf{p}}$.

We may find the time rate of the mátrix element of a dynamical variable $\hat{\Omega}_S$ in the Schrödinger picture:

$$\frac{d}{dt}\langle\alpha_S(t)|\hat{\Omega}_S|\beta_S(t)\rangle = \left[\frac{d}{dt}\langle\alpha_S(t)|\right]\hat{\Omega}_S|\beta_S(t)\rangle$$
$$+ \left\langle\alpha_S(t)\left|\frac{\partial\hat{\Omega}_S}{\partial t}\right|\beta_S(t)\right\rangle$$
$$+ \langle\alpha_S(t)|\hat{\Omega}_S\left[\frac{d}{dt}|\beta_S(t)\rangle\right]$$
$$= \left\langle\alpha_S(t)\left|\frac{\partial\hat{\Omega}_S}{\partial t}\right|\beta_S(t)\right\rangle$$
$$+ \frac{1}{i\hbar}\langle\alpha_S(t)|(\hat{\Omega}_S\hat{H} - \hat{H}\hat{\Omega}_S)|\beta_S(t)\rangle.$$

$$(1.55)$$

The time rate of the matrix element in the Schrödinger picture is made of the expectation value of the time rate of the operator itself and the expectation value of the commutator

$$[\hat{\Omega}_S, \hat{H}] = \hat{\Omega}_S\hat{H} - \hat{H}\hat{\Omega}_S.$$

One interesting feature of Eq. (1.55) appears when $\hat{\Omega}_S$ commutes with $\hat{H}$ and has no explicit time dependence. In this case, all terms in the right-hand side vanish and so all matrix elements of $\hat{\Omega}_S$ are constant in time. This property of the dynamical variable is the definition of a *constant of the motion*.

Let us see what happens if we substitute the Schrödinger ket Eq. (1.54) into the matrix element time rate Eq. (1.55). This gives

$$\frac{d}{dt}\left\langle\alpha_S(0)\left|e^{i\hat{H}t/\hbar}\hat{\Omega}_S e^{-i\hat{H}t/\hbar}\right|\beta_S(0)\right\rangle$$
$$= \left\langle\alpha_S(0)\left|e^{i\hat{H}t/\hbar}\frac{\partial\hat{\Omega}_S}{\partial t}e^{-i\hat{H}t/\hbar}\right|\beta_S(0)\right\rangle$$
$$+ \frac{1}{i\hbar}\left\langle\alpha_S(0)\left|\left[e^{i\hat{H}t/\hbar}\hat{\Omega}_S e^{-i\hat{H}t/\hbar}, \hat{H}\right]\right|\beta_S(0)\right\rangle,$$

where we have made use of the fact that $\hat{H}$ commutes with $e^{\pm i\hat{H}t/\hbar}$. It is convenient to define the time-dependent operator

$$\hat{\Omega}_H \equiv e^{i\hat{H}t/\hbar}\frac{\partial\hat{\Omega}_S}{\partial t}e^{-i\hat{H}t/\hbar}, \quad (1.56)$$

where the subscript H denotes the convention that the operator is presented in the *Heisenberg picture*. Consequently, the Schrödinger ket should be changed to the corresponding Heisenberg ket

$$|\alpha_H(t)\rangle \equiv e^{i\hat{H}t/\hbar}|\alpha_S(t)\rangle = |\alpha_S(0)\rangle. \qquad (1.57)$$

So the Heisenberg ket coincides with the Schrödinger ket at time $t = 0$. Thus, the Heisenberg ket does not depend on time for allowing to write $|\alpha_H(t)\rangle = |\alpha_H\rangle$. In the Heisenberg picture, $\hat{\Omega}_H$ depends on time t no matter how $\hat{\Omega}_S$ depends on time or not, unless $\hat{\Omega}_S$ commutes with $\hat{H}$. The matrix element time rate is then

$$\left\langle \alpha_H \left| \frac{d}{dt}\hat{\Omega}_H \right| \beta_H \right\rangle = \left\langle \alpha_H \left| \frac{\partial \hat{\Omega}_H}{\partial t} \right| \beta_H \right\rangle + \frac{1}{i\hbar}\left\langle \alpha_H \left| [\hat{\Omega}_H, \hat{H}] \right| \beta_H \right\rangle.$$

Since this relation is valid for an arbitrary bra and an arbitrary ket, it is valid for the operators themselves. The resulting operator relation is known as the *Heisenberg equation of motion*:

$$\frac{d\hat{\Omega}_H}{dt} = \frac{\partial \hat{\Omega}_H}{\partial t} + \frac{1}{i\hbar}[\hat{\Omega}_H, \hat{H}]. \qquad (1.58)$$

This Heisenberg equation of motion serves the central role in this book.

The time evolution of a quantum system is governed by the unitary matrix $e^{-i\hat{H}t/\hbar}$; the choice of the Schrödinger picture or the Heisenberg picture is a matter of choice where the unitary matrix is attached to the state vector or the operator, respectively. There is yet another picture to describe the time evolution of a quantum system by concentrating on this unitary matrix; it is the *interaction picture* or *Dirac picture*. The first step to implement the interaction picture is to divide the Hamiltonian into two parts:

$$\hat{H} = \hat{H}_0 + \hat{H}'.$$

The division criterion is rather arbitrary, but $\hat{H}_0$ is commonly chosen not to depend on time and to posses a simple structure. Most common choice is that $\hat{H}_0$ is the kinetic energy, while $\hat{H}'$ is the potential energy. One may choose $\hat{H}_0$ to be the Coulomb field and $\hat{H}'$ to be some external electromagnetic interaction. In general, $\hat{H}$, $\hat{H}_0$, and $\hat{H}'$ do not commute with each other. The interaction picture defines the state vectors and the operators as

$$|\alpha_I(t)\rangle \equiv e^{i\hat{H}_{0S}t/\hbar}|\alpha_S(t)\rangle$$
$$\hat{\Omega}_I(t) \equiv e^{i\hat{H}_{0S}t/\hbar}\hat{\Omega}_S e^{-i\hat{H}_{0S}t/\hbar}, \tag{1.59}$$

where the subscript I indicates the state vectors and the operators are described in the interaction picture.

It is straightforward to show the time evolution of the state vectors and the operators

$$i\hbar\frac{d}{dt}|\alpha_I(t)\rangle = \hat{H}'_I|\alpha_I(t)\rangle$$
$$\frac{d\hat{\Omega}_I}{dt} = \frac{\partial\hat{\Omega}_I}{\partial t} + \frac{1}{i\hbar}\left[\hat{\Omega}_I, \hat{H}_{0I}\right], \tag{1.60}$$

where $\hat{H}'_I = e^{i\hat{H}_{0S}t/\hbar}\hat{H}'_S e^{-i\hat{H}_{0S}t/\hbar}$. In the interaction picture, the time evolution of the state vectors is governed by the interaction Hamiltonian $\hat{H}'_I$, while the time evolution of the operators is governed by the reference Hamiltonian $\hat{H}_{0I}$.

To see the importance of the interaction picture, let us introduce a parameter λ, which varies from 0 to 1. We can modify the Hamiltonian using the parameter λ to

$$\hat{H} = \hat{H}_0 + \lambda\hat{H}'.$$

The parameter λ is tuned to be 0 at time $t = -\infty$ and is increased slowly to 1 as time increases until $t = 0$, when an experiment begins and the state vector and the operator in the Schrödinger picture, the Heisenberg picture, and the interaction picture coincide with each other. The operator time evolution is steady because we have chosen $\hat{H}_{0I}$ to not depend on time. The solution to the operator can be obtained once in any time t, and this solution persists for all time interval from $t = -\infty$ to $t = \infty$. On the other hand, the state vector can be prepared at time $t = -\infty$ with the reference Hamiltonian $\hat{H}_{0I}$. Since $\lambda = 0$ at $t = -\infty$, the state vector and the operators coincide with those of the Heisenberg picture. This state vector prepared at $t = -\infty$ evolves in time to coincide with the state vector in the Schrödinger picture at time $t = 0$ and the further evolution like the Schrödinger picture, but with only the interaction Hamiltonian $\hat{H}'_I$. So the experiment from $t = 0$ to $t = \infty$ observes the effects of the *interaction* $\hat{H}'_I$ to the state vector prepared by the Hamiltonian $\hat{H}_{0I}$. This procedure is of great importance in the equilibrium many-body theory.

An important example in the venue of many-body theory is the problem of a simple harmonic oscillator of which a particle of mass m oscillates in one-dimensional space with a small amplitude around its equilibrium position through Hooke's law with the spring constant K. The Hamiltonian operator of the simple harmonic oscillator is

$$\hat{H} = \frac{\hat{p}^2}{2m} + \frac{1}{2}m\omega^2\hat{x}^2, \tag{1.61}$$

where we find a characteristic oscillation frequency $\omega = \sqrt{K/m}$. The symmetric powers of $\hat{x}$ and $\hat{p}$ operators in the Hamiltonian suggest us that there is no difference whether we describe the simple harmonic oscillator in the position space or in the momentum space. It is convenient to describe the simple harmonic oscillator in another space composed of another set of two operators that are connected by the Hermitian adjoint to each other, say $\hat{a}$ and $\hat{a}^\dagger$. One may find the set of two operators by employing the symmetric and antisymmetric linear combinations of the operators $\hat{x}$ and $\hat{p}$ to be[i]

$$\hat{a} = \sqrt{\frac{m\omega}{2\hbar}}\left(\hat{x} + i\frac{\hat{p}}{m\omega}\right), \quad \hat{a}^\dagger = \sqrt{\frac{m\omega}{2\hbar}}\left(\hat{x} - i\frac{\hat{p}}{m\omega}\right). \tag{1.62}$$

These operators satisfy a commutation relation

$$[\hat{a}, \hat{a}^\dagger] = \frac{1}{2\hbar}\left(-i\,[\hat{x}, \hat{p}] + i\,[\hat{p}, \hat{x}]\right) = 1. \tag{1.63}$$

We introduce a new operator $\hat{N} = \hat{a}^\dagger\hat{a}$, which is obviously Hermitian.[j] This operator is explained by the $\hat{x}$ and $\hat{p}$ operators, from the definitions Eq. (1.62),

$$\hat{N} = \hat{a}^\dagger\hat{a} = \left(\frac{m\omega}{2\hbar}\right)\left(\hat{x}^2 + \frac{1}{m^2\omega^2}\hat{p}^2\right) + \left(\frac{1}{2\hbar}\right)[\hat{x}, \hat{p}].$$

This result suggests us to rewrite the simple harmonic oscillator Hamiltonian operator Eq. (1.61) in terms of the operator $\hat{N}$:

$$\hat{H} = \hbar\omega\left(\hat{N} + \frac{1}{2}\right). \tag{1.64}$$

---

[i]This is nothing more than a linear algebra exercise: $\begin{pmatrix} \hat{a} \\ \hat{a}^\dagger \end{pmatrix} = \begin{pmatrix} c_{xx} & c_{xp} \\ c_{px} & c_{pp} \end{pmatrix}\begin{pmatrix} \hat{x} \\ \hat{p} \end{pmatrix}$.

[j]One can check its Hermitian property by $\hat{N}^\dagger = \left(\hat{a}^\dagger\hat{a}\right)^\dagger = (\hat{a})^\dagger\left(\hat{a}^\dagger\right)^\dagger = \hat{a}^\dagger\hat{a} = \hat{N}$.

Since the Hamiltonian operator $\hat{H}$ is a linear combination of the operator $\hat{N}$, an eigenvector of $\hat{N}$ could be an eigenvector of $\hat{H}$. Let us write the eigenvalue equation of the operator $\hat{N}$ by introducing its eigenvector $|n\rangle$ with the corresponding eigenvalue n:

$$\hat{N}\,|n\rangle = n\,|n\rangle\,, \tag{1.65}$$

so the eigenvalue equation of the Hamiltonian will be

$$\hat{H}\,|n\rangle = \hbar\omega\left(n + \frac{1}{2}\right)|n\rangle\,, \tag{1.66}$$

from which the energy eigenvalue of the Hamiltonian becomes

$$E_n = \hbar\omega\left(n + \frac{1}{2}\right). \tag{1.67}$$

The energy eigenvalue Eq (1.67) has more information than Planck's relation $E = \hbar\omega$ by the factor $\left(n + \frac{1}{2}\right)$. We need to understand what is the meaning of the quantum number n, which is essentially a dimensionless nonnegative number for ensuring the positive definiteness of the harmonic oscillator energy. To this end, let us find the operator relations among $\hat{N}$, $\hat{a}$, and $\hat{a}^\dagger$.

We may first test the commutation relations

$$\begin{aligned}
\left[\hat{N}, \hat{a}\right] &= \left[\hat{a}^\dagger\hat{a}, \hat{a}\right] = \hat{a}^\dagger\left[\hat{a}, \hat{a}\right] + \left[\hat{a}^\dagger, \hat{a}\right]\hat{a} = -\hat{a},\\
\left[\hat{N}, \hat{a}^\dagger\right] &= \left[\hat{a}^\dagger\hat{a}, \hat{a}\right] = \hat{a}^\dagger\left[\hat{a}, \hat{a}^\dagger\right] + \left[\hat{a}^\dagger, \hat{a}^\dagger\right]\hat{a} = \hat{a}^\dagger.
\end{aligned} \tag{1.68}$$

Using these relations, we can investigate the effects of operators $\hat{a}$ and $\hat{a}^\dagger$ on the eigenvector $|n\rangle$. One may show that

$$\hat{N}\hat{a}^\dagger\,|n\rangle = \left(\left[\hat{N}, \hat{a}^\dagger\right] + \hat{a}^\dagger\hat{N}\right)|n\rangle = (n+1)\,\hat{a}^\dagger\,|n\rangle \tag{1.69a}$$

$$\hat{N}\hat{a}\,|n\rangle = \left(\left[\hat{N}, \hat{a}^\dagger\right] + \hat{a}\hat{N}\right)|n\rangle = (n-1)\,\hat{a}\,|n\rangle\,. \tag{1.69b}$$

Hence, one may consider $\hat{a}\,|n\rangle$ or $\hat{a}^\dagger\,|n\rangle$ to be another eigenvector of the operator $\hat{N}$, say $|m\rangle$, to yield

$$\hat{N}\,|m\rangle = (n-1)\,|m\rangle$$

or

$$\hat{N}\,|m\rangle = (n+1)\,|m\rangle\,.$$

So we can say

$$\begin{aligned}
|m\rangle &\to |n-1\rangle \quad \text{for} \quad \hat{a}\\
|m\rangle &\to |n+1\rangle \quad \text{for} \quad \hat{a}^\dagger
\end{aligned}$$

and these relations give the effects of the operators $\hat{a}$ and $\hat{a}^\dagger$ to $|n\rangle$ to be

$$\begin{aligned}
\hat{a}\,|n\rangle &= c\,|n-1\rangle\,, \\
\hat{a}^\dagger\,|n\rangle &= c'\,|n+1\rangle\,,
\end{aligned} \tag{1.70}$$

where c and c' are the appropriate phase factors. The implication of the effects of the operator $\hat{a}$ or $\hat{a}^\dagger$ on the eigenvector $|n\rangle$ of the operator $\hat{N}$ is obvious; besides the corresponding phase factor, $\hat{a}$ decreases the eigenvalue n by 1 or $\hat{a}^\dagger$ increases the eigenvalue n by 1. These properties yield the name of the operators $\hat{a}$ and $\hat{a}^\dagger$ to be the *annihilation* (or *destruction*) operator and the *creation* (or *construction*) operator, respectively.

Let us find the phase factors c and c'. First we consider the square of c and c':

$$\begin{aligned}
|c|^2 &= \langle n-1\,|c^*c|\,n-1\rangle = \langle n\,|\hat{a}^\dagger\hat{a}|\,n\rangle = \langle n\,|\hat{N}|\,n\rangle = n, \\
|c'|^2 &= \langle n+1\,|c'^*c'|\,n+1\rangle = \langle n\,|\hat{a}\hat{a}^\dagger|\,n\rangle \\
&= \langle n\,|(\hat{N}+1)|\,n\rangle = (n+1)\,.
\end{aligned}$$

By convention, we take c and c' to be positive and real to obtain

$$\begin{aligned}
\hat{a}\,|n\rangle &= \sqrt{n}\,|n-1\rangle\,, \\
\hat{a}^\dagger\,|n\rangle &= \sqrt{n+1}\,|n+1\rangle\,.
\end{aligned} \tag{1.71}$$

Now we operate the annihilation operator $\hat{a}$ to the eigenstate $|n\rangle$ of the number operator $\hat{N}$, successively. The sequence will be

$$\begin{aligned}
\hat{a}\,|n\rangle &= \sqrt{n}\,|n-1\rangle\,, \\
\hat{a}^2\,|n\rangle &= \sqrt{n(n-1)}\,|n-2\rangle\,, \\
\hat{a}^3\,|n\rangle &= \sqrt{n(n-1)(n-2)}\,|n-3\rangle\,, \\
&\ \ \vdots
\end{aligned} \tag{1.72}$$

We know the positive definiteness of n as

$$n = \langle n\,|\hat{N}|\,n\rangle = \big((n|\,\hat{a}^\dagger)\cdot(\hat{a}\,|n\rangle)\big) = |c|^2 \geq 0,$$

which proves the physical argument of the positive definiteness of the harmonic oscillator energy $E_n = \hbar\omega\left(n+\frac{1}{2}\right)$. The requirement of the positive definiteness restricts n to be a nonnegative integer; otherwise, the annihilation sequence Eq. (1.72) will never stop, leading to a negative value n. We conclude that $n = 0$ is the minimum eigenvalue with which the Hamiltonian eigenvalue equation yields

$$\hat{H} |0\rangle = \frac{1}{2} \hbar\omega |0\rangle, \qquad (1.73)$$

which constitutes the *ground state* $|0\rangle$ of the simple harmonic oscillator combined with the ground-state energy $E_0 = \frac{1}{2}\hbar\omega$.

We can now generate any simple harmonic oscillator states $|n\rangle$ by applying successively the creation operator $\hat{a}^\dagger$ to this ground state:

$$|1\rangle = \frac{1}{\sqrt{0+1}} \hat{a}^\dagger |0\rangle,$$

$$|2\rangle = \frac{1}{\sqrt{1+1}} \hat{a}^\dagger |1\rangle = \frac{1}{\sqrt{2 \cdot 1}} \left(\hat{a}^\dagger\right)^2 |0\rangle,$$

$$|3\rangle = \frac{1}{\sqrt{2+1}} \hat{a}^\dagger |2\rangle = \frac{1}{\sqrt{3 \cdot 2 \cdot 1}} \left(\hat{a}^\dagger\right)^3 |0\rangle, \qquad (1.74)$$

$$\vdots$$

$$|n\rangle = \frac{1}{\sqrt{n}} \left(\hat{a}^\dagger\right)^n |0\rangle,$$

with the corresponding energy

$$E_n = \hbar\omega \left(n + \frac{1}{2}\right), \qquad n = 0, 1, 2, 3, \cdots \qquad (1.75)$$

We may interpret that $\hat{N}$ operator *counts* the number of oscillators n of frequency ω, so $\hat{N}$ is called the *number operator*. This contrasts sharply to a classical simple harmonic oscillator, which possesses its energy proportional to its amplitude square:

$$E = \frac{1}{2} m\omega^2 A^2, \qquad (1.76)$$

where A is the amplitude and is a *continuous* positive real number. The minimum energy of the classical simple harmonic oscillator is, of course, zero, while the ground state of the quantum simple harmonic oscillator is $\frac{1}{2}\hbar\omega$.

We can obtain the matrix elements for the annihilation operator and the creation operator,

$$\langle n' |\hat{a}| n\rangle = \sqrt{n}\delta_{n',n-1}, \qquad \langle n' |\hat{a}^\dagger| n\rangle = \sqrt{n+1}\delta_{n',n+1}, \qquad (1.77)$$

respectively. It can also represent the position operator $\hat{x}$ and the momentum operator $\hat{p}$ in terms of the annihilation $\hat{a}$ and creation $\hat{a}^\dagger$ operators as

$$\hat{x} = \sqrt{\frac{\hbar}{2m\omega}}\left(\hat{a}^\dagger + \hat{a}\right), \quad \hat{p} = i\sqrt{\frac{\hbar}{2m\omega}}\left(\hat{a}^\dagger - \hat{a}\right),$$

whose matrix elements are

$$\langle n' |\hat{x}| n \rangle = \sqrt{\frac{\hbar}{2m\omega}}\left(\sqrt{n+1}\delta_{n',n+1} + \sqrt{n}\delta_{n',n-1}\right),$$

$$\langle n' |\hat{p}| n \rangle = i\sqrt{\frac{\hbar}{2m\omega}}\left(\sqrt{n+1}\delta_{n',n+1} - \sqrt{n}\delta_{n',n-1}\right). \tag{1.78}$$

1.3 Second Quantization

The quantum mechanics formulation described in the previous sections has shown its success when applied to a system whose interaction can be modeled by a mean property single-particle system. The physical world, which is made of many interacting particles, requires the inclusion of the inter-particle potentials in the many-particle Schrödinger equation. The many-body wavefunction in configuration space contains all possible dynamical information. Since it is impractical to solve the many-body Schrödinger equation directly, one should resort to other techniques: second quantization, quantum-field theory, and the Green's function formalism. The idea of second quantization was introduced by Dirac[k] by applying the concept of the annihilation and creation of particles in order to bypass the mathematical difficulties in relativistic quantum electrodynamics. Soon after Dirac, it has been shown that the second quantization concept greatly simplifies the complicated problems of many identical interacting particles.[l]

---

[k]P. A. M. Dirac, *Proc. R. Soc.* (London) **114A**, 243 (1927).
[l]P. Jordan and O. Klein, *Z. Physik* **45**, 751 (1927); P. Jordan and E. P. Wigner, *Z. Physik* **47**, 631 (1928); V. Fock, *Z. Physik* **75**, 622 (1932). The annotator suggests the reader to refer the descent textbooks Alexander L. Fetter and John Dirk Walecka, *Quantum Theory of Many-Particle Systems* (McGraw-Hill, New York, 1971) and John W. Negele and Henri Orland, *Quantum Many-Particle Systems* (Addison-Wesley, Redwood City, California, 1988).

In relativistic quantum theory spins raise in natural such a way to let a quantum field satisfy the relativistic transformation properties.[m] It is, however, convenient to accept the existence of fermions with spin half-integers and of bosons with spin integers, for non-relativistic quantum many-body theory. The spin degree of freedom is indicated separately, for example σ, and it is attached to some dynamical variables, for example the position eigenvector $|x\rangle = |\mathbf{r}\sigma\rangle$. Further internal degrees of freedom, such as isospins, can be attached in addition to such notation with an appropriate quantum number.

A wavefunction of N identical particles $\Psi_N(x_1, x_2, \ldots, x_N, t)$ represents the probability amplitude for finding particles at the N positions and combined with the corresponding spin states, $x_1, x_2, \ldots, x_N$ at a given time t and has to satisfy the definiteness condition to be written in terms of inner product condition

$$(\Psi_N, \Psi_N) = \int dx_1 dx_2 \cdots dx_N \; |\Psi_N(x_1, x_2, \ldots, x_N, t)|^2 < +\infty.$$

$$(1.79)$$

The N-particle Schrödinger equation, with the Hamiltonian operator

$$\hat{H} = \sum_k^N \hat{K}(x_k) + \frac{1}{2} \sum_{k \neq l}^N \hat{V}(x_k, x_l),$$

$$(1.80)$$

where $\hat{K}$ and $\hat{V}$ are the many-particle kinetic energy and potential energy, respectively, is given by

$$i\hbar \frac{\partial}{\partial t} \Psi_N(x_1, x_2, \ldots, x_N, t) = \hat{H} \Psi_N(x_1, x_2, \ldots, x_N, t)$$

$$(1.81)$$

together with an appropriate set of boundary conditions for the wavefunction Ψ_N.

The most important boundary condition of the many-particle wavefunction Ψ_N would be the permutation properties of indistinguishable identical particles. Let $(P1, P2, \ldots, PN)$ represent any permutation P of the set $(1, 2, \ldots, N)$. We do not know if someone permutes two particles collected from the set of the N identical

---

[m]For a rigorous derivation, one may refer J. M. Jauch and F. Rohrlich, *The Theory of Photons and Electrons, Second Expanded Edition* (Springer-Verlag, New York, 1976), Chapter 1.

particles, but the N-particle wavefunction changes its sign according to

$$\Psi_N \left(x_{P1}, x_{P2}, \cdots, x_{PN}, t \right) = \zeta^P \Psi_N \left(x_1, x_2, \cdots, x_N, t \right), \qquad (1.82)$$

where P is the parity of permutation, and ζ is $+1$ for bosons and -1 for fermions.

Let $\psi_{\alpha_k}(x_k)$ be a *time-independent* single-particle wavefunction to represent an *independent* single-particle quantum state $|\alpha_k\rangle$, which is complete

$$\int^{\Sigma} dk \, |\alpha_k\rangle \langle \alpha_k| = 1.$$

Conveniently, we consider that the infinite set of single-particle quantum number α_k is ordered $(1, 2, 3, \cdots, r, s, t, \cdots, \infty)$ and α_k runs over this set of eigenvalues, so we have a *fixed set of quantum numbers*

$$\alpha_1, \alpha_2, \cdots, \alpha_N. \qquad (1.83)$$

We are going to construct the many-body wavefunction by expanding in terms of the (*independent*) single-particle wavefunctions $\psi_{\alpha_k}(x_k)$:

$$\Psi_N \left(x_1, x_2, \ldots, x_N, t \right)$$
$$= \sum_P \sum_{\alpha_{P1}, \alpha_{P2}, \cdots, \alpha_{PN}} C \left(\alpha_{P1}, \alpha_{P2}, \cdots, \alpha_{PN}, t \right) \psi_{\alpha_{P1}}(x_1) \psi_{\alpha_{P2}}(x_2) \cdots \psi_{\alpha_{PN}}(x_N).$$
$$(1.84)$$

Since the $\psi_\alpha(x)$ are time independent, all the time dependence of the many-particle wavefunction is described in the coefficients $C \left(\alpha_1, \alpha_2, \cdots, \alpha_N, t \right)$, which also follows the permutation property Eq. (1.82) in such a way that

$$C \left(\alpha_{P1}, \alpha_{P2}, \cdots, \alpha_{PN}, t \right) = \zeta^P C \left(\alpha_1, \alpha_2, \cdots, \alpha_N, t \right). \qquad (1.85)$$

To obtain the coefficients $C \left(\alpha_1, \alpha_2, \cdots, \alpha_N, t \right)$ we require the normalization condition

$$\int^{\Sigma} dx_k \, |\Psi_N \left(\{x_k\} \right)|^2 = 1. \qquad (1.86)$$

The orthonormality condition of the single-particle wavefunction yields a condition

$$\sum_{\alpha_1, \alpha_2, \cdots, \alpha_N} |C \left(\alpha_1, \alpha_2, \cdots, \alpha_N, t \right)|^2 = 1. \qquad (1.87)$$

Let us multiply Eq. (1.84) to the left by $\psi_{\alpha_1}^\dagger \psi_{\alpha_2}^\dagger \cdots \psi_{\alpha_N}^\dagger$, which is the product of adjoint wavefunctions corresponding to the fixed order of quantum numbers $\alpha_1 \alpha_2 \cdots \alpha_N$ as appeared in the left-hand side, and then integrate over all the appropriate coordinates. The result becomes

$$
\begin{aligned}
C\left(\alpha_1, \alpha_2, \cdots, \alpha_N, t\right) = & \sum_{\alpha_{P1}, \alpha_{P2}, \cdots, \alpha_{PN}} C\left(\alpha_{P1}, \alpha_{P2}, \cdots, \alpha_{PN}, t\right) \\
& \times \int^\Sigma dx_k \, \psi_{\alpha_1}^\dagger(x_1) \psi_{\alpha_2}^\dagger(x_1) \cdots \psi_{\alpha_N}^\dagger(x_N) \\
& \times \psi_{\alpha_{P1}}(x_{P1}) \psi_{\alpha_{P2}}(x_{P2}) \cdots \psi_{\alpha_{PN}}(x_{PN}) \\
= & \sum_{\alpha_{P1}, \alpha_{P2}, \cdots, \alpha_{PN}} C\left(\alpha_{P1}, \alpha_{P2}, \cdots, \alpha_{PN}, t\right) \\
& \times \sum_P \zeta^P \int dx_1 \psi_{\alpha_1}^\dagger(x_1) \psi_{\alpha_{P1}}(x_{P1}) \\
& \times \int dx_2 \psi_{\alpha_2}^\dagger(x_2) \psi_{\alpha_{P2}}(x_{P2}) \cdots \\
& \times \int dx_N \psi_{\alpha_N}^\dagger(x_N) \psi_{\alpha_{PN}}(x_{PN}).
\end{aligned}
$$

$$(1.88)$$

We know from the Copenhagen interpretation that the integral $\int dx_k \psi_{\alpha_k}^\dagger(x_k) \psi_{\alpha_{Pk}}(x_{Pk})$ measures whether a particle occupies the quantum state $|\alpha_k\rangle$ if $\alpha_{Pk} = \alpha_k$. Let n_k be the count of how many times one has the same permuted states $|\alpha_{Pk}\rangle$ as the state $|\alpha_k\rangle$, so we say that n_k is the occupation number. We also know that a quantum state of fermions cannot accommodate more than one particle because of the exclusion principle. The exclusion principle allows one and only one permutation. On the other hand, bosons can occupy any number of particles. In both cases, the sum of the occupation numbers must be the same as the total number of particles N:

$$N = \sum_k n_k. \qquad (1.89)$$

The factors $\sum_P \zeta^P \int dx_1 \psi_{\alpha_1}^\dagger(x_1) \psi_{\alpha_{P1}}(x_{P1}) \cdots$ in the right-hand side of Eq. (1.88) become $(-1)^P n_1! n_2! n_3! \cdots n_\infty$ for fermions and $n_1! n_2! n_3! \cdots n_\infty!$ for bosons. Here we do not limit the index of occupation numbers because $0! = 1$. It is now available to switch the expansion coefficients $C\left(\alpha_1, \alpha_2, \cdots \alpha_N, t\right)$ to the occupation number coefficients $\bar{C}\left(n_1, n_2, n_3, \cdots n_\infty\right)$, which satisfies the normalization conditions

$$\sum_{n_1, n_2, \cdots n_\infty} \left|\bar{C}\left(n_1, n_2, n_3, \cdots, n_\infty, t\right)\right|^2 \frac{N!}{n_1! n_2! \cdots n_\infty!} = 1 \qquad (1.90a)$$

for bosons and

$$\sum_{n_1, n_2, \cdots, n_\infty} \left|\bar{C}\left(n_1, n_2, n_3, \cdots, n_\infty\right)\right|^2 \frac{N!}{n_1! n_2! \cdots n_\infty!} (-1)^P = 1$$
$$(1.90b)$$

for fermions. Let us define another coefficient

$$f\left(n_1, n_2, \cdots, n_\infty, t\right) \equiv \left(\frac{N!}{n_1! n_2! \cdots n_\infty!}\right)^{1/2} \bar{C}\left(n_1, n_2, \cdots n_\infty, t\right),$$

which satisfies the corresponding normalization condition

$$\sum_{n_1, n_2, \cdots n_\infty} |f\left(n_1, n_2, \cdots, n_\infty, t\right)|^2 = 1.$$

The original N-particle wavefunction is now expressed as

$$\Psi_N\left(x_1, x_1, \cdots x_N, t\right) = \sum_{\alpha_1, \alpha_2, \cdots \alpha_N} C\left(\alpha_1, \alpha_2, \cdots \alpha_N, t\right) \psi_{\alpha_1}\left(x_1\right) \psi_{\alpha_2}\left(x_2\right) \cdots \psi_{\alpha_N}\left(x_N\right)$$

$$= \sum_{\alpha_1, \alpha_2, \cdots \alpha_N} \bar{C}\left(n_1, n_2, \cdots n_N, t\right) \psi_{\alpha_1}\left(x_1\right) \psi_{\alpha_2}\left(x_2\right) \cdots \psi_{\alpha_N}\left(x_N\right)$$

$$= \sum_{n_1, n_2, \cdots n_\infty} f\left(n_1, n_2, \cdots, n_\infty, t\right) \left(\frac{n_1! n_2! \cdots n_\infty!}{N!}\right)^{1/2}$$

$$\times \sum_{P} \zeta^P \sum_{\alpha_{P1}, \alpha_{P2}, \cdots \alpha_{PN}} \psi_{\alpha_{P1}}\left(x_1\right) \psi_{\alpha_{P2}}\left(x_2\right) \cdots \psi_{\alpha_{PN}}\left(x_N\right)$$

$$= \sum_{n_1, n_2, \cdots n_\infty} f\left(n_1, n_2, \cdots, n_\infty, t\right) \Phi_{n_1, n_2, \cdots n_\infty}\left(x_1, x_2, \cdots x_N\right),$$
$$(1.91)$$

where we introduced a *symmetric* $(\zeta = 1)$ or an *antisymmetric* $(\zeta = -1)$ complete orthonormal basis function

$$\Phi_{n_1, n_2, \cdots n_\infty}\left(x_1, x_2, \cdots x_N\right) = \left(\frac{n_1! n_2! \cdots n_\infty!}{N!}\right)^{1/2}$$
$$\times \sum_{P} \zeta^P \sum_{\alpha_{P1}, \alpha_{P2}, \cdots \alpha_{PN}} \psi_{\alpha_{P1}}\left(x_1\right) \psi_{\alpha_{P2}}\left(x_2\right) \cdots \psi_{\alpha_{PN}}\left(x_N\right),$$
$$(1.92)$$

which is *independent of time*. Let us write, as an explicit example, a spinless three-boson $(\zeta = +1)$ wavefunction, which describes two particles occupying the ground state (denoted by the subscript 1) and the rest particle occupying the first excited state (denoted by the subscript 2):

$$\Phi_{210\cdots(n_\infty=0)}(x_1, x_2, x_3)$$
$$= \frac{1}{\sqrt{3}}[\psi_1(1)\psi_1(2)\psi_2(3) + \psi_1(1)\psi_2(2)\psi_1(3)$$
$$+ \psi_2(1)\psi_1(2)\psi_1(3)],$$

where we employ an abbreviation that (1) represents the coordinate x_1. Yet another example of a two-fermion $(\zeta = -1)$ wavefunction, which describes one particle occupying the ground state and the other occupying the first ground state, can be expressed as

$$\Phi_{110\cdots(n_\infty=0)}(x_1, x_2) = \frac{1}{\sqrt{2}}[\psi_1(1)\psi_2(2) - \psi_2(1)\psi_1(2)],$$

in which the single-particle wavefunctions form a *Slater determinant*[n] in their permutations due to the factor $(-1)^P$.

Great conveniences are achieved when the time-independent many-body wavefunctions $\Phi_{n_1 n_2 \cdots n_\infty}(x_1, x_1, \cdots x_\infty)$ are represented by the Dirac notation. We introduce a *time-independent occupation number state vector*

$$|n_1 n_2 \cdots n_\infty\rangle_\zeta,$$

which represents a physical state that the state α_1 is occupied by n_1 particles, the state α_2 is occupied by n_2 particles, etc. This occupation number state satisfies the orthogonality

$$\langle n_1' n_2' \cdots n_\infty' | n_1 n_2 \cdots n_\infty\rangle_\zeta = \delta_{n_1' n_1}\delta_{n_2' n_2}\cdots\delta_{n_\infty' n_\infty} \qquad (1.93a)$$

and the completeness

$$\sum_{n_1, n_2, \cdots n_\infty} |n_1 n_2 \cdots n_\infty\rangle\langle n_1 n_2 \cdots n_\infty|_\zeta = 1. \qquad (1.93b)$$

---

[n]J. C. Slater, *Phys. Rev.* **34**, 1293 (1929). One may write

$$\Phi_{n_1 n_2 \cdots n_\infty}(x_1, x_1, \cdots x_\infty) = \left(\frac{n_1! n_2! \cdots n_\infty!}{N!}\right)^{1/2} \begin{vmatrix} \psi_1(x_1) & \psi_1(x_2) & \cdots & \psi_1(x_N) \\ \psi_2(x_1) & \psi_2(x_2) & \cdots & \psi_2(x_N) \\ & & \vdots & \\ \psi_N(x_1) & \psi_N(x_2) & \cdots & \psi_N(x_N) \end{vmatrix}.$$

We consider that the occupation-number basis states are simply the direct product, or tensor product, of eigenstates of the number operator of each mode[o]

$$|n_1 n_2 \cdots n_\infty\rangle_\zeta = |n_1\rangle_\zeta |n_2\rangle_\zeta \cdots |n_\infty\rangle_\zeta , \qquad (1.94)$$

which forms the so-called *Fock space.*

The single-mode occupation number states $|n_k\rangle$ suggest us to extend the annihilation $\hat{c}_k$ and creation $\hat{c}_k^\dagger$ operators, which satisfy the commutation relations

$$[\hat{c}_k, \hat{c}_l]_{-\zeta} = 0, \qquad \left[\hat{c}_k^\dagger, \hat{c}_l^\dagger\right]_{-\zeta} = 0$$

$$\left[\hat{c}_k, \hat{c}_l^\dagger\right]_{-\zeta} = \delta_{kl}, \qquad (1.95)$$

where we introduced an extended commutator $\left[\hat{A}, \hat{B}\right]_{-\zeta} = \hat{A}\hat{B} - \zeta \hat{B}\hat{A}$. For $\zeta = +1$, the operators $\hat{c}_k = \hat{b}_k$ and $\hat{c}_k^\dagger = \hat{b}_k^\dagger$ are bosonic and their effects when applied to s single-mode state $|n_k\rangle$ are well studied, because they follow the same commutation rules of simple harmonic oscillator:

$$\hat{b}_k |n_k\rangle_+ = \sqrt{n_k} |n_k - 1\rangle_+$$

$$\hat{b}_k^\dagger |n_k\rangle_+ = \sqrt{n_k + 1} |n_k + 1\rangle_+ \qquad (1.96)$$

$$\hat{n}_{k,+} \left(= \hat{b}_k^\dagger \hat{b}_k\right) |n_k\rangle_+ = n_k |n_k\rangle_+ , \qquad n_k = 0, 1, 2, \cdots \infty.$$

The operations of the bosonic operators to a many-body state are simply extended as

$$\hat{b}_k^\dagger \hat{b}_l |n_1 n_2 \cdots n_\infty\rangle_+ = \sqrt{n_k + 1}\sqrt{n_l} |n_1 \cdots n_k + 1 \cdots n_l - 1 \cdots n_\infty\rangle_+.$$
$$(1.97)$$

On the other hand, for $\zeta = -1$, the properties of the fermionic operators $\hat{c}_k = \hat{a}_k$ and $\hat{c}_k^\dagger = \hat{a}_k^\dagger$ are rather different from that of the bosonic ones, due to the *anti*commutation rules

$$\hat{a}_k \hat{a}_l + \hat{a}_l \hat{a}_k = 0, \qquad \hat{a}_k^\dagger \hat{a}_l^\dagger + \hat{a}_l^\dagger \hat{a}_k^\dagger = 0,$$

$$\hat{a}_k \hat{a}_l^\dagger + \hat{a}_l^\dagger \hat{a}_k = \delta_{kl}. \qquad (1.98)$$

---

[o]Some references use the notation of tensor product:

$$|n_1 n_2 \cdots n_\infty|_\zeta = |n_1\rangle_\zeta \otimes |n_2\rangle_\zeta \otimes \cdots \otimes |n_\infty\rangle_\zeta .$$

First of all, they have the most important properties $\hat{a}_k \hat{a}_k = \hat{a}_k^\dagger \hat{a}_k^\dagger = 0$ so that $\hat{a}_k^\dagger \hat{a}_k^\dagger |0\rangle = 0$, which prevents two particles from occupying the same state k. Second, it is easy to show, by omitting the subscripts, that

$$
\begin{aligned}
\left(\hat{a}^\dagger \hat{a}\right)^2 &= 1 - 2\hat{a}\hat{a}^\dagger + \hat{a}\hat{a}^\dagger \hat{a}\hat{a}^\dagger = 1 - 2\hat{a}\hat{a}^\dagger + \hat{a}\left(1 - \hat{a}\hat{a}^\dagger\right)\hat{a}^\dagger \\
&= 1 - \hat{a}\hat{a}^\dagger = \hat{a}^\dagger \hat{a}
\end{aligned}
$$

In other words, the number operator for the kth mode $\hat{n}_{k,-} = \hat{a}_k^\dagger \hat{a}_k$ satisfies the condition that

$$
\hat{n}_{k,-}\left(1 - \hat{n}_{k,-}\right) = 0, \tag{1.99}
$$

which suggests that the number operator has the eigenvalues zero or one. Consequently, for a given state, the properties of the operators $\hat{a}_k$ and $\hat{a}_k^\dagger$ to the state $|n_k\rangle$ are

$$
\begin{aligned}
\hat{a}_k^\dagger |0_k\rangle &= |1_k\rangle, \quad \hat{a}_k |1_k\rangle = |0\rangle, \\
\hat{a}_k^\dagger |1_k\rangle &= 0, \quad\;\; \hat{a}_k |0_k\rangle = 0.
\end{aligned} \tag{1.100}
$$

We define a fermionic many-particle state in the occupation number representation by operating the creation operators to the vacuum state $|0\rangle$:

$$
|n_1 n_2 \cdots n_\infty\rangle_- = \left(\hat{a}_1^\dagger\right)^{n_1} \left(\hat{a}_2^\dagger\right)^{n_2} \cdots \left(\hat{a}_\infty^\dagger\right)^{n_\infty} |0\rangle. \tag{1.101}
$$

The effect of an annihilation operator $\hat{a}_k$ on this state is

$$
\begin{aligned}
\hat{a}_k |n_1 n_2 \cdots n_\infty\rangle_- &= \hat{a}_k \left(\hat{a}_1^\dagger\right)^{n_1} \left(\hat{a}_2^\dagger\right)^{n_2} \cdots \left(\hat{a}_\infty^\dagger\right)^{n_\infty} |0\rangle \\
&= (-1)\hat{a}_1^\dagger \hat{a}_k \left(\hat{a}_1^\dagger\right)^{n_1-1} \left(\hat{a}_2^\dagger\right)^{n_2} \cdots \left(\hat{a}_\infty^\dagger\right)^{n_\infty} |0\rangle \\
&\quad\vdots \\
&= (-1)^{n_1} \left(\hat{a}_1^\dagger\right)^{n_1} \hat{a}_k \left(\hat{a}_2^\dagger\right)^{n_2} \cdots \left(\hat{a}_\infty^\dagger\right)^{n_\infty} |0\rangle \\
&\quad\vdots \\
&= (-1)^{S_k} \left(\hat{a}_1^\dagger\right)^{n_1} \left(\hat{a}_2^\dagger\right)^{n_2} \cdots \left(\hat{a}_k \left(\hat{a}_k^\dagger\right)^{n_k}\right) \\
&\quad\cdots \left(\hat{a}_\infty^\dagger\right)^{n_\infty} |0\rangle,
\end{aligned}
$$

where $S_k = n_1 + n_2 + \cdots + n_{k-1}$. If $n_k = 0$, the operator $\hat{a}_k$ can permute with all the operators at the right without further payments for changing the sign until $\hat{a}_k \left| 0 \right\rangle$, which yields to zero. If $n_k = 1$, on the other hand, the anticommutation rule $\hat{a}_k \hat{a}_k^\dagger = 1 - \hat{a}_k^\dagger \hat{a}_k$ is applied first and then the second term $\hat{a}_k^\dagger \hat{a}_k$ to the vacuum yields to zero. The operations of the fermionic annihilation and creation operators to the many-body state in the occupation number representation are summarized as

$$\hat{a}_k \left| \cdots n_k \cdots \right\rangle_- = \begin{cases} (-1)^{S_k} \sqrt{n_k} \left| \cdots n_k - 1 \cdots \right\rangle_-, & \text{if } n_k = 1 \\ 0, & \text{if } n_k = 0 \end{cases}$$

$$\hat{a}_k^\dagger \left| \cdots n_k \cdots \right\rangle_- = \begin{cases} 0, & \text{if } n_k = 1 \\ (-1)^{S_k} \sqrt{n_k + 1} \left| \cdots n_k + 1 \cdots \right\rangle_-, & \text{if } n_k = 0 \end{cases}$$

$$\hat{n}_{k,-} \left(= \hat{a}_k^\dagger \hat{a}_k \right) \left| \cdots n_k \cdots \right\rangle_- = n_k \left| \cdots n_k \cdots \right\rangle_-, \quad n_k = 0, 1.$$

$$(1.102)$$

Armed with the occupation number representation of many-particle states combined with the single-mode annihilation and creation operators, one defines the *field operators*

$$\hat{\psi} \left(\mathbf{r} \right) \equiv \sum_k \psi_k \left(\mathbf{r} \right) \hat{c}_k,$$

$$\hat{\psi}^\dagger \left(\mathbf{r} \right) \equiv \sum_k \psi_k^\dagger \left(\mathbf{r} \right) \hat{c}_k^\dagger,$$

$$(1.103)$$

where the coefficients are the single-particle wavefunctions at states $\left| \alpha_k \right\rangle$ and the sum is over the complete set of single-particle quantum numbers. It is convenient to split the spinor quantum numbers, for example spin index α, by writing the wavefunctions having two components

$$\psi_k \left(\mathbf{r} \right) = \begin{bmatrix} \psi_k \left(\mathbf{r} \right)_1 \\ \psi_k \left(\mathbf{r} \right)_2 \end{bmatrix} \equiv \psi_k \left(\mathbf{r} \right)_\alpha, \quad \alpha = 1, 2,$$

so we may write the index field operators $\hat{\psi}_\alpha \left(\mathbf{r} \right)$ and $\hat{\psi}_\alpha^\dagger \left(\mathbf{r} \right)$. The field operators satisfy the following quantization conditions:

$$\left[\hat{\psi}_\alpha \left(\mathbf{r} \right), \hat{\psi}_\beta \left(\mathbf{r}' \right) \right]_{-\zeta} = 0, \quad \left[\hat{\psi}_\alpha^\dagger \left(\mathbf{r} \right), \hat{\psi}_\beta^\dagger \left(\mathbf{r}' \right) \right]_{-\zeta} = 0,$$

$$\left[\hat{\psi}_\alpha \left(\mathbf{r} \right), \hat{\psi}_\beta^\dagger \left(\mathbf{r}' \right) \right]_{-\zeta} = \delta_{\alpha\beta} \delta \left(\mathbf{r} - \mathbf{r}' \right).$$

$$(1.104)$$

Equation (1.104) is another form of quantization of dynamical variables, the fields, and it is known as the *second quantization* rules, or the *field quantization* rules, distinguished from the first quantization rules Eq. (1.2), which quantize the position and momentum operators.[p] One of the advantages of the second quantization in many-body quantum theory is that the position vector is treated as a parameter, not as an operator in the first quantization, like the time in the first quantization language. This allows us to treat the quantum mechanical problems of many-body system on the space and time equally.

It would be helpful if the readers verify the following operators in the second quantization language: The Hamiltonian operator can be represented as

$$
\hat{H} = \int d\mathbf{r} \, \hat{\psi}^\dagger(\mathbf{r}) \, \hat{T}(\mathbf{r}) \, \hat{\psi}(\mathbf{r})
$$
$$
+ \frac{1}{2} \iint d\mathbf{r} d\mathbf{r}' \, \hat{\psi}^\dagger(\mathbf{r}) \, \hat{\psi}^\dagger(\mathbf{r}') \, V(\mathbf{r}, \mathbf{r}') \, \hat{\psi}(\mathbf{r}') \, \hat{\psi}(\mathbf{r}')
$$
(1.105)

and the total-number operator

$$
\hat{N} = \int d\mathbf{r} \, \hat{n}(\mathbf{r}) = \int d\mathbf{r} \, \hat{\psi}^\dagger(\mathbf{r}) \, \hat{\psi}(\mathbf{r}).
$$
(1.106)

It is noticeable that the number operator $\hat{N}$ commutes with the Hamiltonian operator Eq. (1.105), which is physically commensurate with the fact that the ordinary Schrödinger Hamiltonian does not change the total number of particles. We infer that $\hat{N}$ is a constant of motion and can be diagonalized simultaneously with the Hamiltonian.

---

[p]Here we do not discuss the quantizations of the other conjugate variables energy and time.

Chapter 2

Mathematical Introduction

2.1 Basic Definitions

The properties of a quantum mechanical system composed of many identical particles are most conveniently described in terms of the second-quantized, Heisenberg representation, particle-creation, and annihilation operators. The creation operator $\hat{\psi}^\dagger(\mathbf{r}, t)$, when acting to the right on a state of the system, adds a particle to the state at the space–time point $\mathbf{r}$, t; the annihilation operator $\hat{\psi}(\mathbf{r}, t)$, the adjoint of the creation operator, acting to the right, removes a particle from the state at the point $\mathbf{r}$, t.

The macroscopic operators of direct physical interest can all be expressed in terms of products of a few ψ's and $\psi^\dagger$'s. For example, the density of particles at the point $\mathbf{r}$, t is

$$\hat{n}(\mathbf{r}, t) = \hat{\psi}^\dagger(\mathbf{r}, t)\,\hat{\psi}(\mathbf{r}, t) \qquad (2.1a)$$

Since the act of removing and then immediately replacing a particle at $\mathbf{r}$, t measures the density of particles at that point, the operator for the total number of particles is

$$\hat{N}(t) = \int d\mathbf{r}\, \hat{\psi}^\dagger(\mathbf{r}, t)\,\hat{\psi}(\mathbf{r}, t) \qquad (2.1b)$$

Similarly, the total energy of a system of particles of mass m interacting through an instantaneous two-body potential $v(r)$ is

Annotations to Quantum Statistical Mechanics
In-Gee Kim
Copyright © 2018 Pan Stanford Publishing Pte. Ltd.
ISBN 978-981-4774-15-4 (Hardcover), 978-1-315-19659-6 (eBook)
www.panstanford.com

given by

$$\hat{H}(t) = \int d\mathbf{r} \frac{\nabla \hat{\psi}^\dagger(\mathbf{r}, t) \cdot \nabla \hat{\psi}(\mathbf{r}, t)}{2m} \tag{2.2}$$

$$+ \frac{1}{2} \int d\mathbf{r} d\mathbf{r}' \hat{\psi}^\dagger(\mathbf{r}, t) \hat{\psi}^\dagger(\mathbf{r}', t) v(|\mathbf{r} - \mathbf{r}'|) \hat{\psi}(\mathbf{r}', t) \hat{\psi}(\mathbf{r}, t)$$

In general, we shall take $\hbar = 1$.

The equation of any operator $\hat{X}(t)$ in the Heisenberg representation is

$$i \frac{\partial \hat{X}(t)}{\partial t} = \left[\hat{X}(t), \hat{H}(t)\right] \tag{2.3}$$

Since $\left[\hat{H}(t), \hat{H}(t)\right] = 0$, we see that the Hamiltonian is independent of time. Also the Hamiltonian does not change the number of particles, $\left[\hat{H}, \hat{N}(t)\right] = 0$; therefore, $\hat{N}(t)$ is also independent of time. Because of the time independence of H, (2.3) may be integrated in the form

$$\hat{X}(t) = e^{i\hat{H}t} \hat{X}(0) e^{-i\hat{H}t} \tag{2.4}$$

Particles may be classified into one of two types: Fermi–Dirac particles, also called fermions, which obey the exclusion principle, and Bose–Einstein particles, or bosons, which do not. The wavefunction of any state of a collection of bosons must be a symmetric function of the coordinates of the particles, whereas, for fermions, the wavefunction must be antisymmetric. One of the main advantages of the second-quantization formalism is that these symmetry requirements are very simply represented in the equal-time commutation relations of the creation and annihilation operators. These commutation relations are

$$\hat{\psi}(\mathbf{r}, t) \hat{\psi}(\mathbf{r}', t) \mp \hat{\psi}(\mathbf{r}', t) \hat{\psi}(\mathbf{r}, t) = 0$$
$$\hat{\psi}^\dagger(\mathbf{r}, t) \hat{\psi}^\dagger(\mathbf{r}', t) \mp \hat{\psi}^\dagger(\mathbf{r}', t) \hat{\psi}^\dagger(\mathbf{r}, t) = 0 \tag{2.5}$$
$$\hat{\psi}(\mathbf{r}, t) \hat{\psi}^\dagger(\mathbf{r}', t) \mp \hat{\psi}^\dagger(\mathbf{r}', t) \hat{\psi}(\mathbf{r}, t) = \delta(\mathbf{r} - \mathbf{r}')$$

where the upper sign refers to Bose–Einstein particles and the lower sign refers to Fermi–Dirac particles. We see, for fermions, that $\hat{\psi}^2(\mathbf{r}, t) = 0$. This is an expression of the exclusion principle in space—it is impossible to find two identical fermions at the same point in space and time.

We shall be interested in describing the behavior of many-particle systems at finite temperature. For a system in thermodynamic equilibrium, the expectation value of any operator $\hat{X}$ may be computed by using the grand-canonical ensemble of statistical mechanics. Thus

$$\langle \hat{X} \rangle = \frac{\sum_i \langle i | \hat{X} | i \rangle \, e^{-\beta(E_i - \mu N_i)}}{\sum_i e^{-\beta(E_i - \mu N_i)}} \qquad (2.6a)$$

Here $|i\rangle$ represents a state of the system, normalized to unity, with energy E_i and number of particles N_i. The sum runs over all states of the system with all possible numbers of particles. A more compact way of writing the average (2.6a) is

$$\langle \hat{X} \rangle = \frac{\operatorname{tr}\left[e^{-\beta(\hat{H} - \mu \hat{N})} \hat{X} \right]}{\operatorname{tr}\left[e^{-\beta(\hat{H} - \mu \hat{N})} \right]} \qquad (2.6b)$$

where tr denotes the trace.

The thermodynamic state of the system is now defined by the parameters μ, the chemical potential, and β, the inverse temperature measured in energy units, i.e., $\beta = 1/k_B T$, where k_B is Boltzmann's constant. Zero temperature, or $\beta \to \infty$, describes the ground state of the system.

Green's functions, which shall form the base of our discussion of many-particle systems, are thermodynamic averages of product of the operators $\hat{\psi}$ (1) and $\hat{\psi}$ (1'). (We use the abbreviated notation 1 to mean $\mathbf{r}_1 t_1$ and 1' to mean $\mathbf{r}_{1'} t_{1'}$, etc.) The one-particle Green's function is defined by

$$G\left(1, 1'\right) = \frac{1}{i} \langle \hat{T} \left(\hat{\psi} \left(1\right) \hat{\psi}^\dagger \left(1'\right) \right) \rangle \qquad (2.7a)$$

while the two-particle Green's function is defined by

$$G_2 \left(12, 1'2'\right) = \frac{1}{i^2} \langle \hat{T} \left(\hat{\psi} \left(1\right) \hat{\psi} \left(2\right) \hat{\psi}^\dagger \left(2'\right) \hat{\psi}^\dagger \left(1'\right) \right) \rangle \qquad (2.7b)$$

In these Green's functions, $\hat{T}$ represents the Wick time-ordering operation. When applied to a product of operators, it arranges them in chronological order with the earliest time appearing on the right and the latest on the left. For bosons, this is the full effect of $\hat{T}$. For fermions, however, it is convenient to define $\hat{T}$ to include an extra factor ± 1, depending on whether the resulting time-ordered product is an even or odd permutation of the original order. Thus, for example,

$$\hat{T}\left(\hat{\psi}\,(1)\,\hat{\psi}^{\dagger}\,(1')\right) = \begin{cases} \hat{\psi}\,(1)\,\hat{\psi}^{\dagger}\,(1') & \text{for} \quad t_1 > t_{1'} \\ \pm\hat{\psi}^{\dagger}\,(1')\,\hat{\psi}\,(1) & \text{for} \quad t_1 < t_{1'} \end{cases}$$

As in (2.5), the upper sign refers to bosons and the lower for fermions. We shall use this sign convention throughout these lectures.

The one-particle Green's function $G\,(1, 1')$ has a direct physical interpretation. It describes the propagation of disturbances in which a single particle is either added or removed from the many-particle equilibrium system. For example, when $t_1 > t_{1'}$, the creation operator acts first, producing a disturbance by adding a particle at the space–time point $\mathbf{r}_{1'}t_{1'}$. This disturbance then propagates to the later time t_1, when a particle is removed at $\mathbf{r}_1$ ending the disturbance and returning the system to its equilibrium state.[a] For $t_1 < t_{1'}$, $\hat{\psi}$ acts first. The disturbance, which is now produced by the removal of a particle at $\mathbf{r}_1 t_1$, propagates to time $t_{1'}$, when it is terminated by the addition of a particle at the point $\mathbf{r}_{1'}$.

Similarly, the two-particle Green's function describes, for the various time orders, disturbances produced by the removal or addition of two particles. For example, when t_1 and t_2 are both later than $t_{1'}$ and $t_{2'}$, $G_2\,(12, 1'2')$ describes the addition of two particles and the subsequent removal of two particles. Yet when t_1 and $t_{1'}$ are later than t_2 and $t_{2'}$, the two-particle Green's function describes the disturbance produced by the addition of one particle and the removal of one particle, and the subsequent return to equilibrium by the removal of a particle and the addition of a particle. We shall make extensive use of this physical interpretation of Green's functions.

In addition to the one-particle Green's function, we define the correlation functions

$$G^{>}\left(1, 1'\right) = \frac{1}{i}\left\langle \hat{\psi}\,(1)\,\hat{\psi}^{\dagger}\,(1')\right\rangle$$

$$G^{<}\left(1, 1'\right) = \pm\frac{1}{i}\left\langle \hat{\psi}^{\dagger}\,(1')\,\hat{\psi}\,(1)\right\rangle \tag{2.8}$$

The notations $>$ and $<$ are intended as a reminder that for $t_1 > t_{1'}$, $G = G^{>}$, while for $t_1 < t_{1'}$, $G = G^{<}$.

---

[a] Here the typographic errors at the subscripts appeared in $t_{1'}$ and $\mathbf{r}_{1'}$ in the original text are corrected.

2.2 The Boundary Condition

The time-development operator $e^{-it\hat{H}}$ bears a strong formal similarity to the weighting factor $e^{\beta\hat{H}}$ that occurs in the grand-canonical average. Indeed for $t = -i\beta$, the two are the same. We can exploit this mathematical similarity to discover identities obeyed by Green's functions. In particular, we shall now derive a fundamental relation between $G^>$ and $G^<$.

Our argument is based on the fact that the time dependence of $\hat{\psi}$ and $\hat{\psi}^\dagger$, given by (2.4), may be used to define the creation and annihilation operators and, therefore, $G^>$ and $G^<$, for complex values of their time arguments. In fact, the function $G^>$, which we may write as

$$G^> (1, 1') = \left(\frac{1}{i}\right) \frac{\text{tr}\left[e^{-\beta(\hat{H}-\mu\hat{N})}e^{it_1\hat{H}}\,\hat{\psi}\,(\mathbf{r}_1, 0)\,e^{-i(t_1-t_{1'})\hat{H}}\,\hat{\psi}^\dagger\,(\mathbf{r}_{1'}, 0)\,e^{-it_{1'}\hat{H}}\right]}{\text{tr}\left[e^{-\beta(\hat{H}-\mu\hat{N})}\right]}$$

is an analytic function for complex values of the time arguments in the region $0 > \Im (t_1 - t_{1'}) > -\beta$. This analyticity follows directly from the assumption that the $e^{-\beta(\hat{H}-\mu\hat{N})}$ factor is sufficient to guarantee the absolute convergence of the trace for real time. Similarly, $G^< (1, 1')$ is an analytic function in the region $0 < \Im (t_1 - t_{1'}) < \beta$.

To derive the relation between $G^>$ and $G^<$, we notice that the expression

$$G^< (1, 1')\big|_{t_1=0} = \left(\pm\frac{1}{i}\right) \frac{\text{tr}\left[e^{-\beta(\hat{H}-\mu\hat{N})}\hat{\psi}^\dagger\,(\mathbf{r}_{1'}, t_{1'})\,\hat{\psi}\,(\mathbf{r}_1, 0)\right]}{\text{tr}\left[e^{-\beta(\hat{H}-\mu\hat{N})}\right]}$$

may be rearranged, using the cyclic invariance of the trace ($\text{tr}\,\hat{A}\,\hat{B} = \text{tr}\,\hat{B}\,\hat{A}$), to become

$$G^< (1, 1')\big|_{t_1=0}$$

$$= \pm\frac{1}{i} \frac{\text{tr}\left\{e^{-\beta(\hat{H}-\mu\hat{N})}\left[e^{\beta(\hat{H}-\mu\hat{N})}\hat{\psi}\,(\mathbf{r}_1, 0)\,e^{-\beta(\hat{H}-\mu\hat{N})}\hat{\psi}^\dagger\,(\mathbf{r}_{1'}, t_{1'})\right]\right\}}{\text{tr}\left[e^{-\beta(\hat{H}-\mu\hat{N})}\right]}$$

$$= \pm\left(\frac{1}{i}\right) \left\langle e^{\beta(\hat{H}-\mu\hat{N})}\hat{\psi}\,(\mathbf{r}_1, 0)\,e^{-\beta(\hat{H}-\mu\hat{N})}\hat{\psi}^\dagger\,(\mathbf{r}_{1'}, t_{1'})\right\rangle$$

Because $\hat{\psi}\,(\mathbf{r}_1, 0)$ removes a particle, we have

$$\hat{\psi}\,(\mathbf{r}_1, 0)\,\mathfrak{f}\,(\hat{N}) = \mathfrak{f}\,(\hat{N} + 1)\,\hat{\psi}\,(\mathbf{r}_1, 0)$$

where $f\left(\hat{N}\right)$ is any function of the number operator $\hat{N}$. In particular,

$$e^{-\beta\mu\hat{N}}\,\hat{\psi}\,(\mathbf{r}_1,\,0)\,e^{\beta\mu\hat{N}} = e^{\beta\mu}\,\hat{\psi}\,(\mathbf{r}_1,\,0)$$

and from (2.7a) it follows that

$$e^{\beta\hat{H}}\,\hat{\psi}\,(\mathbf{r}_1,\,0)\,e^{-\beta\hat{H}} = \hat{\psi}\,(\mathbf{r}_1,\,-i\beta)$$

Thus,

$$
\begin{aligned}
G^<\left(1,\,1'\right)\big|_{t_1=0} &= \pm\left(\frac{1}{i}\right)\langle\hat{\psi}\,(\mathbf{r}_1,\,-i\beta)\,\hat{\psi}^{\,\dagger}\left(1'\right)\rangle\,e^{\beta\mu}\\
&= \pm e^{\beta\mu}\,G^>\left(1,\,1'\right)\big|_{t_1=-i\beta}
\end{aligned}
\tag{2.9}
$$

This relationship is crucial to all our Green's function analysis.

Notice that Eq. (2.9) follows directly from the cyclic invariance of the trace and the structure of the time dependence of $\hat{\psi}\,(1)$. Since G_2 is also defined as a trace, we can go through an entirely similar analysis for it, splitting it into several non-time-ordered expectation values of $\hat{\psi}$'s and $\hat{\psi}^{\dagger}$'s and proving a set of relations similar to Eq. (2.9). However, this analysis is much too complicated because G_2 is composed of too many different analytic pieces, corresponding to all the different possible time orderings of its four times variables.

We employ the following simple device to exhibit a relation like Eq. (2.9) for G_2. We consider the time variable to be restricted to the interval

$$0 \le it \le \beta$$

Equation (2.4) defines the field operators and, therefore, Green's functions for imaginary times. To complete the definition of Green's functions in this time domain, we extend the definition of the time-ordering symbol $\hat{T}$ to mean "$i \times t$" ordering when the times are imaginary. The further down the imaginary axis a time is, the "later" it is. Then Green's functions are well defined in the interval $0 \le it \le \beta$. For example, the one-particle Green's function is

$$
G\left(1,\,1'\right) = \begin{cases} G^>\left(1,\,1'\right) & \text{for} \quad it_1 > it_{1'}\\ G^<\left(1,\,1'\right) & \text{for} \quad it_1 < it_{1'} \end{cases}
$$

For $0 < it_{1'} < \beta$, we have

$$G\left(1,\,1'\right)\big|_{t_1=0} = G^<\left(1,\,1'\right)\big|_{t_1=0} \quad (\text{since } 0 = it_1 < it_{1'} \text{ for all } t_{1'})$$

and

$$G\left(1, 1'\right)\big|_{t_1=-i\beta} = G^>\left(1, 1'\right)\big|_{t_1=-i\beta} \quad \text{(since } \beta = it_1 > it_{1'} \text{ for all } t_{1'})$$

Therefore, Eq. (2.9) can be restated as a relation between the values of $G\left(1, 1'\right)$ at the boundaries of the imaginary time domain:

$$G\left(1, 1'\right)\big|_{t_1=0} = \pm e^{\beta\mu}\, G\left(1, 1'\right)\big|_{t_1=-i\beta} \tag{2.10}$$

Moreover, we can see immediately that G_2 on the imaginary time axis obeys exactly this same boundary condition.

$$G_2\left(12, 1'2'\right)\big|_{t_1=0} = \pm e^{\beta\mu}\, G\left(12, 1'2'\right)\big|_{t_1=-i\beta} \tag{2.11a}$$

and also

$$G_2\left(12, 1'2'\right)\big|_{t_{1'}=0} = \pm e^{-\beta\mu}\, G\left(12, 1'2'\right)\big|_{t_{1'}=-i\beta} \tag{2.11b}$$

These boundary conditions on G and G_2 will be used over and over again in the subsequent analysis.

It is only at a later stage that we shall need the imaginary-time Green's functions. Now we shall restrict our attention to the one-particle function, for which Eq. (2.9) is a suitable representation of the boundary condition.

Because of the translational and rotational invariance of the Hamiltonian Eq. (2.2) in space and its translational invariance in time, $G^>$ and $G^<$ depend only on $|\mathbf{r}_1 - \mathbf{r}_{1'}|$ and $t_1 - t_{1'}$. When we want to emphasize that these functions depend only on the difference variables, we shall write them as $G^{>(<)}\left(1 - 1'\right)$ or as $G^{>(<)}\left(|\mathbf{r}_1 - \mathbf{r}_{1'}|, t_1 - t_{1'}\right)$. In terms of the difference variables, Eq. (2.9) is

$$G^<\left(\mathbf{r}, t\right) = \pm e^{i\beta}G^>\left(\mathbf{r}, t - i\beta\right)$$

We now introduce the Fourier transformations of $G^>$ and $G^<$, defined by

$$G^>\left(\mathbf{p}, \omega\right) = i \int d\mathbf{r} \int_{-\infty}^{\infty} dt\, e^{-i\mathbf{p}\cdot\mathbf{r}+i\omega t}G^>\left(\mathbf{r}, t\right)$$
$$G^<\left(\mathbf{p}, \omega\right) = \pm i \int d\mathbf{r} \int_{-\infty}^{\infty} dt\, e^{-i\mathbf{p}\cdot\mathbf{r}+i\omega t}G^<\left(\mathbf{r}, t\right) \tag{2.12}$$

Note the explicit factors of i and $\pm i$ that we have included here to make $G^>\left(\mathbf{p}, \omega\right)$ and $G^<\left(\mathbf{p}, \omega\right)$ real nonnegative quantities. Equation (2.9) then becomes the simpler relationship

$$G^<\left(\mathbf{p}, \omega\right) = e^{-\beta(\omega-\mu)}G^>\left(\mathbf{p}, \omega\right) \tag{2.13}$$

It is useful to introduce the "spectral function" $A(\mathbf{p}, \omega)$ defined by

$$A(\mathbf{p}, \omega) = G^{>}(\mathbf{p}, \omega) \mp G^{<}(\mathbf{p}, \omega) \tag{2.14}$$

The boundary condition on G can then be represented by writing

$$G^{>}(\mathbf{p}, \omega) = [1 \pm f(\omega)] A(\mathbf{p}, \omega)$$
$$G^{<}(\mathbf{p}, \omega) = f(\omega) A(\mathbf{p}, \omega) \tag{2.15}$$

where

$$f(\omega) = \frac{1}{e^{\beta(\omega-\mu)} \mp 1} \tag{2.16}$$

The term f can be recognized as the average occupation number in the grand-canonical ensemble of a mode with energy ω.

[The statement is, more precisely, that if the Hamiltonian can be diagonalized to the form $\sum_\lambda \epsilon_\lambda \hat{\psi}_\lambda^\dagger \hat{\psi}_\lambda$, then $\hat{\psi}_\lambda^\dagger$ is a creation operator for a mode of the system with energy ϵ_λ. The average occupation number of the mode λ is $\langle \hat{\psi}_\lambda^\dagger \hat{\psi}_\lambda \rangle = f(\epsilon_\lambda)$.]

From the definitions of $G^{>}$ and $G^{<}$, it follows that

$$A(\mathbf{p}, \omega) = \int d\mathbf{r} \int_{-\infty}^{\infty} dt\, e^{-i\mathbf{p}\cdot\mathbf{r}+i\omega t} \left\langle \left[\hat{\psi}(\mathbf{r}, t) \hat{\psi}^\dagger(\mathbf{0}, 0) \mp \hat{\psi}^\dagger(\mathbf{0}, 0) \hat{\psi}(\mathbf{r}, t)\right] \right\rangle$$

Thus, as a consequence of the equal-time commutation relation Eq. (2.5), A satisfies the sum rule

$$\int \frac{d\omega}{2\pi} A(\mathbf{p}, \omega) = \int d\mathbf{r} e^{-i\mathbf{p}\cdot\mathbf{r}} \left\langle \left[\hat{\psi}(\mathbf{r}, 0) \hat{\psi}^\dagger(\mathbf{0}, 0) \mp \hat{\psi}^\dagger(\mathbf{0}, 0) \hat{\psi}(\mathbf{r}, 0)\right] \right\rangle$$
$$= \int d\mathbf{r}\delta(\mathbf{r}) = 1 \tag{2.17}$$

We can use the relations that we have just derived to find G for the trivial case of free particles, for which the Hamiltonian is

$$\hat{H}_0 = \int d\mathbf{r} \frac{\nabla \hat{\psi}^\dagger(\mathbf{r}, t) \cdot \nabla \hat{\psi}(\mathbf{r}, t)}{2m}$$

We notice that

$$G^{<}(\mathbf{p}, \omega) = \int dt \frac{e^{i\omega t}}{\Omega} \left\langle \hat{\psi}^\dagger(\mathbf{p}, 0) \hat{\psi}(\mathbf{p}, t) \right\rangle$$

where Ω is the volume of the system and $\hat{\psi}(\mathbf{p}, t)$ is the spatial Fourier transform of $\hat{\psi}(\mathbf{r}, t)$. Since $\hat{\psi}(\mathbf{p}, 0)$ removes a free particle with momentum $\mathbf{p}$, it must remove energy $p^2/2m$ from the system. Thus,

$$\hat{\psi}(\mathbf{p}, t) = e^{i\hat{H}t} \hat{\psi}(\mathbf{p}, 0) e^{-i\hat{H}t} = e^{-i\left(\frac{p^2}{2m}\right)t} \hat{\psi}(\mathbf{p}, 0)$$

so that

$$G^< (\mathbf{p}, \omega) = \left(\frac{2\pi}{\Omega} \right) \delta \left(\omega - \frac{p^2}{2m} \right) \langle \hat{\psi}^\dagger (\mathbf{p}, 0) \psi (\mathbf{p}, 0) \rangle$$

Hence, $A (\mathbf{p}, \omega)$ is proportional to $\delta \left(\omega - \frac{p^2}{2m} \right)$, and the constant of proportionality is determined from the sum rule Eq. (2.17) to be 2π. Thus, for free particles,

$$A (\mathbf{p}, \omega) = A_0 (\mathbf{p}, \omega) = 2\pi \delta \left(\omega - \frac{p^2}{2m} \right) \qquad (2.18)$$

$$G_0^> (\mathbf{r}, t) = \int \frac{d\mathbf{p}}{(2\pi)^3} e^{i \mathbf{p} \cdot \mathbf{r} - i \left(\frac{p^2}{2m} \right) t} \left(\frac{1 \pm f \left(\frac{p^2}{2m} \right)}{i} \right)$$

$$G_0^< (\mathbf{r}, t) = \int \frac{d\mathbf{p}}{(2\pi)^3} e^{i \mathbf{p} \cdot \mathbf{r} - i \left(\frac{p^2}{2m} \right) t} \left(\frac{f \left(\frac{p^2}{2m} \right)}{i} \right) \qquad (2.19)$$

Since $\hat{\psi}^\dagger (\mathbf{p}, 0) \hat{\psi} (\mathbf{p}, 0)$ is the operator representing the density of particles with momentum $\mathbf{p}$, it follows that for free particles, the average number of particles with momentum $\mathbf{p}$ is

$$\langle n (\mathbf{p}) \rangle = \frac{\langle \hat{\psi}^\dagger (\mathbf{p}, 0) \hat{\psi} (\mathbf{p}, 0) \rangle}{\Omega} = f \left(\frac{p^2}{2m} \right)$$

$$= \frac{1}{e^{\beta \left(\frac{p^2}{2m} - \mu \right)} \mp 1} \qquad (2.20)$$

This is a result familiar from elementary statistical mechanics.

Chapter 3

Information Contained in $G^>$ and $G^<$

3.1 Dynamical Information

Now that we have set down the preliminaries, we shall try to gain some insight into $G^>$ and $G^<$.

The Fourier transformation of the field operator $\hat{\psi}\,(\mathbf{r},\,t)$, given by

$$\hat{\psi}\,(\mathbf{p},\,\omega) = \int d\mathbf{r} \int dt\, e^{-i\mathbf{p}\cdot\mathbf{r}+i\omega t}\hat{\psi}\,(\mathbf{r},\,t)$$

is an operator that annihilates a particle with momentum $\mathbf{p}$ and energy ω. Thus, $G^<\,(\mathbf{p},\,\omega)$ can be identified as the average density of particles in the system with momentum $\mathbf{p}$ and energy ω:

$$G^<\,(\mathbf{p},\,\omega) = \langle n\,(\mathbf{p},\,\omega)\rangle = A\,(\mathbf{p},\,\omega)\,f\,(\omega) \tag{3.1}$$

The interpretation of this result is evident. As we have pointed out, $f\,(\omega)$ is the average occupation number of a mode with energy ω; the spectral function $A\,(\mathbf{p},\,\omega)$ is a weighting function with total weight unity, which, whenever it is nonzero, defines the spectrum of possible energies ω, for a particle with momentum $\mathbf{p}$ in the medium.

To check this result, we may note that the density of particles,

$$\langle n\,(\mathbf{r},\,t)\rangle = \left\langle \hat{\psi}^\dagger\,(\mathbf{r},\,t)\,\hat{\psi}\,(\mathbf{r},\,t)\right\rangle = \pm iG^<\,(\mathbf{r}t,\,\mathbf{r}t)$$

$$= \int \frac{d\omega}{2\pi}\,\frac{d\mathbf{p}}{(2\pi)^3}\,G^<\,(\mathbf{p},\,\omega) \tag{3.2}$$

Annotations to Quantum Statistical Mechanics
In-Gee Kim

This says that the total density of particles is equal to the integral over all $\mathbf{p}$ and ω of the density of particles with momentum $\mathbf{p}$ and energy ω. Since $\langle n(\mathbf{r}, t)\rangle$ is independent of $\mathbf{r}$ and t, we shall represent it simply by the symbol n.

As an example, for a system of free particles,

$$A_0(\mathbf{p}, \omega) = 2\pi\delta\left(\omega - \frac{p^2}{2m}\right)$$

Hence, $A_0(\mathbf{p}, \omega)$ is non-vanishing only when $\omega = \frac{p^2}{2m}$. This says that the only possible energy value for a free particle with momentum $\mathbf{p}$ is $\frac{p^2}{2m}$. The total density of particles with momentum $\mathbf{p}$ is

$$\langle n(\mathbf{p})\rangle = \int\frac{d\omega}{2\pi}\langle n(\mathbf{p}, \omega)\rangle = f\left(\frac{p^2}{2m}\right) = \frac{1}{e^{\beta\left(\frac{p^2}{2m}-\mu\right)} \mp 1} \tag{3.3}$$

To see what happens in the classical limit, we explicitly write the factors of $\hbar$ in the expression of the density:

$$n = \int\frac{d\mathbf{p}}{(2\pi\hbar)^3}\frac{1}{e^{\beta\left(\frac{p^2}{2m}-\mu\right)} \mp 1} \tag{3.4}$$

In order that at a fixed temperature, the density does not diverge as $\hbar \to 0$, the factor $e^{-\beta\mu}$ must become very large. Thus, the classical limit is given by $\beta\mu \to -\infty$. We may then neglect the ∓ 1 in the denominator of Eq. (3.4), so that the momentum distribution becomes the familiar Maxwell–Boltzmann distribution

$$\langle n(\mathbf{p})\rangle = (\text{const})\, e^{-\beta\left(\frac{p^2}{2m}\right)}$$

Equation (3.4) indicates that $\beta\mu \to -\infty$ is also the low-density limit.

On the other hand, for a highly degenerate (i.e., high-density) Fermi gas, $\beta\mu$ becomes very large and positive. Defining the Fermi momentum[a] p_F by $\mu = \frac{p_F^2}{2m}$, we find

$$\langle n(\mathbf{p})\rangle \simeq \begin{cases} 0 & \text{for} \quad p > p_F \\ 1 & \text{for} \quad p < p_F \end{cases}$$

All states with momentum $p < p_F$ are filled, and all states with $p > p_F$ are empty.

---

[a]The symbol for the Fermi momentum was originally p_f instead of p_F.

For a Bose system, μ cannot become positive, but instead it approaches zero as the density increases. Then the total density of particles with nonzero momentum cannot become arbitrarily large, but it is instead limited by

$$\int \frac{d\mathbf{p}}{(2\pi)^3} \frac{1}{e^{\beta\left(\frac{p^2}{2m}\right)} - 1} = \frac{1}{2\pi^2} \left(\frac{2m}{\beta}\right)^{\frac{3}{2}} \int_0^\infty \frac{x^2}{e^{x^2} - 1} dx$$

In order to reach a higher density, the system puts a macroscopic number of particles into the mode $p = 0$. The mathematical possibility of this occurrence is the fact that $\mu = 0$, $f(0) = \infty$. This phenomenon, called the Bose–Einstein condensation, is reflected in the physical world as the phase transition of He^4 to the superfluid state.

When there is an interaction between the particles, $A(\mathbf{p}, \omega)$ will not be a single delta function. To see the detailed structure of A, let us compute $G^>(\mathbf{p}, \omega)$ by explicitly introducing sums over states. Then $G^>(\mathbf{p}, \omega)$ is

$$G^>(\mathbf{p}, \omega) = A(\mathbf{p}, \omega)\,[1 \pm f(\omega)]$$

$$= \int_{-\infty}^\infty dt \frac{e^{i\omega t}}{\Omega} \sum_i e^{-\beta(E_i - \mu N_i)} \frac{\langle i | \hat{\psi}(\mathbf{p})\, e^{-i\hat{H}t} \hat{\psi}^\dagger(\mathbf{p}) | i \rangle}{\mathrm{tr}\left[e^{-\beta(\hat{H} - \mu\hat{N})}\right]}$$

$$= \frac{1}{\Omega} \sum_{i,j} e^{-\beta(E_i - \mu N_i)} \left| \langle i | \hat{\psi}^\dagger(\mathbf{p}) | j \rangle \right|^2 \frac{2\pi\delta(\omega + E_i - E_j)}{\mathrm{tr}\left[e^{-\beta(\hat{H} - \mu\hat{N})}\right]}$$

$$(3.5)$$

It is clear then that the values of ω for which $A(\mathbf{p}, \omega)$ is non-vanishing are just the possible energy differences that result from adding a single particle of momentum $\mathbf{p}$ to the system. Almost always the energy spectrum of the system is sufficiently complex so that $A(\mathbf{p}, \omega)$ finally appears to have no delta functions in it but is instead a continuous function of ω. However, there are often sharp peaks in A. These sharp peaks represent coherent and long-lived excitations, which behave in many ways like free or weakly interacting particles. These excitations are usually called quasi-particles.

We can notice from Eq. (3.5) that $G^>(\mathbf{p}, \omega)$ is proportional to the averaged transition probability for processes in which an extra particle with momentum **p**, when added to the system, increases the energy of the system by ω. This transition probability measures the density of states available for added particles. Therefore, $G^>$ (**p**, ω) is the density of states available for an extra particle with momentum **p** and energy ω.

Similarly, $G^>$ (**p**, ω) is proportional to the averaged transition probability for processes involving the removal of a particle with momentum **p**, and leading to a decrease in the energy of the system by ω. Since the transition probability for the removal of a particle is just a measure of the density of particles, we again see that $G^>$ (**p**, ω) is the density of particles with momentum **p** and energy ω. The interpretation of $G^>$ as a density of states and $G^<$ as a density of particles will be used many times in our further work.

In terms of these two transition probabilities,[b] the boundary condition Eq. (2.13)[c] is

$$\frac{\mathfrak{P} \text{ (adding } \mathbf{p}, \omega)}{\mathfrak{P} \text{ (removing } \mathbf{p}, \omega)} = \frac{A\left(1 \pm f\left(\omega\right)\right)}{Af\left(\omega\right)} = e^{\beta(\omega-\mu)} \tag{3.6}$$

This statement, called the "detailed balancing condition," is a direct consequence of the use of an equilibrium ensemble.

3.2 Statistical Mechanical Information Contained in *G*

In addition to the detailed dynamical information, G contains all possible information about the statistical mechanics of the system.

We have already seen how we can write the expectation value of the density of particles in terms of $G^<$. Similarly, we can express the total energy, i.e., the expectation value of the Hamiltonian Eq. (2.2), in terms of $G^<$. To do this, we must make use of the equations of motion for $\hat{\psi}$ and $\hat{\psi}^\dagger$. Using the equation of motion Eq. (2.3) and the

---

[b]Here we introduce the symbol $\mathfrak{P}$ for representing the transition probability. In the original text, the symbol was T.P.
[c]The equation number was originally (2.12), but this is just definition of the Fourier transformations of $G^>$ and $G^<$.

commutation relations, Eq. (2.5), we see that

$$\left(i\frac{\partial}{\partial t} + \frac{\nabla^2}{2m}\right)\hat{\psi}\,(\mathbf{r}, t) = \int d\bar{\mathbf{r}}\, v\,(\mathbf{r} - \bar{\mathbf{r}})\,\hat{\psi}^{\dagger}\,(\bar{\mathbf{r}}, t)\,\hat{\psi}\,(\bar{\mathbf{r}}, t)\,\hat{\psi}\,(\mathbf{r}, t)$$

$$(3.7a)$$

and

$$\left(-i\frac{\partial}{\partial t'} + \frac{\nabla'^2}{2m}\right)\hat{\psi}^{\dagger}\,(\mathbf{r}', t') = \hat{\psi}^{\dagger}\,(\bar{\mathbf{r}}', t')\int d\bar{\mathbf{r}}'\, v\,(\mathbf{r}' - \bar{\mathbf{r}}')\,\hat{\psi}\,(\bar{\mathbf{r}}', t')\,\hat{\psi}\,(\bar{\mathbf{r}}', t')$$

$$(3.7b)$$

Therefore, it follows that

$$
\frac{1}{4}\int d\mathbf{r}\,\left[\left(i\frac{\partial}{\partial t} - i\frac{\partial}{\partial t'}\right)\hat{\psi}^{\dagger}\,(\mathbf{r}, t')\,\hat{\psi}\,(\mathbf{r}, t)\right]_{t'=t}
$$
$$
= \frac{1}{4}\int d\mathbf{r}\,\left[\left(-\frac{\nabla^2}{2m} - \frac{\nabla'^2}{2m}\right)\hat{\psi}^{\dagger}\,(\mathbf{r}, t')\,\hat{\psi}\,(\mathbf{r}, t)\right]_{\mathbf{r}'=\mathbf{r}} \qquad (3.8)
$$
$$
+ \frac{1}{2}\int d\mathbf{r}d\bar{\mathbf{r}}\hat{\psi}^{\dagger}\,(\mathbf{r}, t)\,\hat{\psi}^{\dagger}\,(\bar{\mathbf{r}}, t)\,v\,(\mathbf{r} - \bar{\mathbf{r}})\,\hat{\psi}\,(\bar{\mathbf{r}}, t)\,\hat{\psi}\,(\mathbf{r}, t)
$$

The right side of Eq. (3.8) is half the kinetic energy put all the potential energy. When we add the other half of the kinetic energy, we find that

$$
\langle\hat{H}\rangle = \frac{1}{4}\int d\mathbf{r}\,\left[\left(i\frac{\partial}{\partial t} - i\frac{\partial}{\partial t'} + \frac{\nabla\cdot\nabla'}{m}\right)\langle\hat{\psi}^{\dagger}\,(\mathbf{r}', t')\,\hat{\psi}\,(\mathbf{r}, t)\rangle\right]_{\mathbf{r}'=\mathbf{r},\,t'=t}
$$
$$
= \pm\frac{i}{4}\int d\mathbf{r}\,\left[\left(i\frac{\partial}{\partial t} - i\frac{\partial}{\partial t'} + \frac{\nabla\cdot\nabla'}{m}\right)G^{<}\,(\mathbf{r}t, \mathbf{r}'t')\right]_{\mathbf{r}'=\mathbf{r},\,t'=t}
$$
$$
= \Omega\int\frac{d\mathbf{p}}{(2\pi)^3}\frac{d\omega}{2\pi}\frac{\omega + \left(\frac{p^2}{2m}\right)}{2}f\,(\omega)\,A\,(\mathbf{p}, \omega) \qquad (3.9)
$$

where Ω is the volume of the system. Equation (3.9) is very useful for evaluating ground-state energies, specific heats, etc.

All statistical–mechanical information can be obtained from the grand partition function[d]

$$\Xi = \mathrm{tr}\left[e^{-\beta\left(\hat{H}-\mu\hat{N}\right)}\right] \qquad (3.10a)$$

We shall now show how we can find Ξ from G. Statistical mechanics tells us that in the limit of large volume, the grand partition function

---

[d] The symbol for the grand partition function was Z_g in the original text.

is related to the pressure P by

$$\Xi = e^{\beta P \Omega} \tag{3.10b}$$

Differentiating the logarithm of Ξ with respect to μ at fixed β and Ω, we find

$$\beta \Omega \left. \frac{\partial P}{\partial \mu} \right|_{\beta \Omega} = \frac{\partial}{\partial \mu} \ln \Xi = \frac{\partial}{\partial \mu} \ln \text{tr} \left[e^{-\beta (\hat{H} - \mu \hat{N})} \right]$$

$$= \beta \frac{\text{tr} \left[e^{-\beta (\hat{H} - \mu \hat{N})} \right] \hat{N}}{\text{tr} \left[e^{-\beta (\hat{H} - \mu \hat{N})} \right]}$$

$$= \beta \langle \hat{N} \rangle$$

so that the density of particles is given by

$$n = \left. \frac{\partial P}{\partial \mu} \right|_{\beta \Omega} \tag{3.11}$$

This is a very commonly used thermodynamic identity. Since we know that, in the limit $\mu \to -\infty$, the density and the pressure both go to zero, we can integrate Eq. (3.11) to obtain

$$P(\beta, \mu) = \int_{-\infty}^{\beta} d\mu' \, n(\beta, \mu') \tag{3.12}$$

Consequently if, or for a given β, we know Green's function as a function of μ, we can calculate P and hence the partition function.

Unfortunately, the integral in Eq. (3.12) can rarely be performed explicitly. One of the few cases for which a moderately simple result emerges is for a free gas. Here

$$n(\beta, \mu) = \int \frac{d\mathbf{p}}{(2\pi)^3} \frac{1}{e^{\beta \left[\left(\frac{p^2}{2m} \right) - \mu \right]} + 1} \tag{3.13a}$$

and hence

$$P(\beta, \mu) = \mp \frac{1}{\beta} \int \frac{d\mathbf{p}}{(2\pi)^3} \ln \left\{ 1 \mp e^{\beta \left[\left(\frac{p^2}{2m} \right) - \mu \right]} \right\} \tag{3.13b}$$

In the classical limit, $\beta \mu \to -\infty$. Then we see that

$$n = \int \frac{d\mathbf{p}}{(2\pi)^3} e^{-\beta \left[\left(\frac{p^2}{2m} \right) - \mu \right]}$$

and

$$P = \beta^{-1} \int \frac{d\mathbf{p}}{(2\pi)^3} e^{-\beta \left[\left(\frac{p^2}{2m} \right) - \mu \right]}$$

so that $P = \beta^{-1} n = n k_B T$. This is the well-known equation of state of an ideal gas.

There is, however, another method of constructing the grand partition function, which is very useful in practice. Let us write a coupling constant λ in front of the potential energy term in Eq. (2.2). Then

$$\hat{H} = \hat{H}_0 + \lambda \hat{V}$$

where $\hat{H}_0$ is the kinetic energy and $\hat{V}$ is the potential energy operator,

$$\hat{V} = \frac{1}{2} \int d\mathbf{r} d\bar{\mathbf{r}} \, \hat{\psi}^\dagger (\mathbf{r}) \, \hat{\psi}^\dagger (\bar{\mathbf{r}}) \, v \, (|\mathbf{r} - \bar{\mathbf{r}}|) \, \hat{\psi} (\bar{\mathbf{r}}) \, \hat{\psi} (\mathbf{r})$$

When we differentiate $\ln \Xi$ with respect to λ, at fixed β, μ, and Ω, we find

$$\frac{\partial}{\partial \lambda} \ln \Xi = \frac{1}{\Xi} \mathrm{tr} \left[\frac{\partial}{\partial \lambda} e^{-\beta \left(\hat{H}_0 + \lambda \hat{V} - \mu \hat{N} \right)} \right] = -\beta \left\langle \hat{V} \right\rangle \tag{3.14}$$

(We do not have to worry about the noncommutatibility of $\hat{V}$ with $\hat{H}_0 - \mu \hat{N}$ because of the cyclic invariance of the trace.) Integrating both sides of Eq. (3.14) with respect to λ, from $\lambda = 0$ to $\lambda = 1$, we find

$$[\ln \Xi]_{\lambda=1} - [\ln \Xi]_{\lambda=0} = -\beta \int_0^1 \frac{d\lambda}{\lambda} \left\langle \lambda \hat{V} \right\rangle_\lambda \tag{3.15}$$

Now $\left\langle \lambda \hat{V} \right\rangle_\lambda$ is the expectation value of the potential energy, for coupling strength λ. It may be expressed in terms of $G^<$ by subtracting from Eq. (3.8) half the kinetic energy. Then

$$\left\langle \lambda \hat{V} \right\rangle_\lambda = \Omega \int \frac{d\mathbf{p}}{(2\pi)^3} \frac{d\omega}{2\pi} \frac{\omega - \left(\frac{p^2}{2m} \right)}{2} A_\lambda (\mathbf{p}, \omega) f (\omega) \tag{3.16}$$

so that

$$\beta P \Omega = [\ln \Xi]_{\lambda=1}$$

$$= [\ln \Xi]_{\lambda=0} - \beta \Omega \int_0^1 \frac{d\lambda}{\lambda} \int \frac{d\mathbf{p}}{(2\pi)^3} \frac{d\omega}{2\pi} \frac{\omega - \left(\frac{p^2}{2m} \right)}{2} A_\lambda (\mathbf{p}, \omega) f (\omega)$$

$$\tag{3.17}$$

The constant term $[\ln \Xi]_{\lambda=0}$ is just $\beta P \Omega$ for free particles, which we have evaluated in Eq. (3.13b).

Chapter 4

The Hartree and Hartree–Fock Approximations

4.1 Equations of Motion

We have seen that the one-particle Green's function contains very useful dynamic and thermodynamic information. However, to extract this information, we must first develop techniques for determining G.

Our methods will be based on the equation of motion satisfied by the one-particle Green's function. This equation of motion is derived from the equation of motion (3.7a) for $\hat{\psi}\,(1)$. From Eq. (3.7a), it follows that[a]

$$\left(\frac{1}{i}\right)\left\langle \hat{T}\left[\left(i\frac{\partial}{\partial t_1} + \frac{\nabla_1^2}{2m}\right)\hat{\psi}\,(1)\,\hat{\psi}^\dagger\,(1')\right]\right\rangle$$
$$= \pm\left(\frac{1}{i}\right)\int d\mathbf{r}_2\, v\,(\mathbf{r}_1 - \mathbf{r}_2)$$
$$\times \left\langle \hat{T}\left(\hat{\psi}\,(1)\,\hat{\psi}\,(2)\,\hat{\psi}^\dagger\,(2^+)\,\hat{\psi}^\dagger\,(1')\right)\right\rangle\Big|_{t_2=t_1}$$
$$= \pm i\int d\mathbf{r}_2\, v\,(\mathbf{r}_1 - \mathbf{r}_2)\, G_2\,(12; 1'2^+)\Big|_{t_2=t_1}$$

(4.1)

---

[a]The equation number (4.1) was omitted in the original text.

Annotations to Quantum Statistical Mechanics
In-Gee Kim
Copyright © 2018 Pan Stanford Publishing Pte. Ltd.
ISBN 978-981-4774-15-4 (Hardcover), 978-1-315-19659-6 (eBook)
www.panstanford.com

Here, the notation 2^+ is intended to serve as a reminder that the time argument of $\hat{\psi}^\dagger$ (2) must be chosen to be infinitesimally larger than the time arguments of the $\hat{\psi}$'s in order that the time ordering in G_2 reproduces the order of factors that appear in Eq. (3.7a). [Since $\hat{\psi}$'s commute (or anti-commute) at equal times, we do not have to worry about the time ordering of $\hat{\psi}$ (1) and $\hat{\psi}$ (2).]

To convert Eq. (4.1) into an equation for G, we must take the time derivatives outside the $\hat{T}$-ordering symbol. The spatial derivatives commute with the time-ordering operation, but the time derivative does not. Since $\hat{T}$ changes the time ordering when $t_1 = t_{1'}$, the difference

$$\frac{\partial}{\partial t_1} \left\langle \hat{T} \left(\hat{\psi} \left(1 \right) \hat{\psi}^\dagger \left(1' \right) \right) \right\rangle - \left\langle \hat{T} \left(\frac{\partial}{\partial t_1} \hat{\psi} \left(1 \right) \hat{\psi}^\dagger \left(1' \right) \right) \right\rangle$$

must be proportional to a delta function of $t_1 - t_{1'}$. The constant of proportionality is the discontinuity of $\left\langle \hat{T} \left(\hat{\psi} \left(1 \right) \hat{\psi}^\dagger \left(1' \right) \right) \right\rangle$ as t_1 passes through $t_{1'}$, i.e.,

$$\frac{\partial}{\partial t_1} \left\langle \hat{T} \left(\hat{\psi} \left(1 \right) \hat{\psi}^\dagger \left(1' \right) \right) \right\rangle - \left\langle \hat{T} \left(\frac{\partial}{\partial t_1} \hat{\psi} \left(1 \right) \hat{\psi}^\dagger \left(1' \right) \right) \right\rangle$$
$$= \delta \left(t_1 - t_{1'} \right) \left\langle \left(\hat{\psi} \left(1 \right) \hat{\psi}^\dagger \left(1' \right) \mp \hat{\psi}^\dagger \left(1' \right) \hat{\psi} \left(1 \right) \right) \right\rangle$$
$$= \delta \left(t_1 - t_{1'} \right) \delta \left(\mathbf{r}_1 - \mathbf{r}_{1'} \right) = \delta \left(1 - 1' \right)$$

In this way, we find that Eq. (4.1) becomes an equation of motion for G:

$$\left(i \frac{\partial}{\partial t_1} + \frac{\nabla_1^2}{2m} \right) G \left(1, 1' \right) = \delta \left(1 - 1' \right)$$
$$\pm i \int d\mathbf{r}_2 v \left(\mathbf{r}_1 - \mathbf{r}_2 \right) G_2 \left(12; 1'2^+ \right) \Big|_{t_2 = t_1}$$

$$(4.2a)$$

In a similar fashion, we can also write an equation of motion for G_2 involving G_3, one for G_3 involving G_4, and so on. As we will have no need for these equations, we shall not write them down.

Starting from the equation of motion of $\hat{\psi}^\dagger$ (1'), we also derive the adjoint equation of motion,

$$\left(-i \frac{\partial}{\partial t_{1'}} + \frac{\nabla_{1'}^2}{2m} \right) G \left(1, 1' \right) = \delta \left(1 - 1' \right)$$
$$\pm i \int d\mathbf{r}_2 G_2 \left(12^-; 1'2 \right) \Big|_{t_2 = t_1} v \left(\mathbf{r}_2 - \mathbf{r}_{1'} \right)$$

$$(4.2b)$$

Equations (4.2) are equally valid for the real-time and the imaginary-time Green's function. The only difference between the two cases is that for imaginary times, one has to interpret the delta function in time as being defined with respect to integrations along the imaginary time axis.

Equations (4.2a) and (4.2b) both determine G in terms of G_2. It is in general impossible to know G_2 exactly. We shall find G by making approximations for G_2 in the equations of motion (4.2).

However, even if G_2 were precisely known, Eq. (4.2) would not be sufficient to determine G unambiguously. These equations are first-order differential equations in time, and thus a single supplementary boundary condition is required to fix their solution precisely. The necessary boundary condition is, of course, Eq. (2.10):

$$G\left(1, 1'\right)\big|_{t_1=0} = \pm e^{\beta\mu}\, G\left(1, 1'\right)\big|_{t_1=-i\beta} \qquad (2.10)$$

A very natural representation of G, which automatically takes the quasi-periodic boundary condition into account, is to express G as a Fourier series, which we write in momentum space as

$$G\left(p, t - t'\right) = \frac{1}{-i\beta} \sum_\nu e^{-iz_\nu(t-t')} G\left(p, z_\nu\right) \quad \text{for} \quad \begin{matrix} 0 \le it \le \beta \\ 0 \le it' \le \beta \end{matrix}$$

$$(4.3)$$

where $z_\nu = \left(\frac{\pi\nu}{-i\beta}\right) + \mu$. The sum is taken to run over all even integers for Bose statistics and over all odd integers for Fermi statistics in order to reproduce correctly the $\pm$ in the boundary conditions.

The equation of motion directly determines the Fourier coefficient $G\left[\left(\frac{\pi\nu}{-i\beta}\right) + \mu\right]$. However, we want to know the spectral weight function A. To relate G to A, we invert the Fourier series (4.3):

$$G\left(p, z_\nu\right) = \int_0^{-i\beta} dt\, e^{i\left[\left(\frac{\pi\nu}{-i\beta}\right)+\mu\right](t-t')} G\left(p, t - t'\right)$$

This integral must be independent of t' and is most simply evaluated by taking $t' = 0$. Then

$$G\left(p, t\right) = G^{>}\left(p, t\right) = \int \frac{d\omega}{2\pi i} e^{-i\omega t} \frac{A\left(p, \omega\right)}{1 \mp e^{-\beta(\omega-\mu)}}$$

and we find

$$G\left(p, z_\nu\right) = \int_{-\infty}^{\infty} \frac{d\omega}{2\pi i} \int_0^{-i\beta} dt\, \left[e^{i\left[\left(\frac{\pi\nu}{-i\beta}\right)+\mu-\omega\right]t}\right] \frac{A\left(p, \omega\right)}{1 \mp e^{-\beta(\omega-\mu)}}$$

$$= \int \frac{d\omega}{2\pi} \frac{A\left(p, \omega\right)}{z_\nu - \omega} \qquad (4.4)$$

Thus, the Fourier coefficient is just the analytic function

$$G\left(p,\, z\right) = \int \frac{d\omega}{2\pi}\, \frac{A\left(p,\, \omega\right)}{z - \omega} \tag{4.5}$$

evaluated at $z = z_v = \left(\frac{\pi v}{-i\beta}\right) + \mu$. The procedure for finding A from the Fourier coefficients is then very simple. One merely continues the Fourier coefficients—a function defined on the points $z = \left(\frac{\pi v}{-i\beta}\right) + \mu$—to an analytic function for all (nonreal) z. The unique continuation that has no essential singularity at $z = \infty$ is the function Eq. (4.5). Then, $A\left(p,\, \omega\right)$ is given by the discontinuity of $G\left(p,\, z\right)$ across the real axis, i.e.,

$$A\left(p,\, \omega\right) = i\left[G\left(p,\, \omega + i\epsilon\right) - G\left(p,\, \omega - i\epsilon\right)\right] \tag{4.6}$$

since

$$\frac{1}{\omega - \omega' + i\epsilon} = \wp\, \frac{1}{\omega - \omega'} - \pi i \delta\left(\omega - \omega'\right)$$

where $\wp$ denotes the principal value integral and ϵ is an infinitesimal positive number.

The three concepts—equations of motion, boundary conditions, and analytic continuations—form the mathematical basis of all our techniques for determining Green's function.

4.2 Free Particles

Let us illustrate these methods by considering some very simple approximations for G. The most trivial example is that of free particles. Since $v = 0$, the equation of motion (4.2a) is simply

$$\left(i\frac{\partial}{\partial t_1} + \frac{\nabla_1^2}{2m}\right) G\left(1,\, 1'\right) = \delta\left(1 - 1'\right) \tag{4.7}$$

We multiply this equation by

$$\exp\left[-i\mathbf{p}\cdot\left(\mathbf{r}_1 - \mathbf{r}_{1'}\right) + i\left(\frac{\pi v}{-i\beta} + \mu\right)\left(t_1 - t_{1'}\right)\right]$$

Integrate over all $\mathbf{r}_1$ and all t_1 in the interval 0 to $-i\beta$. Then Eq. (4.7) becomes an equation for the Fourier coefficient,

$$\left(z_v - \frac{p^2}{2m}\right) G\left(p,\, z_v\right) = 1$$

Therefore,

$$G(p, z_\nu) = \frac{1}{z_\nu - \left(\frac{p^2}{2m}\right)} \tag{4.8a}$$

The analytic continuation of this formula is

$$G(p, z) = \frac{1}{z - \left(\frac{p^2}{2m}\right)} \tag{4.8b}$$

This analytic continuation involves nothing more than replacing $\left(\frac{\pi\nu}{-i\beta}\right) + \mu$ by the general complex variable z. The analytic continuations we shall perform will never be more complicated than this. We see directly from Eqs. (4.6) and (4.8b) that

$$A_0(p, \omega) = 2\pi\delta\left[\omega - \left(\frac{p^2}{2m}\right)\right]$$

This by-now-familiar result expresses the fact that a free particle with momentum p can only have energy $\frac{p^2}{2m}$. Once we know A, we know $G^>$ and $G^<$.

4.3 Hartree Approximation

To determine G when $\nu \neq 0$, we must approximate the G_2 that appears in Eq. (4.2a). Approximations to G_2 can be physically motivated by the propagator interpretation of $G(1, 1')$ and $G_2(12; 1'2')$.

The one-particle Green's function, $G(1, 1')$, represents the propagation of a particle added to the medium at $1'$ and removed at 1. We can represent this pictorially by a line going from $1'$ to 1:

$$G(1, 1') = \quad 1' \longrightarrow 1$$

Notice that this line represents propagation through the medium, and not free-particle propagation. Similarly,

$$G_2(12; 1'2') =$$

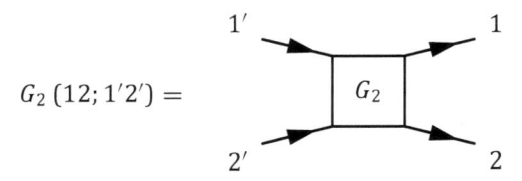

describes the propagation of two particles added to the medium at 1′ and 2′ and removed at 1 and 2. In general, the motion of the particles is correlated because the added particles interact with each other, either directly or intermediately through other particles in the system.

However, as a first approximation, we may neglect this correlation and assume that the added particles propagate through the medium completely independent of each other. That is, we use the approximation[b]

$$G_2\left(12; 1'2'\right) \simeq = G\left(1, 1'\right) G\left(2, 2'\right) \tag{4.9}$$

If we then substitute Eq. (4.9) into the equation of motion (4.2a), we obtain the approximate equation for G:

$$\left[i\frac{\partial}{\partial t_1} + \frac{\nabla_1^2}{2m} \mp i \int d\mathbf{r}_2\, v\left(\mathbf{r}_1 - \mathbf{r}_2\right) G\left(2, 2^+\right)\right] G\left(1, 1'\right)$$

$$= \left[i\frac{\partial}{\partial t_1} + \frac{\nabla_1^2}{2m} - \int d\mathbf{r}_2\, v\left(\mathbf{r}_1 - \mathbf{r}_2\right) \langle n\left(\mathbf{r}_2\right)\rangle\right] G\left(1, 1'\right) \tag{4.10}$$

$$= \delta\left(1 - 1'\right)$$

Equation (4.10) is a Green's function statement of the well-known Hartree approximation. It is the same equation as we would have obtained had we considered a set of independent particles moving through the potential field

$$V\left(\mathbf{r}_1\right) = \int d\mathbf{r}_2\, v\left(\mathbf{r}_1 - \mathbf{r}_2\right) \langle n\left(\mathbf{r}_2\right)\rangle \tag{4.11}$$

The potential field Eq. (4.11), called the self-consistent Hartree field, is the average field generated by all the other particles in the system. Thus, we see that the Hartree approximation describes the many-particle system as a set of independent particles, each

---

[b]The symbol $\simeq$ in the second line was omitted in the original text.

particle, however, moving through the average field produced by all the particles.

For a translationally invariant system, Eq. (4.10) is quite trivial. Since $\langle n (\mathbf{r}_2) \rangle$ is independent of the position $\mathbf{r}_2$, the average potential is also constant. Letting $v = \int d\mathbf{r} v (r)$, we may write

$$V = nv$$

Then, by just the same procedure as in the free-particle case, we find from Eq. (4.10) the equation for the Fourier constant:

$$\left[z_\nu - \left(\frac{p^2}{2m} \right) - nv \right] G (p, z_\nu) = 1$$

The continuation from z_ν to all complex z of the Fourier coefficient is, therefore,

$$G (p, z) = \frac{1}{z - \left(\frac{p^2}{2m} \right) - nv} \tag{4.12}$$

so that in the Hartree approximation

$$A (p, \omega) = 2\pi \delta \left[\omega - \left(\frac{p^2}{2m} \right) - nv \right] \tag{4.13}$$

Thus, the particles move as free particles, except that they each have the added energy nv.

To complete the solution to the Hartree approximation, we must solve for the density of particles in terms of μ, or vice versa. This can be computed from Eq. (3.2):

$$n = \pm iG^< (\mathbf{r}t, \mathbf{r}t) = \int \frac{d\mathbf{p}}{(2\pi)^3} \frac{d\omega}{2\pi} A (p, \omega) f(\omega) \tag{4.14}$$

which for the Hartree approximation becomes

$$n = \int \frac{d\mathbf{p}}{(2\pi)^3} \frac{d\omega}{2\pi} \frac{1}{e^{\beta \left[\left(\frac{p^2}{2m} \right) + nv - \mu \right]} \mp 1} \tag{4.15}$$

Similarly, we find the energy per unit volume from Eq. (3.9):

$$\frac{\langle H \rangle}{\Omega} = \int \frac{d\mathbf{p}}{(2\pi)^3} \left(\frac{p^2}{2m} + \frac{nv}{2} \right) \frac{1}{e^{\beta \left[\left(\frac{p^2}{2m} \right) + nv - \mu \right]} \mp 1}$$

$$= \left(\frac{1}{2} \right) n^2 v + \int \frac{d\mathbf{p}}{(2\pi)^3} \frac{\frac{p^2}{2m}}{e^{\beta \left[\left(\frac{p^2}{2m} \right) + nv - \mu \right]} \mp 1} \tag{4.16}$$

Finally, we may obtain the equation of state of a gas in the Hartree approximation. We do this in the low-density limit for simplicity. We start out by considering the effect of changing the chemical potential by an infinitesimal amount $d\mu$ at fixed temperature. Then the familiar thermodynamic identity,

$$dP = nd\mu \tag{4.17}$$

gives the change in the pressure. When Eq. (4.15) is taken in the low-density limit ($\beta\mu \to \infty$), it becomes

$$n = e^{\beta(\mu - nv)} \int \frac{d\mathbf{p}}{(2\pi)^3} e^{-\beta\left(\frac{p^2}{2m}\right)}$$

Hence at fixed β,

$$dn = \beta n \left(d\mu - vdn\right)$$

Thus, from Eq. (4.17),

$$dP = \left(\frac{1}{\beta}\right) dn + vndn = k_B T dn + \left(\frac{1}{2}\right) vd\left(n^2\right)$$

Since the pressure vanishes at $n = 0$, we find

$$P - \left(\frac{1}{2}\right) n^2 v = nk_B T \tag{4.18}$$

This is in the form of a van der Waals equation,

$$\left(P - an^2\right)\left(\Omega - \Omega_{\text{exc}}\right) = nk_B T$$

but without the volume-exclusion effect. For an interacting whose long-range part is attractive, v is negative, and quite reasonably the pressure is reduced from its free-particle value.

We could never hope to discover a volume-exclusion term from the Hartree approximation. Such a term arises because the particles can never penetrate each other's hard cores. However, in deriving the Hartree approximation, we have said that the particles move independently; therefore, this correlation effect has been completely left out. In order to treat hard-core interactions, it is necessary to include in the approximation for G_2 the fact that the motion of one particle depends on the detailed positions of the other particles in the medium.

The Hartree approximation is much less trivial when the particles are sitting in an external potential $U(\mathbf{r})$. The system for which Hartree originated his approximation was that of electrons in an atom, under the influence of the central potential of the nucleus.

The equation of motion for G in the presence of an external potential is[c]

$$\left[i\frac{\partial}{\partial t_1} + \frac{\nabla_1^2}{2m} - U(\mathbf{r}_1)\right] G(1, 1')$$

$$= \delta(1 - 1') \pm i \int d\mathbf{r}_2 \, v(\mathbf{r}_1 - \mathbf{r}_2) \, G_2(12; 1'2^+)\Big|_{t_2=t_1}$$

and in the Hartree approximation, this reduced to

$$\left[i\frac{\partial}{\partial t_1} + \frac{\nabla_1^2}{2m} - U(\mathbf{r}_1) - \int d\mathbf{r}_2 v(\mathbf{r}_1\mathbf{r}_2) \langle n(\mathbf{r}_2)\rangle\right] G(1, 1') = \delta(1 - 1')$$

$$(4.19)$$

Again this equation is the same as we would have obtained had we considered independent particles in the effective potential field

$$U_{\text{eff}}(\mathbf{r}_1) = U(\mathbf{r}_1) + \int d\mathbf{r}_2 \, v(\mathbf{r}_1 - \mathbf{r}_2) \langle n(\mathbf{r}_2)\rangle \qquad (4.20)$$

Since the system is no longer translationally invariant, we cannot consider $\langle n \rangle$ or U_{eff} to be independent of position, and the equation cannot be diagonalized by Fourier transforming in space. It can, however, be diagonalized on the basis of normalized eigenfunctions, $\varphi_i(\mathbf{r})$, of the effective single-particle Hamiltonian, $\hat{H}_1(\mathbf{r}) = \left(-\frac{\nabla^2}{2m}\right) + U_{\text{eff}}(\mathbf{r})$:

$$\hat{H}_1(\mathbf{r}) \varphi_i(\mathbf{r}) = E_i \varphi_i(\mathbf{r}) \qquad (4.21)$$

The procedure for solving the equation is to first take Fourier coefficients of the equation of motion, finding

$$\left[z_\nu - \hat{H}_1(\mathbf{r})\right] G(\mathbf{r}, \mathbf{r}'; z_\nu) = \delta(\mathbf{r} - \mathbf{r}') \qquad (4.22)$$

so that in terms of the φ_i,

$$G(\mathbf{r}, \mathbf{r}'; z_\nu) = \sum_i \frac{\varphi_i(\mathbf{r}) \varphi_i^*(\mathbf{r}')}{z_\nu - E_i}$$

Hence

$$A(\mathbf{r}, \mathbf{r}', \omega) = 2\pi \sum_i \varphi_i(\mathbf{r}) \varphi_i^*(\mathbf{r}') \delta(\omega - E_i) \qquad (4.23)$$

---

[c]The integral variable symbol $d\mathbf{r}_2$ was omitted in the original text.

We see that the single-particle Hamiltonian $\hat{H}_1$ defines both the single-particle energies and wavefunctions of the particles in the system.

Once more, to complete the solution, we have to compute the density, since this determines U_{Jeff}. We have

$$\langle n\,(\mathbf{r},\,t)\rangle = \int \frac{d\omega}{2\pi}\,A\,(\mathbf{r},\,\mathbf{r},\,\omega)\,f\,(\omega)$$

$$= \sum_i |\varphi_i\,(\mathbf{r})|\,f\,(E_i) \tag{4.24}$$

The term $f\,(E_i)$ gives the average occupation of the i-th single-particle level, while $|\varphi_i\,(\mathbf{r})|^2$ is obviously the probability of observing at $\mathbf{r}$ a particle in the i-th level.

Notice that to determine $\phi_i\,(\mathbf{r})$, it is necessary to solve a nonlinear equation, since $\hat{H}_1\,(\mathbf{r})$ itself depends on all the φ_i through its dependence on the density. The process of solving this nonlinear equation is called obtaining a "self-consistent" Hartree solution.

4.4 Hartree–Fock Approximation

The Hartree approximation (4.9) for the two-particle Green's function does not take into account the identity of the particles. Since the particles are identical, we cannot distinguish processes in which the particle added at $1'$ appears at 1 from a process in which it appears at 2. These processes contribute coherently. To include this possibility of exchange, we can write[d]

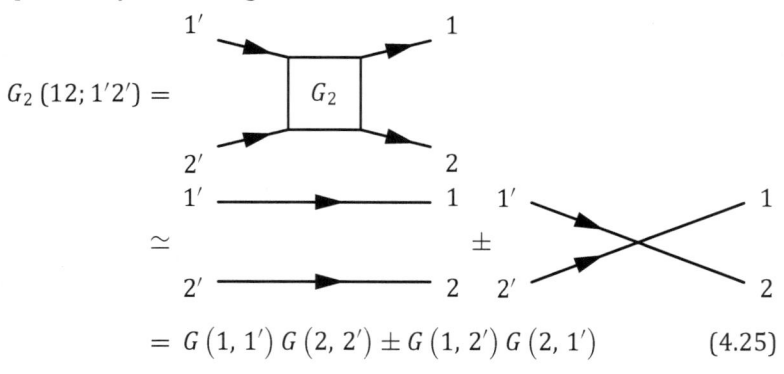

$$= G\,(1,\,1')\,G\,(2,\,2') \pm G\,(1,\,2')\,G\,(2,\,1') \tag{4.25}$$

---

[d]The symbol $\simeq$ in the second line was $=$ in the original text.

This approximation to G_2 leads to the Hartree–Fock approximation. In fixing the relative signs of the two terms in Eq. (4.25), we use the fact that $G_2 (12; 1'2') = \pm G_2 (21; 1'2')$. This symmetry can be verified directly from the definition of G_2, Eq. (2.7b).

The approximate equation of G resulting from substituting Eq. (4.25) into Eq. (4.2a) takes the form

$$\left(i\frac{\partial}{\partial t_1} + \frac{\nabla_1^2}{2m} \right) G (1, 1') + \int d\mathbf{r}_2 \, \langle \mathbf{r}_1 | \mathcal{V} | \mathbf{r}_2 \rangle \, G (2, 1')\big|_{t_2=t_1} = \delta (1 - 1')$$

$$(4.26)$$

where

$$\langle \mathbf{r}_1 | \mathcal{V} | \mathbf{r}_2 \rangle = \delta (\mathbf{r}_1 - \mathbf{r}_2) \int d\mathbf{r})_3 v (\mathbf{r}_1 - \mathbf{r}_3) \langle n (\mathbf{r}_3) \rangle$$
$$+ iv (\mathbf{r}_1 - \mathbf{r}_2) G^< (1, 2)\big|_{t_2=t_1}$$

$$(4.27)$$

again has the interpretation of an average, self-consistent potential field through which the particles move. However, with the inclusion of exchange, $\mathcal{V}$ becomes nonlocal in space.

In the case of translationally invariant system, we can Fourier transform Eqs. (4.26) and (4.27) in space to obtain

$$\left[i\frac{\partial}{\partial t_1} - E (p) \right] G (p, t_1 - t_{1'}) = \delta (t_1 - t_{1'})$$

$$(4.28)$$

and

$$E (p) = \frac{p^2}{2m} + nv \pm \int \frac{d\mathbf{p}'}{(2\pi)^3} \, v (\mathbf{p} - \mathbf{p}') \langle n (\mathbf{p}) \rangle$$

$$(4.29)$$

where $v (\mathbf{p}) = \int d\mathbf{r} \; e^{-i\mathbf{p}\cdot\mathbf{r}} v(r)$ is the Fourier transform of the potential $v (\mathbf{r})$. Just as before

$$A (p, \omega) = 2\pi \delta (\omega - E (p))$$

$$(4.30)$$

so that

$$\langle n (\mathbf{p}) \rangle = f (E (p)) = \frac{1}{e^{\beta [E (p) - \mu]} \mp 1}$$

$$(4.31)$$

The Hartree–Fock single-particle energy $E (p)$ must then be obtained as the solution to Eqs. (4.29) and (4.31).

To sum up: Both the Hartree and the Hartree–Fock approximations are derived by assuming that there is no correlation between the motion of two particles added to the medium. Thus, these approximations describe the particles as moving independently through an average potential field. The particles then find themselves in perfectly stable single-particle states. There is no possibility for collisions and indeed no mechanism at all for particles moving from one single-particle state to another.

In Chapter 5, we describe a way of introducing the effect of collisions into our Green's function analysis.

Chapter 5

Effects of Collisions on G

5.1 Lifetime of Single-Particle States

The Hartree and Hartree–Fock approximations have the characteristic feature that A has the form

$$A(p, \omega) = 2\pi \delta(\omega - E(p))$$

so that there is just a single possible energy for each momentum.
This result is physically quite unreasonable. The interaction between the particles should result in the existence of a spread in these
possible energies. Perhaps the best way of seeing the necessity of
this spread is to consider

$$\frac{1}{\Omega^2} \left| \langle \hat{\psi}(\mathbf{p}, t)\, \hat{\psi}^\dagger(\mathbf{p}, t') \rangle \right|^2 = \left| G^>(p, t - t') \right|^2$$

$$= \left| \int \frac{d\omega}{2\pi}\, A(p, \omega)\, [1 \pm f(\omega)]\, e^{-i\omega(t - t')} \right|^2$$

$$\tag{5.1}$$

If the expectation value in Eq. (5.1) involved only a single state,
Eq. (5.1) would be the probability that one could add a particle
with momentum p to this state at the time t', remove a particle at
the time t, and then come back to the very same state as in the

Annotations to Quantum Statistical Mechanics
In-Gee Kim
Copyright © 2018 Pan Stanford Publishing Pte. Ltd.
ISBN 978-981-4774-15-4 (Hardcover), 978-1-315-19659-6 (eBook)
www.panstanford.com

beginning. Clearly, as the addition and removal processes become very separated in time, i.e., $|t - t'| \rightarrow \infty$, this probability should decrease. The expectation value in Eq. (5.1) actually contains a sum of many different states. This sum should lead to a result decreasing even more strongly in time.

However, in the Hartree and Hartree–Fock approximations, the right-hand side of Eq. (5.1) is independent of time. Therefore, this approximation predicts an infinite lifetime for any state produced by adding a single particle to the system. Thus, we must look for better approximations if we are to have an understanding of the lifetime of single-particle excited states.

It is possible to estimate this lifetime for a classical gas without doing any calculation. If we first add a particle and then remove a particle with the same momentum, we should come back to the same state only if, in the intervening time, the added particle has not collided with any of the other particles in the gas. Therefore, we should expect that the probability Eq. (5.1) should decay as $e^{-\Gamma(p)|t-t'|}$, where $\Gamma(p)$ is the collision rate for the added particle. This collision rate can be estimated as

$$\Gamma(p) \sim \langle n \rangle \, \sigma \left(\frac{\bar{p}}{m} \right) \tag{5.2}$$

where σ is an average collision cross section, and $\frac{\bar{p}}{m}$ is an average relative velocity of the added particle with respect to the other particles in the medium.

This decay of single-particle excited states is an exceedingly important feature of many-particle systems. It is responsible for the return of the system to thermodynamic equilibrium after a disturbance.

It is very easy to find a form for A that will lead to a proper decay of the probability Eq. (5.1). No A, which is a sum of a finite number of delta functions, will lead to exponential decay in Eq. (5.1). But any continuously varying A will lead to rapid decay. Consider, for example, the Lorentzian line shape

$$A(p, \omega) = \frac{\Gamma(p)}{[\omega - E(p)]^2 + \left[\frac{\Gamma(p)}{2} \right]^2} \tag{5.3}$$

When the dispersion in energy $\Gamma(p)$ is much less than β, we can perform the integral in Eq. (5.1) by replacing $f(\omega)$ by $f(E(p))$. Then the probability does indeed decay as $e^{-\Gamma(p)|t_1 - t_{1'}|}$. Thus, $\Gamma(p)$ represents both the energy dispersion and decay rate of the single-particle excited state with momentum p. The average energy of the added particle is $E(p)$.

5.2 Born Approximation Collisions

We now want to describe an approximation that includes the simplest effects of collisions. We have already noticed that if one just takes into account independent particle propagation in G_2, i.e.,

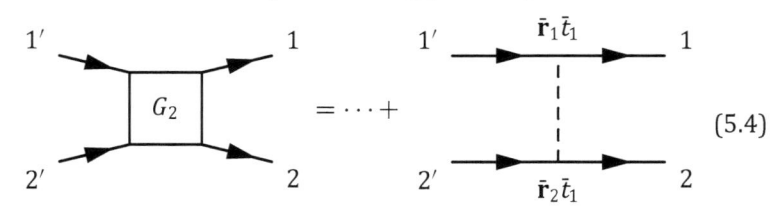

then no lifetime appears.[a] The simplest type of process that can lead to a lifetime is one in which the two particles added at $1'$ and $2'$ propagate to the spatial points $\bar{\mathbf{r}}_1$ and $\bar{\mathbf{r}}_2$; at the time $\bar{t}_1$, when the particles are at these spatial points, the potential acts between the particles, scattering them. Then the particles propagate to the points 1 and 2, where they are removed from the system. We can represent the contribution of this process to G_2 pictorially as

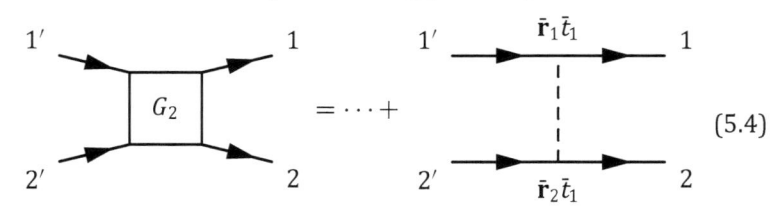

$$= \cdots + \qquad (5.4)$$

where the dashed line represents $v(\bar{\mathbf{r}}_1 - \bar{\mathbf{r}}_2)$.

At first sight, it appears quite easy to write down Green's functions that correspond to our physical picture (5.4). We replace each line by a propagator and integrate over all possible points at which the intermediate interaction could occur. Then we find that

---

[a] The symbol $\pm$ was missing in the original text.

the value of this picture is

$$(?) \times \int d\bar{\mathbf{r}}_1 \int d\bar{\mathbf{r}}_2 \int_{(?)}^{(?)} d\bar{t}_2 \left[G\left(1, \bar{1}\right) G\left(\bar{1}, 1'\right) v\left(\bar{\mathbf{r}}_1 - \bar{\mathbf{r}}_2\right) G\left(2, \bar{2}\right) G\left(\bar{2}, 2'\right) \right]_{\bar{t}_2 = \bar{t}_1}$$

$$(5.5)$$

The three question marks in Eq. (5.5) represent the quantities that we cannot fix by a physical argument alone. First, there is the numerical factor in front of the entire expression. We shall see in Chapter 6 that it should be i. More important is the ambiguity of the limits on the $\bar{t}_1$ integration. Should this integral run over all times? Over all times after the particles have been added? Or when? This question is very hard to settle on the basis of physical arguments alone. To remove this latter ambiguity, we consider Green's functions defined in the pure imaginary time domain, $0 < it < \beta$. There G and G_2 must satisfy the boundary conditions

$$G\left(1, 1'\right)\big|_{t_1=0} = \pm e^{\beta\mu} \, G\left(1, 1'\right)\big|_{t_1=-i\beta} \qquad (2.10)$$

$$G_2\left(12, 1'2'\right)\big|_{t_1=0} = \pm e^{\beta\mu} \, G\left(12, 1'2'\right)\big|_{t_1=-i\beta} \qquad (2.11)$$

Notice that the G_2 we used to define the Hartree–Fock approximation certainly satisfies Eq. (2.11), since

$$\left[G\left(1, 1'\right) G\left(2, 2'\right) \pm G\left(1, 2'\right) G\left(2, 1'\right) \right]_{\bar{t}_1=0}$$

$$= \pm e^{\beta\mu} \left[G\left(1, 1'\right) G\left(2, 2'\right) \pm G\left(1, 2'\right) G\left(2, 1'\right) \right]_{\bar{t}_1=-i\beta}$$

Expression (5.5) will also satisfy Eq. (2.11) if the $\bar{t}_1$ integral is taken to run from 0 to $-i\beta$. In that case, Eq. (5.5) is of the form

$$F\left(1, \ldots\right) = \int_0^{-i\beta} d\bar{t}_1 \int d\mathbf{r}_1 \, G\left(1, \bar{1}\right) \cdots$$

so that

$$F\left(1, \ldots\right)\big|_{t_1=0} = \int G\left(1, \bar{1}\right)\big|_{t_1=0} \cdots = \pm e^{\beta\mu} \int G\left(1, \bar{1}\right)\big|_{t_1=-i\beta} \cdots$$

$$= \pm e^{\beta\mu} \, F\left(1, \ldots\right)\big|_{t_1=-i\beta}$$

All the above is just an elaborate justification for approximating G_2 by

$$G_2 (12, 1'2') =$$

$$= G(1, 1') G(2, 2') \pm G(1, 2') G(2, 1')$$
$$+ i \int_0^{-i\beta} d\bar{t}_1 d\bar{\mathbf{r}}_1 d\bar{\mathbf{r}}_2 \, v(\bar{\mathbf{r}}_1 - \bar{\mathbf{r}}_2)$$
$$\times \{ G(1, \bar{1}) G(\bar{1}, 1') G(2, \bar{2}) G(\bar{2}, 2')$$
$$\pm G(1, \bar{1}) G(\bar{1}, 2') G(2, \bar{2}) G(\bar{2}, 1') \}_{\bar{t}_2 = \bar{t}_1} \qquad (5.6)$$

This approximation describes the two particles added to the system as either propagating independently or scattering through single interaction. Both direct and exchange processes are included. Since only the first-order terms in v are included in describing the scattering, clearly Eq. (5.6) gives no better a picture of the scattering than the first Born approximation of conventional scattering theory. For that reason, we shall call Eq. (5.6) and the resulting approximation for G the Born scattering or collision approximation.

In Chapter 6, this approximation will be shown to be the first two terms in an expansion of G_2 in power series in G and v.

When Eq. (5.6) is substituted into the equation of motion for G, Eq. (4.2a), the Born scattering approximation takes the form

$$\left[i \frac{\partial}{\partial t_1} + \frac{\nabla_1^2}{2m} \right] G(1, 1') - \int_0^{-i\beta} d\bar{t}_1 d\bar{\mathbf{r}}_1 \, \Sigma(1, \bar{1}) G(\bar{1}, 1')$$
$$= \delta(1 - 1') \quad \text{for} \quad \begin{array}{c} 0 \le it \le \beta \\ 0 \le it' \le \beta \end{array} \qquad (5.7)$$

where $\Sigma\,(1, 1')$, which is usually called the self-energy, can be split into two parts,

$$\Sigma\,(1, 1') = \Sigma_{HF}\,(1, 1') + \Sigma_c\,(1, 1') \tag{5.8}$$

The Hartree–Fock part of Σ, whose effects we have already treated in detail, is

$$\Sigma_{HF}\,(1, 1') = \delta\,(t_1 - t_{1'})\left\{\delta\,(\mathbf{r}_1 - \mathbf{r}_{1'})\int d\mathbf{r}_2\,v\,(\mathbf{r}_1 - \mathbf{r}_2)\,\langle n\,(\mathbf{r}_2)\rangle\right.$$
$$\left.+\,iv\,(\mathbf{r}_1 - \mathbf{r}_2)\,G^<\,(1, 2)\big|_{t_1=t_2}\right\} \tag{5.9}$$

while the part of the self-energy due to collisions is, in the Born scattering approximation,

$$\Sigma_c\,(1, 1') = \pm\,i^2\int d\mathbf{r}_2 d\mathbf{r}_{2'}\,v\,(\mathbf{r}_1 - \mathbf{r}_2)\,v\,(\mathbf{r}_{1'} - \mathbf{r}_{2'})$$
$$\times\,[G\,(1, 1')\,G\,(2, 2')\,G\,(2', 2) \tag{5.10}$$
$$\pm\,G\,(1, 2')\,G\,(2, 1')\,G\,(2', 2)]_{t_2=t_1,\,t_{2'}=t_{1'}}$$

As a first step in solving Eq. (5.7), we Fourier-transform it in space and find

$$\left[i\frac{\partial}{\partial t_1} - E\,(p)\right]G\,(p, t - t') - \int_0^{-i\beta} d\bar{t}\,\Sigma\,(p, t - \bar{t})$$
$$\times\,G\,(p, \bar{t} - t') = \delta\,(t - t') \tag{5.11}$$

where $E\,(p)$ is just the Hartree–Fock single-particle energy defined by Eq. (4.28). The Fourier transform of the collisional part of the self-energy is, from Eq. (5.10),[b]

$$\Sigma_c\,(p, t - t') = \pm\,i^2\int\frac{d\mathbf{p}'}{(2\pi)^3}\frac{d\bar{\mathbf{p}}}{(2\pi)^3}\frac{d\bar{\mathbf{p}}'}{(2\pi)^3}$$
$$\times\left(\frac{1}{2}\right)\left(v\,(\mathbf{p} - \bar{\mathbf{p}}) \pm v\,(\mathbf{p} - \bar{\mathbf{p}}')\right)^2(2\pi)^3 \tag{5.12}$$
$$\times\,\delta\,(\mathbf{p} + \mathbf{p}' - \bar{\mathbf{p}} - \bar{\mathbf{p}}')$$
$$\times\,G\,(p', t' - t)\,G\,(\bar{p}, t - t')\,G\,(\bar{p}', t - t')$$

In our later analysis, we shall see in detail that the integrand in Eq. (5.12) describes processes in which particles with momentum $\mathbf{p}$ and $\mathbf{p}'$ scatter into states with momentum $\bar{\mathbf{p}}$ and $\bar{\mathbf{p}}'$ as well as

---

[b]The integral variables in the original text have to be corrected.

the inverse processes in which the barred momenta go into the unbarred ones. In either cases, we recognize that the momentum delta function in Eq. (5.12) represents the conservation of momentum, while the combination $\left(\frac{1}{2}\right)(v\,(\mathbf{p}-\bar{\mathbf{p}}) \pm v\,(\mathbf{p}-\bar{\mathbf{p}}'))^2$ represents the first Born approximation collision cross section with exchange included.

We now have to solve Eq. (5.11) and Eq. (5.12) to obtain the physically interesting functions $G^>$, $G^<$, and A. However, it is convenient for us to obtain the solution to these equations by using properties of Σ_c, which are generally valid. Hence we turn to a discussion of these general properties.[c]

5.3 Structure of Σ_c and A

From Eq. (5.12), we notice that Σ_c, like G, is composed of two analytic functions:

$$\Sigma_c\,(p, t - t') = \begin{cases} \Sigma^>\,(p, t - t') & \text{for} \quad it > it' \\ \Sigma^<\,(p, t - t') & \text{for} \quad it < it' \end{cases} \tag{5.13}$$

where

$$\Sigma^>\,(p, t - t') = \int \cdots G^<\,(p', t' - t)\, G^>\,(\bar{p}, t - t')\, G^>\,(\bar{p}', t - t')$$

$$\Sigma^<\,(p, t - t') = \int \cdots G^>\,(p', t' - t)\, G^<\,(\bar{p}, t - t')\, G^<\,(\bar{p}', t - t')$$

$$\tag{5.14}$$

In fact, it is true in general that Σ_c is composed of two analytic functions, as indicated in Eq. (5.13).

It is in general convenient to represent the functions $\Sigma^>$ and $\Sigma^<$ as Fourier integral analogous to Eq. (2.12):

$$\Sigma^>\,(p, t) = \int_{-\infty}^{\infty} \frac{d\omega}{2\pi i}\, \Sigma^>\,(p, \omega)\, e^{-i\omega t}$$

$$\Sigma^<\,(p, t) = \pm \int_{-\infty}^{\infty} \frac{d\omega}{2\pi i}\, \Sigma^<\,(p, \omega)\, e^{-i\omega t}$$

$$\tag{5.15}$$

---

[c]*(Original)* ‡The general properties discussed below are not all valid when dealing with hard-core interactions. This will be taken up in Chapter 14.

We have again written the explicit factors of i and $\pm i$ so that the functions $\Sigma^> (p, \omega)$ and $\Sigma^< (p, \omega)$ will turn out to be real and nonnegative.

In particular approximation Ref. (5.14)

$$\Sigma^> (p, \omega) = \int \frac{d\mathbf{p}'d\omega'}{(2\pi)^4} \frac{d\bar{\mathbf{p}}d\bar{\omega}}{(2\pi)^4} \frac{d\bar{\mathbf{p}}'d\bar{\omega}'}{(2\pi)^4}$$
$$\times (2\pi)^4 \, \delta \left(\mathbf{p} + \bar{\mathbf{p}}' - \bar{\mathbf{p}} - \mathbf{p}' \right) \delta \left(\omega + \omega' - \bar{\omega} - \bar{\omega}' \right)$$
$$\times \left(v \left(\mathbf{p} - \bar{\mathbf{p}} \right) \pm v \left(\mathbf{p} - \bar{\mathbf{p}}' \right) \right)^2$$
$$\times G^< \left(p', \omega' \right) G^> \left(\bar{p}, \bar{\omega} \right) G^> \left(\bar{p}', \bar{\omega}' \right)$$

$$\Sigma^< (p, \omega) = \int \cdots G^> \left(p', \omega' \right) G^< \left(\bar{p}, \bar{\omega}' \right) G^< \left(\bar{p}', \bar{\omega}' \right) \qquad (5.16)$$

The second important property of $\Sigma_c (t - t')$ is that it satisfies the same boundary condition (2.10) as G. This is derived from the fact that G_2 satisfies the boundary condition (2.11). Thus, for $0 < it_{1'} < \beta$,

$$\Sigma_c \left(1, 1' \right)\big|_{t_1=0} = \pm e^{\beta \mu} \, \Sigma_c \left(1, 1' \right)\big|_{t_1=-i\beta}$$

or[d]

$$\Sigma^< \left(1, 1' \right)\big|_{t_1=0} = \pm e^{\beta \mu} \, \Sigma^> \left(1, 1' \right)\big|_{t_1=-i\beta} \qquad (5.17)$$

Therefore, $\Sigma^> (p, \omega)$ and $\Sigma^< (p, \omega)$ are related in exactly the same way as $G^> (p, \omega)$ and $G^< (p, \omega)$. In analogy to A, we define

$$\Gamma (p, \omega) = \Sigma^> (p, \omega) \pm \Sigma^< (p, \omega) \qquad (5.18)$$

so that in analogy with Eq. (2.15),

$$\Sigma^> (p, \omega) = \Gamma (p, \omega) \left[1 \pm f(\omega) \right]$$
$$\Sigma^< (p, \omega) = \Gamma (p, \omega) f(\omega)$$

Since Σ_c obeys the quasi-periodicity condition (5.17), it too may be expanded in a Fourier series like (4.3) in the imaginary time interval, with the Fourier coefficients given by

$$\Sigma_c (p, z_\nu) = \int_{-\infty}^{\infty} \frac{d\omega}{2\pi} \frac{\Gamma (p, \omega)}{z_\nu - \omega} \qquad (5.19)$$

where $z_\nu = \left(\frac{\pi \nu}{-i\beta} \right) + \mu$.

---

[d]The subscript c appeared in the original text is omitted.

Now we can see quite directly how to solve Eq. (5.11) for G. We take Fourier coefficients of both sides of this equation by multiplying by $e^{iz_\nu(t-t')}$ and integrating over all t from 0 to $-i\beta$. Then we find

$$[z_\nu - E(p) - \Sigma_c(p, z_\nu)] G(p, z_\nu) = 1$$

This is a relation between the functions $G(p, z)$ and $\Sigma(p, z)$ on the set of points z_ν, and it must, therefore, hold for all complex z. Thus

$$G(p, z) = \frac{1}{z - E(p) - \Sigma_c(p, z)}$$

$$= \frac{1}{z - E(p) - \int \frac{d\omega'}{2\pi} \frac{\Gamma(p, \omega')}{z - \omega'}} \tag{5.20}$$

We recall that A is given in terms of G by the discontinuity of G across the real axis. Hence

$$A(p, \omega) = \frac{i}{\omega + i\epsilon - E(p) - \int \frac{d\omega'}{2\pi} \frac{\Gamma(p, \omega')}{\omega + i\epsilon - \omega'}}$$

$$- \frac{i}{\omega - i\epsilon - E(p) - \int \frac{d\omega'}{2\pi} \frac{\Gamma(p, \omega')}{\omega - i\epsilon - \omega'}}$$

Since

$$\frac{1}{x + i\epsilon} = \wp \frac{1}{x} - i\delta(x)$$

we may write

$$A(p, \omega) = \frac{i}{\omega - E(p) - \Re\Sigma_c(p, \omega) + \left(\frac{i}{2}\right)\Gamma(p, \omega)}$$

$$- \frac{i}{\omega - E(p) - \Re\Sigma_c(p, \omega) - \left(\frac{i}{2}\right)\Gamma(p, \omega)} \tag{5.21}$$

where

$$\Re\Sigma_c(p, \omega) = \wp \int \frac{d\omega'}{2\pi} \frac{\Gamma(p, \omega')}{\omega - \omega'}$$

Finally, we find A in terms of Γ as

$$A(p, \omega) = \frac{\Gamma(p, \omega)}{[\omega - E(p) - \Re\Sigma_c(p, \omega)]^2 + \left[\frac{\Gamma(p, \omega)}{2}\right]^2} \tag{5.22}$$

This equation is an entirely general result.

Notice that A is of the same form as we used in our discussion of the lifetime of single-particle excited states, except that $\Re\Sigma_c$ and Γ depend on frequency. If these are slowly varying functions of the frequency, we can still think of Γ as a lifetime of the single-particle excited state with momentum p. $\Re\Sigma_c$ can clearly be interpreted as the average energy gained by a particle of momentum **p** in virtue of its correlations with all the other particles in the system. Notice that the line shift, $\Re\Sigma_c$, and the line width, Γ, are not independent: They are connected by the dispersion function (5.21). This kind of dispersion relation occurs again and again in many-particle physics.

5.4 Interpretation of the Born Collision Approximation

The above arguments do not depend in the slightest on the use of the Born collision approximation. The result (5.22) is quite generally valid. To gain a more detailed understanding of this result, let us study the lifetime that emerges from the Born collision approximation.

We recall that

$$
\begin{aligned}
\Sigma^> (p, \omega) = \int & \frac{d\mathbf{p}'d\omega'}{(2\pi)^4} \frac{d\bar{\mathbf{p}}d\bar{\omega}}{(2\pi)^4} \frac{d\bar{\mathbf{p}}'d\bar{\omega}'}{(2\pi)^4} \\
& \times (2\pi)^4 \delta \left(\mathbf{p} + \bar{\mathbf{p}}' - \bar{\mathbf{p}} - \bar{\mathbf{p}}' \right) \delta \left(\omega + \omega' - \bar{\omega} - \bar{\omega}' \right) \\
& \times \left(v \left(\mathbf{p} - \bar{\mathbf{p}} \right) \pm v \left(\mathbf{p} - \bar{\mathbf{p}}' \right) \right)^2 \\
& \times G^< \left(p', \omega' \right) G^> \left(\bar{p}, \bar{\omega} \right) G^> \left(\bar{p}', \bar{\omega}' \right) \quad (5.16a)
\end{aligned}
$$

$$
\begin{aligned}
\Sigma^< (p, \omega) = \int & \frac{d\mathbf{p}'d\omega'}{(2\pi)^4} \frac{d\bar{\mathbf{p}}d\bar{\omega}}{(2\pi)^4} \frac{d\bar{\mathbf{p}}'d\bar{\omega}'}{(2\pi)^4} \\
& \times (2\pi)^4 \delta \left(\mathbf{p} + \bar{\mathbf{p}}' - \bar{\mathbf{p}} - \bar{\mathbf{p}}' \right) \delta \left(\omega + \omega' - \bar{\omega} - \bar{\omega}' \right) \\
& \times \left(v \left(\mathbf{p} - \bar{\mathbf{p}} \right) \pm v \left(\mathbf{p} - \bar{\mathbf{p}}' \right) \right)^2 \\
& \times G^> \left(p', \omega' \right) G^< \left(\bar{p}, \bar{\omega}' \right) G^< \left(\bar{p}', \bar{\omega}' \right) \quad (5.16b)
\end{aligned}
$$

Equations (5.16) look rather horrible, but actually they are quite easy to understand. Γ is related to the decay of the probability that when we add a particle with momentum $\mathbf{p}$ to a system at time t' and then remove a particle with this momentum at time t, we return to the same state. In fact when A is a Lorentzian line shape (5.3), this probability is

$$
\left| \langle \hat{\psi} \left(\mathbf{p}t \right) \hat{\psi}^\dagger \left(\mathbf{p}t' \right) \rangle \right|^2 \sim e^{-\Gamma(p)|t-t'|}
$$

Now we do not expect the system to return to the same state if the added particle disturbs the system in any way. In particular, if the particle collides with other particles, this will prevent the system from returning to its initial state. We may interpret $\Sigma^> (p, \omega)$ as the collision rate of the added particles. To see this, consider a collision in which a particle with momentum $\mathbf{p}$ and energy ω scatters off a particle with[e] momentum $\mathbf{p}'$ and energy ω' and the two particles end up in states $\bar{\mathbf{p}}, \omega$ and $\bar{\mathbf{p}}', \bar{\omega}'$:[f]

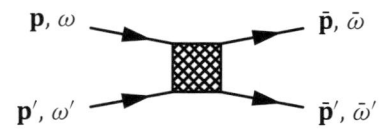

In the Born approximation, the differential cross section for such a process is proportional to $[v (\mathbf{p} - \bar{\mathbf{p}}) \pm v (\mathbf{p} - \bar{\mathbf{p}}')]^2$ times delta functions representing the conservation of energy and momentum in the collision. We can recognize these factors in Eq. (5.16a). To get the collision rate, we must multiply by the density of scatterers, $G^< (p', \omega') = A (p', \omega') f (\omega')$ and by the density of available final states, $G^> (\bar{p}, \bar{\omega}) G^> (\bar{p}', \bar{\omega}') = A (\bar{p}, \bar{\omega}) A (\bar{p}', \bar{\omega}') \times [1 \pm f (\bar{\omega})] [1 \pm f (\bar{\omega}')]$. Thus, we see that Eq. (5.16a) is indeed the total collision rate of the added particle.

In a low-density system, e.g., any classical system, $\Sigma^> (p, \omega)$ represents the entire lifetime. This follows because the boundary condition implies $\Sigma^< (p, \omega) = e^{-\beta(\omega-\mu)} \Sigma^> (p, \omega)$. However, in a low-density system $\beta\mu \to -\infty$, so that $\Sigma^<$ is negligible in comparison with $\Sigma^>$.

In a highly degenerate system, however, $\Sigma^<$ is just as large as $\Sigma^>$. By just the same arguments as we have just gone through, we can see that $\Sigma^<$ is the total collision rate into the configuration $\mathbf{p}, \omega$, assuming that $\mathbf{p}, \omega$ is initially empty. Hence we must conclude that, for fermions, the total decay rate, $\Gamma (\mathbf{p}, \omega)$, is the sum of the rates for scattering in and scattering out, whereas for bosons this total rate is the difference between these two rates. How can this result be understood physically.

---

[e]The typographic errors of the appearance of the symbols $\mathbf{p}'$ and ω' in the original text are fixed.

[f]The hatched circle of the diagram in the original text is replaced by the hatched square.

We said that the system would not come back to the same state whenever the interaction between the added particle and the particles originally present changed the configuration of the system. The added particle has two effects. First, this particle itself undergoes collisions, $\mathbf{p}, \omega + \mathbf{p}', \omega' \rightarrow \bar{\mathbf{p}}, \bar{\omega} + \bar{\mathbf{p}}', \bar{\omega}'$, as represented in $\Sigma^>$. Second, the added particle changes the rate of occurrence of the inverse process, $\bar{\mathbf{p}}, \bar{\omega} + \bar{\mathbf{p}}', \bar{\omega}' \rightarrow \mathbf{p}, \omega + \mathbf{p}', \omega'$, as represented in $\Sigma^<$. For a fermion system, these inverse processes are inhibited because the exclusion principle prevents a scattering from sending a particle into the state $\mathbf{p}, \omega$. Then the net effect of $\Sigma^>$ and $\Sigma^<$ is that extra particles pile up in the configurations $\bar{\mathbf{p}}, \omega$ and $\bar{\mathbf{p}}', \bar{\omega}'$. Thus, for fermions, $\Sigma^>$ and $\Sigma^<$ contribute additively to the lifetime.

On the other hand, for bosons the presence of an extra particle in the state $\mathbf{p}, \omega$ increases the probability of a scattering into that state, since it increases the density available of final state. Now the processes represented in $\Sigma^<$ will tend to decrease the occupation of the configurations $\bar{\mathbf{p}}, \bar{\omega}$ and $\bar{\mathbf{p}}', \bar{\omega}'$, whereas the processes represented by $\Sigma^>$ will tend to increase the occupation of these configurations. Therefore, for bosons, $\Gamma = \Sigma^> - \Sigma^<$.

In a zero-temperature fermion system, it is quite convenient to interpret $\Sigma^>$ and $\Sigma^<$ in the language of "holes and particles." Here, $\Sigma^> (p, \omega)$ is the lifetime of a particle state and vanishes for $\omega \leq \mu$, while $\Sigma^< (p, \omega)$ is the lifetime of a hole state, and it vanishes for $\omega \geq \mu$. When our model is specialized to zero temperature, $\Gamma (p, \omega) = 0$ at $\omega = \mu$. This result, which is true in all order of perturbation theory, enables us to define long-lived single-particle states near the edge of the Fermi sea.

After all this talk about the meaning of the result we have obtained, it is important to notice that we really do not have a solution for A. Equations (5.16) and (5.22) represent a horribly complex set of integral equations for A. To get detailed numerical answers, it is necessary to solve these equations. For example, if Γ is small, to 0-th order, we can take $A (p, \omega) = 2\pi \delta (\omega - E(p))$. To first order, we could substitute this form for A into Eq. (5.16) and obtain the lowest-order results for $\Sigma^>$ and $\Sigma^<$. Then we would substitute these approximations for $\Sigma^>$ and $\Sigma^<$ into Eq. (5.22) and find the first-order solution result for A. And so forth.

5.5 Boltzmann Equation Interpretation

We have just been considering the response of a system, initially in equilibrium, to a disturbance that adds a particle with momentum **p** to the system. A perhaps more familiar way of describing the behavior of a system after a disturbance is by means of the Boltzmann equation. Now we shall indicate how the lifetime obtained in the previous section may also be derived from a Boltzmann equation.

The Boltzmann equation is only valid in cases in which Γ, the dispersion in energy, is small, so that a particle with momentum $\mathbf{p}_1$ can be considered to have the energy $E(p_1)$. Then we can describe the system after the disturbance in terms of $\overline{n(\mathbf{p}_1, T)}$, the average density of particles with momentum $\mathbf{p}_1$ and time T. The Boltzmann equation expresses the time derivative of $\overline{n(\mathbf{p}_1, T)}$ as the rate of scattering of particle into the state with momentum $\mathbf{p}_1$ minus the rate of scattering out of momentum $\mathbf{p}_1$. If we use Born approximation cross section, we find, as the Boltzmann equation,

$$
\begin{aligned}
\frac{\partial}{\partial T}\overline{n(\mathbf{p}_1, T)} = -\int & \frac{d\mathbf{p}'}{(2\pi)^3}\frac{d\bar{\mathbf{p}}}{(2\pi)^3}\frac{d\bar{\mathbf{p}}'}{(2\pi)^3} \\
& \times 2\pi\delta\left[E(p_1) + E(p') - E(\bar{p}) - E(\bar{p}')\right] \\
& \times (2\pi)^3\delta\left(\mathbf{p}_1 + \mathbf{p}' - \bar{\mathbf{p}} - \bar{\mathbf{p}}'\right)\left(\frac{1}{2}\right) \\
& \times \left[v(\mathbf{p}_1 - \bar{\mathbf{p}}) \mp v(\mathbf{p}_1 - \bar{\mathbf{p}}')\right]^2 \\
& \times \left\{\overline{n(\mathbf{p}_1, T)}\,\overline{n(\mathbf{p}', T)}\left[1 \pm \overline{n(\bar{\mathbf{p}}, T)}\right]\left[1 \pm \overline{n(\bar{\mathbf{p}}', T)}\right]\right. \\
& \left. -[1 \pm \overline{n(\mathbf{p}_1, T)}][1 \pm \overline{n(\mathbf{p}', T)}]\overline{n(\bar{\mathbf{p}}, T)}\,\overline{n(\bar{\mathbf{p}}', T)}\right\}
\end{aligned}
$$

(5.23)

After adding at time $T = 0$ a particle with momentum **p** to a system in equilibrium, $\overline{n(\mathbf{p}_1, 0)}$ is given by

$$
\overline{n(\mathbf{p}_1, 0)} = f(E(p_1)) = \frac{1}{e^{\beta(E(p_1)-\mu)} \mp 1} \quad \text{for} \quad \mathbf{p}_1 \neq \mathbf{p} \quad (5.24)
$$

However, $\overline{n(\mathbf{p}, T)}$ is initially not given by its equilibrium value but is instead $\overline{n(\mathbf{p}_1, 0)}$. Now, $\overline{n(\mathbf{p}_1, T)}$ for $\mathbf{p}_1 \neq \mathbf{p}$ will never change appreciably from its equilibrium value. Therefore, for this initial condition, the Boltzmann equation (5.23) reduced to the simple result

$$\frac{\partial}{\partial T}\overline{n\left(\mathbf{p},\,T\right)} = -\overline{n\left(\mathbf{p},\,T\right)}\Sigma^{>}\left(p\right) + \left[1 \pm \overline{n\left(\mathbf{p},\,T\right)}\right]\Sigma^{<}\left(p\right) \quad (5.25)$$

Notice that $\Sigma^{>}\left(p\right)$ and $\Sigma^{<}\left(p\right)$ are precisely the values of $\Sigma^{>}\left(p,\,\omega = E\left(p\right)\right)$ and $\Sigma^{<}\left(p,\,\omega = E\left(p\right)\right)$, which emerges when A is approximated by $2\pi\delta\left(\omega - E\left(p\right)\right)$. Equation (5.25) has the solution

$$\overline{n\left(\mathbf{p},\,T\right)} = f\left(E\left(p\right)\right) + e^{-\Gamma\left(p\right)T}\left[\overline{n\left(p,\,0\right)} - f\left(E\left(p\right)\right)\right]$$

where

$$\Gamma\left(p\right) = \Sigma^{>}\left(p\right) \mp \Sigma^{<}\left(p\right) \quad (5.26)$$

This result indicates a close correspondence between our Born collision approximation and the results of an analysis based on a Boltzmann equation with Born approximation collision cross sections. We shall later use a generalization of the Born collision approximation for G to derive this Boltzmann equation.

Chapter 6

A Technique for Deriving Green's Function Approximations

Up to now we have written approximation for G by relying on the propagator interpretations of G and of the G_2 that appears in the equations of motion for G. We have thus been able to write a few simple approximations for G_2 in terms of the processes that we wish to consider. However, physical intuition can take us just so far. The use of purely imaginary times makes a direct interpretation of these equations difficult. Furthermore, it is hard to find physical ways of determining the numerical factors that appear in front of the various terms in the expansion of G_2. We, therefore, seek a systematic way of deriving approximations for G.

As a purely formal device, we define a generalization of the one-particle Green's function in the imaginary time interval $[0, -i\beta]$:

$$G\left(1, 1'; U\right) = \frac{1}{i} \frac{\left\langle \hat{T}\left[\hat{S}\hat{\psi}\left(1\right)\hat{\psi}^{\dagger}\left(1'\right)\right]\right\rangle}{\left\langle \hat{T}\left[\hat{S}\right]\right\rangle} \tag{6.1}$$

Here $\hat{T}$ means imaginary time ordering and the operator $\hat{S}$ is given by

$$\hat{S} = \exp\left[-i\int_0^{-i\beta} d2\, U\left(2\right)\hat{n}(2)\right] \tag{6.2}$$

Annotations to Quantum Statistical Mechanics
In-Gee Kim
Copyright © 2018 Pan Stanford Publishing Pte. Ltd.
ISBN 978-981-4774-15-4 (Hardcover), 978-1-315-19659-6 (eBook)
www.panstanford.com

$\hat{n}(2) = \hat{\psi}^\dagger(2)\hat{\psi}(2)$ and $U(2)$ is a function of space and times in the interval[a] $[0, -i\beta]$.

One reason that the Green's function (6.1) is convenient to use is that it satisfies the same boundary condition,

$$G\left(1, 1'; U\right)\big|_{t_1=0} = \pm e^{\beta\mu}\, G\left(1, 1'; U\right)\big|_{t_1=-i\beta} \qquad (6.3)$$

as the equilibrium Green's function. The derivation of this boundary condition for $G(U)$ is essentially the same as for the equilibrium functions. The time 0 is the earliest possible time, so that

$$G\left(1, 1'; U\right)\big|_{t_1=0} = \pm \frac{1}{i} \frac{\left\langle \hat{T}\left[\hat{S}\hat{\psi}^\dagger\left(1'\right)\right]\hat{\psi}\left(\mathbf{r}_1, 0\right)\right\rangle}{\left\langle \hat{T}\left[\hat{S}\right]\right\rangle}$$

Since the time $-i\beta$ is the latest possible time,

$$G\left(1, 1'; U\right)\big|_{t_1=-i\beta} = \frac{1}{i} \frac{\left\langle \hat{\psi}\left(\mathbf{r}_1, -i\beta\right)\hat{T}\left[\hat{S}\hat{\psi}^\dagger\left(1'\right)\right]\right\rangle}{\left\langle \hat{T}\left[\hat{S}\right]\right\rangle}$$

The cyclic invariance of the trace that defines the expectation value then implies Eq. (6.3).

Another reason this Green's function is convenient is that it obeys equations of motion quite similar to those obeyed by the equilibrium function G. These are

$$\left[i\frac{\partial}{\partial t_1} + \frac{\nabla_1^2}{2m} - U(1)\right]G\left(1, 1'; U\right) = \delta\left(1 - 1'\right)$$

$$\pm i \int d\mathbf{r}_2\, v\left(\mathbf{r}_1 - \mathbf{r}_2\right) G_2\left(12, 1'2^+; U\right)\big|_{t_2=t_2} \qquad (6.4\mathrm{a})$$

and

$$\left[i\frac{\partial}{\partial t_{1'}} + \frac{\nabla_{1'}^2}{2m} - U\left(1'\right)\right]G\left(1, 1'; U\right) = \delta\left(1 - 1'\right)$$

$$\mp i \int d\mathbf{r}_2\, v\left(\mathbf{r}_2 - \mathbf{r}_{1'}\right) G_2\left(12^-, 1'2; U\right)\big|_{t_2=t_2} \qquad (6.4\mathrm{b})$$

where

$$G_2\left(12, 1'2'; U\right) = \left(\frac{1}{i}\right)^2 \frac{\left\langle \hat{T}\left[\hat{S}\hat{\psi}(1)\hat{\psi}(2)\hat{\psi}^\dagger(2')\hat{\psi}^\dagger(1')\right]\right\rangle}{\left\langle \hat{T}\left[\hat{S}\right]\right\rangle} \qquad (6.5)$$

---

[a] *(Original)* ‡We may regard $G\left(1, 1'; U\right)$ as a one-particle Green's function, written in the interaction reapresentation, for the system developing in imaginary time in the presence of the scalar potential U. This potential is represented by adding a term $\int dr\, U(r, t)\hat{n}(r, t)$ to the Hamiltonian. In the interaction representation, all the U dependence is explicit int the $\hat{S}$ factor, and the field operators are the same as in the absence of the potential.

We derive Eq. (6.4) in exactly the same way as the equations of motion for the equilibrium function $G(1-1')$. The only new feature is the appearance of the terms UG. To see the origin of these terms, consider, for example,

$$\hat{T}\left[\hat{S}\hat{\psi}(1)\right] = \hat{T}\left\{\exp\left[i\int_{t_1}^{-i\beta} d2\, U(2)\hat{n}(2)\right]\right\}$$
$$\times \hat{\psi}(1)\hat{T}\left\{\exp\left[-i\int_0^{t_1} d2\, U(2)\hat{n}(2)\right]\right\}$$

Then

$$i\frac{\partial}{\partial t_1}\hat{T}\left[\hat{S}\hat{\psi}(1)\right] = \hat{T}\left\{\exp\left[-i\int_{t_1}^{-i\beta} d2\, U(2)\hat{n}(2)\right]\right\}$$
$$\times \left\{i\frac{\partial\hat{\psi}(1)}{\partial t_1} + \int d\mathbf{r}_2\, U(\mathbf{r}_2, t_1)\left[\hat{\psi}(1), \hat{n}(\mathbf{r}_2, t_1)\right]\right\}$$
$$\times \hat{T}\left\{\exp\left[i\int_{t_1}^{-i\beta} d2\, U(2)\hat{n}(2)\right]\right\}$$

Since from Eq. (2.5)

$$\left[\hat{\psi}(\mathbf{r}_1, t_1), \hat{n}(\mathbf{r}_2, t_1)\right] = \delta(\mathbf{r}_1 - \mathbf{r}_2)\psi(\mathbf{r}_1, t_1)$$

it follows that

$$i\frac{\partial}{\partial t_1}\left[\hat{T}(\hat{S}\hat{\psi}(1))\right] = \hat{T}\left[\hat{S}i\frac{\partial\hat{\psi}(1)}{\partial t_1}\right] + \hat{T}\left[\hat{S}\hat{\psi}(1)\right] U(1) \qquad (6.6)$$

Such a calculation is the source of the UG term in Eq. (6.4a).

So far we have only succeeded in making things more complicated. We shall learn something by considering the change in $G(U)$ resulting from an infinitesimal change in U. We let

$$U(2) \rightarrow U(2) + \delta U(2) \qquad (6.7)$$

The change in G resulting from this change in U is

$$\delta G(1, 1'; U) = \delta\left\{\frac{1}{i}\frac{\langle\hat{T}[\hat{S}\hat{\psi}(1)\hat{\psi}^\dagger(1')]\rangle}{\langle\hat{T}[\hat{S}]\rangle}\right\}$$
$$= \frac{1}{i}\left[\frac{\langle\hat{T}[\delta\hat{S}\hat{\psi}(1)\hat{\psi}^\dagger(1')]\rangle}{\langle\hat{T}[\hat{S}]\rangle} - \frac{\langle\hat{T}[\delta\hat{S}]\rangle}{\langle\hat{T}[\hat{S}]\rangle}\frac{\langle\hat{T}[\hat{S}\hat{\psi}(1)\hat{\psi}^\dagger(1')]\rangle}{\langle\hat{T}[\hat{S}]\rangle}\right]$$

$$(6.8)$$

When $\delta\hat{S}$ appears in a time-ordered product, it can be evaluated as

$$\delta\hat{S} = \delta\left\{\exp\left[-i\int_0^{-i\beta} d2\, U(2)\hat{n}(2)\right]\right\} = \hat{S}\frac{1}{i}\int_0^{-i\beta} d2\, \delta U(2)\hat{n}(2)$$

$$(6.9)$$

since the $\hat{T}$'s automatically provide the proper (imaginary) time ordering. On substituting Eq. (6.9) into Eq. (6.8), we find

$$\delta G(1, 1'; U) = \int_0^{-i\beta} d2 \left\{ \frac{\langle \hat{T}\, [\hat{S}\hat{\psi}(1)\hat{\psi}^\dagger(1')\,\hat{n}(2)]\rangle}{i^2\,\langle \hat{T}\,[\hat{S}]\rangle} \right.$$

$$\left. - \frac{\langle \hat{T}\,[\hat{S}\hat{\psi}(1)\hat{\psi}^\dagger(1')]\rangle\,\langle \hat{T}\,[\hat{S}\hat{n}(2)]\rangle}{i\,\langle \hat{T}\,[\hat{S}]\rangle} \frac{}{i\,\langle \hat{T}\,[\hat{S}]\rangle} \right\}\delta U(2)$$

$$= \pm\int_0^{-i\beta} d2[G_2(12, 1'2^+; U)$$

$$- G(1, 1'; U)\, G(2, 2^+; U)]\delta U(2) \qquad (6.10)$$

Since this calculation of δG is just a generalization of the method by which one obtains an ordinary derivative, we call the coefficient of $\delta U(2)$ in Eq. (6.10) the functional derivative, or variational derivative, of $G(1, 1'; U)$ with respect to $U(2)$. It is denoted by $\delta G(1, 1'; U)/\delta U(2)$, so that

$$\frac{\delta G(1, 1'; U)}{\delta U(2)} = \pm\left[G_2(12, 1'2^+; U) - G(1, 1'; U)\, G(2, 2^+; U)\right]$$

$$(6.11)$$

We may, therefore, express the G_2 that appears in the equation of motion Eq. (6.4) for G in terms of $\delta G/\delta U$. This equation then becomes

$$\left\{ i\frac{\partial}{\partial t_1} + \frac{\nabla_1^2}{2m} - U(1) \mp i\int d\mathbf{r}_2 v(\mathbf{r}_1 - \mathbf{r}_2) \right.$$

$$\left. \times \left[G(\mathbf{r}_2 t_1, \mathbf{r}_2 t_1^+; U) + \frac{\delta}{\delta U(\mathbf{r}_2, t_1^+)} \right] \right\} G(1, 1'; U) = \delta(1 - 1')$$

$$(6.12)$$

The Green's function $G(U)$ is thus determined by a single functional differential equation.

Unfortunately, there exist no practical techniques for solving such functional differential equations exactly. Equation (6.12) may be used, however, to generate approximate equations for G. We shall begin our discussion by using Eq. (6.12) to derive the beginning of a perturbative expansion of $G(U)$ in a power series in v.

6.1 Ordinary Perturbation Theory

If there is no interaction between the particles, $G(U)$ is determined by the equation

$$\left[i \frac{\partial}{\partial t_1} + \frac{\nabla_1^2}{2m} - U(1) \right] G_0\left(1, 1'; U\right) = \delta\left(1 - 1'\right) \tag{6.13}$$

together with the boundary condition (6.3). The function $G_0(1, 1'; U)$ may be used to convert Eq. (6.12) into an integral equation:[b]

$$G\left(1, 1'; U\right) = G_0\left(1, 1'; U\right) \pm i \int_0^{-i\beta} d\bar{1} d\bar{2}\, G_0\left(1, \bar{1}; U\right) V\left(\bar{1} - \bar{2}\right)$$

$$\times \left[G\left(\bar{2}, \bar{2}^+; U\right) + \frac{\delta}{\delta U\left(\bar{2}\right)} \right] G\left(\bar{1}, 1'; U\right) \tag{6.14}$$

We have introduced the notation

$$V\left(1 - 1'\right) = v\left(|r_1 - r_{1'}|\right) \delta\left(t_1 - t_{1'}\right) \tag{6.15}$$

By applying $\left[i \frac{\partial}{\partial t_1} + \frac{\nabla_1^2}{2m} - U(1) \right]$ to Eq. (6.14), one can verify that Eq. (6.14) is a solution to Eq. (6.12). To see that it satisfies the boundary condition (6.3), we observe that

$$G\left(1, 1'; U\right)\big|_{t_1=0} = G_0\left(1, 1'; U\right)\big|_{t_1=0} + \int_0^{-i\beta} d\bar{1}\, G_0\left(1, 1'; U\right)\big|_{t_1=0} \cdots$$

$$= \pm e^{\beta\mu} \left[G_0\left(1, 1'; U\right)\big|_{t_1=-i\beta} \right.$$

$$+ \int_0^{-i\beta} d\bar{1}\, G_0\left(1, 1'; U\right)\big|_{t_1=-i\beta} \cdots \right]$$

$$= \pm e^{\beta\mu}\, G\left(1, 1'; U\right)\big|_{t_1=-i\beta}$$

Notice that Eq. (6.14) contains time integrals from 0 to $-i\beta$. This is the ultimate origin of the appearance of such integrals in the Born collision approximation.

To expand $G(U)$ in a power series in V, we need only successively iterate Eq. (6.14). To zeroth order $G = G_0$: The first-order term is

---

[b]There was a typographic error in the equation number of the original text.

obtained by substituting $G = G_0$ into the right side of Eq. (6.14). Then to first order in V:

$$G\left(1, 1'; U\right) = G_0\left(1, 1'; U\right) \pm i \int_0^{-i\beta} d\bar{1}d\bar{2}\, G_0\left(1, \bar{1}; U\right) V\left(\bar{1} - \bar{2}\right)$$

$$\times \left[G_0\left(\bar{2}, \bar{2}^+; U\right) + \frac{\delta}{\delta U\left(\bar{2}\right)}\right] G_0\left(\bar{1}, 1'; U\right) \quad (6.16)$$

We then must compute $\frac{\delta}{\delta U(2)} G_0\left(1, 1'; U\right)$. Perhaps the simplest way of finding this derivative is to regard $G_0\left(1, 1'; U\right)$ as a matrix in the variable 1 and $1'$. The inverse of this matrix, defined by

$$\int_0^{-i\beta} d\bar{1}\, G_0\left(1, \bar{1}; U\right) G_0\left(\bar{1}, 1'; U\right) = \delta\left(1 - 1'\right)$$

is, from Eq. (6.13), just

$$G_0^{-1}\left(1, 1'; U\right) = \left[i\frac{\partial}{\partial t_1} + \frac{\nabla_1^2}{2m} - U\left(1\right)\right]\delta\left(1 - 1'\right) \quad (6.17)$$

Varying both sides of the matrix equation $G_0^{-1}G_0 = 1$ with respect to U implies

$$\delta\left[G_0^{-1}G_0\right] = \delta G_0^{-1}G_0 + G_0^{-1}\delta G_0 = 0$$

or

$$\delta G_0 = -G_0\delta G_0^{-1}G_0$$

Thus

$$\frac{\delta G_0\left(1, 1'\right)}{\delta U\left(2\right)} = -\int_0^{-i\beta} d3d3'\, G_0\left(1, 3\right)\left[\frac{\delta G_0^{-1}\left(3, 3'; 0\right)}{\delta U\left(2\right)}\right] G_0\left(3', 1'\right)$$

$$= \int_0^{-i\beta} d3\, G_0\left(1, 3\right)\frac{\delta U\left(3\right)}{\delta U\left(2\right)}G_0\left(3, 1'\right)$$

$$= G_0\left(1, 2\right) G_0\left(2, 1'\right) \quad (6.18)$$

since $\frac{\delta U(3)}{\delta U(2)} = \delta(3 - 2)$. Substituting Eq. (6.18) into Eq. (6.16), we find that to first order in V,

$$G\left(1, 1'; U\right) = G_0\left(1, 1'; U\right) \pm i \int_0^{-i\beta} d\bar{1}d\bar{2}\, G_0\left(1, \bar{1}; U\right) V\left(\bar{1} - \bar{2}\right)$$

$$\times\left[G_0\left(\bar{2}, \bar{2}^+; U\right) G_0\left(\bar{1}, 1'; U\right)\right.$$

$$\left.\pm\, G_0\left(\bar{1}, \bar{2}^+; U\right) G_0\left(\bar{2}, 1'; U\right)\right] \tag{6.19}$$

We represent this pictorially as

$$G\left(1, 1'; U\right) = \quad 1' \longrightarrow 1 + 1' \longrightarrow 1$$

$$\times \quad 1' \longrightarrow 1$$

where the lines signify G_0. When U is set equal to zero, we have the expansion of $G\left(1 - 1'\right)$ to first order in V.

It is instructive to compare this first-order result with the Hartree–Fock approximation, which may be written as

$$\left(i\frac{\partial}{\partial t_1} + \frac{\nabla_1^2}{2m}\right) G\left(1 - 1'\right) = \delta\left(1 - 1'\right) \pm i \int_0^{-i\beta} d\bar{2}\, V\left(\bar{1} - \bar{2}\right)$$

$$\times\left[G\left(\bar{2} - \bar{2}^+\right) G\left(1 - 1'\right)\right.$$

$$\left.\pm\, G\left(1 - \bar{2}^2\right) G\left(\bar{2} - 1'\right)\right]$$

Then

$$G\left(1 - 1'\right) = G_0\left(1 - 1'\right) \pm i \int_0^{-i\beta} d\bar{1}d\bar{2}\, G_0\left(1 - \bar{1}\right) V\left(\bar{1} - \bar{2}\right)$$

$$\times\left[G\left(\bar{2} - \bar{2}^+\right) G\left(1 - 1'\right) \pm G\left(1 - \bar{2}^2\right) G\left(\bar{2} - 1'\right)\right] \tag{6.20}$$

The first-order solution (6.19) is equivalent to the Hartree–Fock solution expanded to first order in V.

To obtain higher-order terms in V, we substitute Eq. (6.19) back into Eq. (6.14), and again use Eq. (6.18). The GG term gives the

second-order contributions

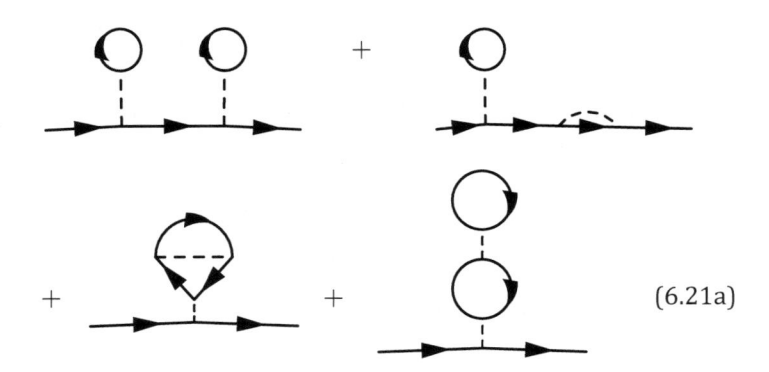

$$\tag{6.21a}$$

while the second-order contribution from $\frac{\delta G}{\delta U}$ is

$$i \int_0^{-i\beta} d\bar{1}d\bar{2}\, G_0\left(1, \bar{1}\, U\right) V\left(\bar{1} - \bar{2}\right) \tag{6.21b}$$

$$\times \frac{\delta}{\delta U\left(\bar{2}\right)} \left\{ \pm i \int d\bar{3}d\bar{4} G_0\left(\bar{1}, \bar{3}; U\right) V\left(\bar{3} - \bar{4}\right) \right.$$

$$\left. \times \left[G_0\left(\bar{4}, \bar{4}^+; U\right) G_0\left(\bar{3}, 1'; U\right) \pm G_0\left(\bar{3}, \bar{4}^+; u\right) G_0\left(\bar{3}, 1'; U\right) \right] \right\}$$

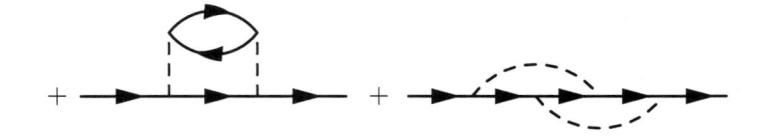

All the terms in Eq. (6.21a) and the first four terms in Eq. (6.21b) arise from an iteration of the Hartree–Fock approximation. However, the last two terms do not appear in the Hartree–Fock approximation, but are instead the lowest-order contributions of the collision terms in the Born collision approximation. In the appendix, we consider this expansion in more detail.

One can iterate further and expand G to arbitrarily high order in V. The general structure of G is given by drawing all topologically different connected diagrams.

We should point out that there are very few situations in which this expansion converges rapidly. Usually, the potential is sufficiently large so that the first few orders of perturbation theory give a very poor answer. Furthermore, physical effects such as the $e^{-\Gamma(t-t')}$ behavior of G and the single-particle energy shift cannot appear in finite order in this expansion. Instead, one would find $e^{-\Gamma(t-t')}$ replaced by its power-series expansion

$$1 - \Gamma\left(t - t'\right) + \left(\frac{1}{2}\right)\Gamma^2\left(t - t'\right)^2 + \cdots$$

which converges slowly for large time differences.

6.2 Expansion of Σ in V and G_0

The difficulties of the expansion of G in powers of V may be avoided by either infinite classes of terms in the expansion, or equivalently by expanding the self-energy $\Sigma\left(1, 1'; U\right)$ in terms of V. We recall that Σ is defined by

$$\left(i\frac{\partial}{\partial t_1} + \frac{\nabla_1^2}{2m}\right)G\left(1 - 1'\right) - \int_0^{-i\beta} d\bar{1}\Sigma\left(1 - \bar{1}\right)G\left(\bar{1} - 1'\right) = \delta\left(1 - 1'\right)$$

$$(6.22)$$

in equilibrium case. In the presence of U, we define Σ by the equation

$$\int_0^{-i\beta} d\bar{1}\left[G_0^{-1}\left(1, \bar{1}; U\right) - \Sigma\left(1, \bar{1}; U\right)\right]G\left(\bar{1}, 1'; U\right) = \delta\left(1 - 1'\right)$$

$$(6.23)$$

If we define the matrix inverse of G by the equation

$$\int_0^{-i\beta} d\bar{1}\, G^{-1}\left(1, \bar{1}; U\right)G\left(\bar{1}, 1'; U\right) = \delta\left(1 - 1'\right)$$

it is clear that

$$G^{-1}\left(1, 1'; U\right) = G_0^{-1}\left(1, 1'; U\right) - \Sigma\left(1, 1'; U\right) \tag{6.24}$$

To find $\Sigma(U)$, we matrix multiply Eq. (6.12) on the right by G^{-1}. Then

$$G^{-1}(1, 1'; U) = G_0^{-1}(1, 1'; U) \mp i \int_0^{-i\beta} d\bar{2} V(\bar{1} - \bar{2}) G(\bar{2}, \bar{2}^+) \delta(1 - 1')$$

$$- i \int_0^{-i\beta} d\bar{2} d\bar{1} V(1 - \bar{2}) \left[\frac{\delta G(1, \bar{1}; U)}{\delta U(2)} \right] G^{-1}(\bar{1}, 1'; U) \tag{6.24a}$$

so that

$$\Sigma\left(1, 1'; U\right) = \pm i \int_0^{-i\beta} d\bar{2} V\left(\bar{1} - \bar{2}\right) G\left(\bar{2}, \bar{2}^+\right) \delta\left(1 - 1'\right)$$

$$+ i \int_0^{-i\beta} d\bar{2} d\bar{1} V(1 - \bar{2}) \left[\frac{\delta G(1, \bar{1}; U)}{\delta U(2)} \right] G^{-1}(\bar{1}, 1'; U) \tag{6.25a}$$

Using $\delta G \cdot G^{-1} + G \delta G^{-1} = 0$, we find

$$\int_0^{-i\beta} d\bar{1} \left[\frac{\delta G\left(1, \bar{1}; U\right)}{\delta U(2)} \right] G^{-1}\left(\bar{1}, 1'; U\right)$$

$$= -\int d\bar{1}\, G\left(1, \bar{1}; U\right) \frac{\delta}{\delta U(2)} \left[G_0^{-1}\left(\bar{1}, 1'; U\right) - \Sigma\left(\bar{1}, 1'; U\right) \right]$$

$$= G\left(1, 1'\right) \delta\left(2 - 1'\right) + \int_0^{-i\beta} d\bar{1}\, G\left(1, \bar{1}; U\right) \frac{\delta \Sigma\left(\bar{1}, 1'; U\right)}{\delta U(2)}$$

Hence Eq. (6.25a) for Σ becomes

$$\Sigma\left(1, 1'; U\right) = \delta\left(1 - 1'\right) \left[\pm i \int d\bar{2}\, V\left(1 - \bar{2}\right) G\left(\bar{2}, \bar{2}^+; U\right) \right]$$

$$+ iV\left(1 - 1'\right) G\left(1, 1'; U\right)$$

$$+ i \int d\bar{1} d\bar{2}\, V\left(1 - \bar{2}\right) G\left(1, \bar{1}; U\right) \frac{\delta \Sigma\left(\bar{1}, 1'; U\right)}{\delta U(\bar{2})} \tag{6.25b}$$

This latter equation is very useful for deriving the expansion of Σ in a power series in G_0 and V. To lowest order in V,

$$\Sigma\left(1, 1'; U\right) = \pm i\delta\left(1 - 1'\right) \int d\bar{2}\, V\left(1 - \bar{2}\right) G_0\left(\bar{2}, \bar{2}^+; U\right)$$

$$+ iV\left(1 - 1'\right) G_0\left(1, 1'; U\right) \tag{6.26}$$

This is clearly just the lowest-order approximation to the Hartree–Fock self-energy. The second-order result for Σ is obtained by taking the Hartree–Fock terms in Eq. (6.25b) to first order in G, using Eq. (6.19). The more interesting second-order terms in Σ result from $\frac{\delta\Sigma}{\delta U}$. To lowest order, these terms are

$$+ \qquad\qquad \tag{6.27}$$

where the lines signify G_0's. Expression (6.27) is just the lowest-order evaluation of the collision term in the Born collision approximation self-energy.

6.3 Expansion of Σ in V and G

In the calculations in previous chapters, we have expanded Σ in V and G instead of V and G_0. The primary reason for doing this is that G has a simple physical interpretation, while the physical significance of G_0 in an interacting system is far from clear. We shall, therefore, indicate how successive iteration of Eq. (6.25b) leads to such an expansion in G and V.

The Hartree approximation is derived by neglecting $\frac{\delta G}{\delta U}$ in Eq. (6.25a). This approximation is the first term in the systematic expansion of Σ in a series in V and G:

$$\Sigma_{\mathrm{HF}}\left(1, 1'; U\right) = \pm i \int d\bar{2}\, V\left(1 - 2\right) G\left(\bar{2}, \bar{2}^+; U\right) \delta\left(1 - 1'\right)$$
$$+ i V\left(1 - 1'\right) G\left(1, 1'; U\right)$$

The next term comes from approximating $\frac{\delta\Sigma}{\delta U}$ by $\frac{\delta\Sigma_{\mathrm{HF}}}{\delta U}$ in Eq. (6.25b). Then Eq. (6.25b) becomes

$$\Sigma\left(1, 1'; U\right) = \Sigma_{\mathrm{HF}}\left(1, 1'; U\right) \pm i^2 \int_0^{-i\beta} d\bar{1}d\bar{2}\, V\left(1 - \bar{2}\right) G\left(1, \bar{1}; U\right)$$
$$\times \frac{\delta}{\delta U\left(\bar{2}\right)}\left[\int d\bar{3}\, V\left(\bar{1} - \bar{3}\right) G\left(\bar{3}, \bar{3}^+; U\right) \delta\left(\bar{1} - 1'\right)\right.$$
$$\left. \pm V\left(\bar{1} - 1'\right) G\left(\bar{1}, 1'; U\right)\right]$$

However, $\delta G = -G \cdot \delta G^{-1} \cdot G$, so that to lowest order,

$$\frac{\delta G\,(1,\,1';U\,)}{\delta U\,(2)} = G\,(1,\,2)\,G\,(2,\,1')$$

Therefore, we find to second order in V,

$$\Sigma\,(1,\,1';U\,) - \Sigma_{\text{HF}}\,(1,\,1';U\,)$$
$$= \pm i^2 \int d\bar{2}d\bar{3}\; V\,(1 - \bar{2})\,V\,(\bar{3} - 1)$$
$$\times [G(1,\,1';U\,)G(\bar{3},\,\bar{2};U\,)G(\bar{2},\,\bar{3};U\,)$$
$$\pm G(1,\,\bar{3};U\,)G(\bar{3},\,2;U\,)G(\bar{2},\,1';U\,)]$$

$$= \qquad\qquad + \qquad\qquad \text{(6.28)}$$

where the lines represent G's. Equation (6.28), when U is set equal to zero, is the Born collision approximation.

Chapter 7

Transport Phenomena

So far we have studied many-body systems by considering the effect of adding or removing one or more particles. From the one particle Green's function $G(1 - 1')$, we were able to determine the energy spectrum and decay times of the single-particle excited states.

We indicate that G_2 can be used to describe the scattering of two particles added to the medium. Higher-order Green's functions, defined similarly to G and G_2, describe the effects of adding or removing more than two particles.

However, there exists a class of disturbances that are not conveniently described in terms of these equilibrium Green's functions. Consider, for example, a disturbance produced by the externally applied force field, $\mathbf{F}(\mathbf{r}, t) = -\nabla U(\mathbf{r}, t)$.[a] This force field may be represented by the addition of the term[b]

$$\hat{H}'(t) = \int d\mathbf{r}\, U(\mathbf{r}, t)\, \hat{n}(\mathbf{r}, t) \tag{7.1}$$

---

[a] In this chapter, the original text uses capital letters for the position vector $\mathbf{R}$ and the time T. Here, we switch the capital letters to the lower cases, $\mathbf{r}$ and t, for avoiding confusions. In Section 7.4, a new coordinate system has been introduced, but it is not necessary at this stage.

[b] The typographic error on assigning the equation number (7.1) in the original text is fixed.

Annotations to Quantum Statistical Mechanics
In-Gee Kim
Copyright © 2018 Pan Stanford Publishing Pte. Ltd.
ISBN 978-981-4774-15-4 (Hardcover), 978-1-315-19659-6 (eBook)
www.panstanford.com

where

$$\hat{n}\left(\mathbf{r}, t\right) = \hat{\psi}^{\dagger}\left(\mathbf{r}, t\right) \hat{\psi}\left(\mathbf{r}, t\right)$$

to the Hamiltonian of the system. One example in which this kind of disturbance is particularly important is a system of charged particles, perturbed from equilibrium by a longitudinal electric field. Then, the external force is the electric field, times e, the charge on each particle, while $e^{-1}U\left(\mathbf{r}, t\right)$ is the scalar potential for the applied electric field.

Other types of external disturbances, e.g., general electromagnetic fields, can be represented by other terms added to the Hamiltonian. These extra terms cause no additional conceptual difficulties. However, for the sake of simplicity, we shall restrict ourselves to the disturbance (7.1).

Many interesting physical phenomena appear as the response of systems to external disturbances of this kind. For example, in an ordinary gas, a slowly varying $U\left(\mathbf{r}, t\right)$ will produce sound waves. A longitudinal electric field, applied to a charged system, will lead to a flow of current. Both processes will be accompanied by the flow of heat. Each of these processes involves the flow of macroscopically observable quantities—momentum (in a sound wave), charge, and energy—and are, therefore, known as transport processes.

Preparatory to developing a Green's function theory of transport, we shall review the conventional approach based on the Boltzmann equation in order to see, on the one hand, its shortcomings and, one the other hand, the features that must be retained in any correct theory.

7.1 Boltzmann Equation Approach to Transport

The conventional Boltzmann equation is an equation of motion for $f\left(\mathbf{p}, \mathbf{r}, t\right)$, the average density of particles with momentum $\mathbf{p}$ at the space–time point[c] $\mathbf{r}, t$. The time derivative of f is computed by

---

[c](*Original*) ‡The reader may argue that it is unreasonable to define an $f\left(\mathbf{p}, \mathbf{r}, t\right)$ quantum mechanically because the uncertainty principle makes it impossible to simultaneously specify the position and momentum of a particle. However, we are not interested in specifying the position of any particle with accuracy much greater than the wavelength of the disturbance. Therefore, when the disturbance varies only over macroscopic distances we can specify the momentum of the particles with macroscopic accuracy.

taking into account the following effects:

(1) Particles with momentum $\mathbf{p}$ continually drift into and out of the volume element of space about $\mathbf{r}$.
(2) Owing to the average forces acting on the particles, the momenta of the particles in this volume are gradually changed.
(3) Collisions that take place in this volume suddenly change the particle momenta. Collision rates are computed by using the free-particle collision cross sections, correcting the collision rates for the density of final states in the many-body system. For fermions, the exclusion principle requires that particles cannot scatter into occupied states; bosons, on the other hand, prefer to scatter into occupied states.

Thus, the Boltzmann equation is

$$\left\{ \frac{\partial}{\partial t} + \frac{\mathbf{p} \cdot \nabla_{\mathbf{r}}}{m} - [\nabla U\left(\mathbf{r}, t\right)] \cdot \nabla_{\mathbf{p}} \right\} f\left(\mathbf{p}, \mathbf{r}, t\right) = \left(\frac{\partial f}{\partial t} \right)_{\text{collision}}$$

(7.2)

where, in terms of Born approximation collision cross sections,

$$\left(\frac{\partial f}{\partial t} \right)_{\text{collision}} = \int \frac{d\mathbf{p}'}{(2\pi)^3} \frac{d\bar{\mathbf{p}}}{(2\pi)^3} \frac{d\bar{\mathbf{p}}'}{(2\pi)^3} \left(\frac{1}{2} \right) \left[v\left(\mathbf{p} - \bar{\mathbf{p}}\right) \pm v\left(\mathbf{p} - \bar{\mathbf{p}}'\right) \right]^2$$
$$\times (2\pi)^3 \, \delta\left(\mathbf{p} + \mathbf{p}' - \bar{\mathbf{p}} - \bar{\mathbf{p}}'\right)$$
$$\times 2\pi \delta \left(\frac{p^2}{2m} + \frac{(p')^2}{2m} - \frac{\bar{p}^2}{2m} - \frac{(\bar{p}')^2}{2m} \right)$$
$$\times \left[(1 \pm f)\left(1 \pm f'\right) \bar{f}\bar{f}' - ff'\left(1 \pm \bar{f}\right)\left(1 \pm \bar{f}'\right) \right]$$

(7.3)

Here

$$f = f\left(\mathbf{p}, \mathbf{r}, t\right), \quad f' = f\left(\mathbf{p}', \mathbf{r}, t\right), \quad \text{etc.}$$

This Boltzmann equation is appropriate only for systems with weak, short-ranged forces.

When particles interact through the Coulomb force, $v(r) = \frac{e^2}{r}$, the force is so long-ranged that the whole picture of instantaneous local collisions breaks down completely. For long-ranged forces, Eq. (7.2) is almost certainly wrong. It is much better to leave out the collision term entirely and consider the particles move independently through an average potential field. This effective field is the sum of the applied field and the averaged field produced by all the particles in the system:

$$U_{\text{eff}} (\mathbf{r}, t) = U (\mathbf{r}, t) + \int d\mathbf{r}' \, v (\mathbf{r} - \mathbf{r}') \int \frac{d\mathbf{p}'}{(2pi)^3} \, f (\mathbf{p}', \mathbf{r}, t) \quad (7.4)$$

Then the Boltzmann equation becomes

$$\left\{ \frac{\partial}{\partial t} + \frac{\mathbf{p} \cdot \nabla_{\mathbf{r}}}{m} - [\nabla U_{\text{eff}} (\mathbf{r}, t)] \cdot \nabla_{\mathbf{p}} \right\} f (\mathbf{p}, \mathbf{r}, t) = 0 \quad (7.5)$$

This equation is often called the Vlasov–Landau equation. It is nonlinear because U_{Jeff} depends on f. We shall defer the discussion of the collisions Boltzmann equation to Chapter 8.

In the absence of a U, Eq. (7.2) has the solution

$$f (\mathbf{p}, \mathbf{r}, t) = \frac{1}{e^{\beta \left[\frac{(\mathbf{p} - m v)^2}{2m} - \mu \right]} \pm 1} \quad (7.6)$$

the readers should check for themselves that, in fact, the collision term vanishes for this choice of f. This solution represents thermodynamic equilibrium. The parameters β, μ, and $\mathbf{v}$ are the five parameters ($\mathbf{v}$ is a vector) necessary to specify the thermodynamic state of the system. The new parameter here is $\mathbf{v}$, the average velocity of the system. Notice that the solution (7.6) is the distribution function in thermodynamic equilibrium for a set of independent particles. Therefore, the Boltzmann equation ignores the change in the equilibrium distribution caused by the inter-particle potential. Our more general theory will overcome this limitation.

Now we use the Boltzmann equation, (7.2), to derive the existence of ordinary sound waves. This derivation indicates the way in which the Boltzmann equation describes transport phenomena. Sound waves appear in the limit in which the disturbance $U (\mathbf{r}, t)$ varies very slowly in space and time.

When $U (\mathbf{r}, t)$ has this slow variation, $f (\mathbf{p}, \mathbf{r}, t)$ must be slowly varying. Then the left-hand side of Eq. (7.2) must be very small, since it is proportional to space or time derivatives. Hence the collision term in Eq. (7.2) must also be small. For an arbitrary choice of f, the collision term is on the order of Γf, where Γ is the typical collision rate. By hypothesis, we are considering a slowly varying disturbance, so that Γf is much greater than $\frac{\partial f}{\partial t}$ or $\left(\frac{\mathbf{p}}{m} \right) \cdot \nabla_{\mathbf{r}} f$. Therefore, the condition that the collision term be small is a strong requirement on the solution f. To lowest order we can determine f by demanding that

$$\left(\frac{\partial f}{\partial t}\right)_{\text{collision}} = 0 \tag{7.7}$$

The solution to Eq. (7.7) must be of the form

$$f(\mathbf{p}, \mathbf{r}, t) = \left\{\exp\left[\beta(\mathbf{r}, t)\frac{(\mathbf{p} - m\mathbf{v}(\mathbf{r}, t))^2}{2m} - \beta(\mathbf{r}, t)\mu(\mathbf{r}, t)\right] \mp 1\right\}^{-1}$$

$$\tag{7.8}$$

The f represented by Eq. (7.8) describes the system as being in local thermodynamic equilibrium. However, the system is not in complete thermodynamic equilibrium since the temperature $\beta(\mathbf{r}, t)$, the chemical potential $\mu(\mathbf{r}, t)$, and the average local velocity of the particles, $\mathbf{v}(\mathbf{r}, t)$, vary from point to point.

Notice that in obtaining Eq. (7.8), we are really thinking the collision term to be a dominant part of the Boltzmann equation. It is the collisions that are responsible for keeping the system in this local thermodynamic equilibrium.

To complete the lowest-order solution, we must determine the five unknown functions, $\beta(\mathbf{r}, t)$, $\mu(\mathbf{r}, t)$, and $\mathbf{v}(\mathbf{r}, t)$, which appear in Eq. (7.8). We can determine these by making use of the five conservation laws for the number of particles, momentum, and energy.

These five conservation laws are obtained by multiplying Eq. (7.2) by 1, $\frac{p^2}{2m}$, or $\mathbf{p}$, and then integrating the resulting equations over all $\mathbf{p}$. In all three cases, the integrals of the collision terms vanish and we find the differential conservation laws:

Number conservation:

$$\frac{\partial}{\partial t}n(\mathbf{r}, t) + \nabla \cdot \mathbf{j}(\mathbf{r}, t) = 0 \tag{7.9a}$$

where

$$n(\mathbf{r}, t) = \int \frac{d\mathbf{p}}{(2\pi)^3} f(\mathbf{p}, \mathbf{r}, t)$$

$$\mathbf{j}(\mathbf{r}, t) = \int \frac{d\mathbf{p}}{(2\pi)^3}\frac{\mathbf{p}}{m} f(\mathbf{p}, \mathbf{r}, t)$$

Energy conservation:

$$\frac{\partial \mathcal{E}(\mathbf{r}, t)}{\partial t} + \nabla_{\mathbf{r}} \cdot \int \frac{d\mathbf{p}}{(2\pi)^3} \frac{\mathbf{p}}{m} \frac{p^2}{2m} f(\mathbf{p}, \mathbf{r}, t) = -\mathbf{j}(\mathbf{r}, t) \cdot \nabla_{\mathbf{r}} U(\mathbf{r}, t)$$

$$(7.9b)$$

where

$$\mathcal{E}(\mathbf{r}, t) = \int \frac{d\mathbf{p}}{(2\pi)^3} \frac{p^2}{2m} f(\mathbf{p}, \mathbf{r}, t)$$

Momentum conservation:

$$m\frac{\partial}{\partial t}\mathbf{j}(\mathbf{r}, t) + \int \frac{d\mathbf{p}}{(2\pi)^3} (\mathbf{p} \cdot \nabla_{\mathbf{r}}) \left[\frac{\mathbf{p}}{m} f(\mathbf{p}, \mathbf{r}, t)\right] = -n(\mathbf{r}, t) \nabla U(\mathbf{r}, t) \quad (7.9c)$$

The number conservation law expresses the result that the time derivative of the density of particles must be equal to the negative divergence of the current. This is also called the equation of continuity. Similarly, the time derivative of the energy density is the negative divergence of the energy current, plus the density of the power added at the point in question. Finally, the time derivative of the momentum density is the negative divergence of the momentum current, plus the applied force density.

The conservation laws (7.9) are exact consequences of the Boltzmann equation. They do not depend in any way on the use of approximation (7.8) for f. However, we can substitute the approximate f into these equations and thereby determine parameters in Eq. (7.8) in terms of U.

To simplify this analysis, we shall consider only the low-density limit $(\beta\mu \rightarrow -\infty)$, in which f has the simpler form

$$f(\mathbf{p}, \mathbf{r}, t) = \exp\left\{-\beta(\mathbf{r}, t)\left[\frac{(\mathbf{p} - m\mathbf{v}(\mathbf{r}, t))^2}{2m} - \mu(\mathbf{r}, t)\right]\right\} \quad (7.8a)$$

As a further simplification, we consider $U(\mathbf{r}, t)$ to be a small perturbation of the system from an initial equilibrium configuration at time $-\infty$ in which $\mathbf{v}(\mathbf{r}, -\infty) = 0$, $\beta(\mathbf{r}, -\infty) = \beta$, and $\mu(\mathbf{r}, -\infty) = \mu$. This enables us to write the conservation laws in a linearized form. These linearized conservation laws are derived by substituting Eq. (7.8a) into Eq. (7.9). Since terms like $\mathbf{v}\cdot\nabla$, v^2, etc., are all of second or higher order, the linearized conservation laws are

$$\frac{\partial n\left(\mathbf{r}, t\right)}{\partial t} = -\nabla \cdot \left(\mathbf{v}\left(\mathbf{r}, t\right) n\left(\mathbf{r}, t\right)\right) \simeq -n\nabla \cdot \mathbf{v}\left(\mathbf{r}, t\right)$$

$$\frac{\partial \mathcal{E}\left(\mathbf{r}, t\right)}{\partial t} = -\nabla_{\mathbf{r}} \left\{ \int \frac{d\mathbf{p}}{\left(2\pi\right)^3} \frac{\mathbf{p}}{m} \frac{p^2}{2m} \right\} e^{-\beta(\mathbf{r},t)\left[\frac{(\mathbf{p}-m\mathbf{v}(\mathbf{r},t))^2}{2m} - \mu(\mathbf{r},t)\right]} \tag{7.10a}$$

so that

$$\frac{\partial \mathcal{E}\left(\mathbf{r}, t\right)}{\partial t} \approx -\left(\mathcal{E} + P\right) \nabla \cdot \mathbf{v}\left(\mathbf{r}, t\right) \tag{7.10b}$$

and

$$mn\frac{\partial \mathbf{v}\left(\mathbf{r}, t\right)}{\partial t} = -\nabla P\left(\mathbf{r}, t\right) - n\nabla U\left(\mathbf{r}, t\right) \tag{7.10c}$$

In Eq. (7.10)

$$n\left(\mathbf{r}, t\right) = \int \frac{d\mathbf{p}}{\left(2\pi\right)^3} \exp\left\{-\beta\left(\mathbf{r}, t\right)\left[\frac{p^2}{2m} - \mu\left(\mathbf{r}, t\right)\right]\right\}$$

$$\mathcal{E}\left(\mathbf{r}, t\right) = \left(\frac{3}{2}\right) n\left(\mathbf{r}, t\right) \beta^{-1}\left(\mathbf{r}, t\right) \tag{7.11}$$

and

$$P\left(\mathbf{r}, t\right) = n\left(\mathbf{r}, t\right) \beta^{-1}\left(\mathbf{r}, t\right)$$

are the particle density, energy density, and pressure of a free low-density gas, expressed as functions of $\beta\left(\mathbf{r}, t\right)$ and $\mu\left(\mathbf{r}, t\right)$. Also n, $\mathcal{E}$, and P are the values of these quantities at time $-\infty$. The linearized hydrodynamic Eqs. (7.10) can be derived for all ordinary fluids. However, Eqs. (7.11) are not always true, since they are the thermodynamic relations for a perfect gas. In a more general discussion of sound propagation, one must use more accurate thermodynamic relations than Eq. (7.11). These cannot be derived from a Boltzmann equation.

We now eliminate $U\left(\mathbf{r}, t\right)$ from Eq. (7.10). If we take m times the time derivative of Eq. (7.10a) and subtract from it the divergence of Eq. (7.10c), we find

$$m\frac{\partial^2 n\left(\mathbf{r}, t\right)}{\partial t^2} - \nabla^2 P\left(\mathbf{r}, t\right) = n\nabla^2 U\left(\mathbf{r}, t\right) \tag{7.12}$$

and from Eqs. (7.10b) and (7.10a), we find

$$\frac{1}{n}\frac{\partial}{\partial t}n\left(\mathbf{r}, t\right) = -\nabla \cdot \mathbf{v}\left(\mathbf{r}, t\right) = \frac{1}{\mathcal{E}+P}\frac{\partial \mathcal{E}\left(\mathbf{r}, t\right)}{\partial t}$$

$$= \frac{3}{5}\frac{1}{P}\frac{\partial P\left(\mathbf{r}, t\right)}{\partial t} \tag{7.13}$$

This last equation is a restriction on the possible changes in $\beta\left(\mathbf{r}, t\right)$ and $\mu\left(\mathbf{r}, t\right)$, and we may use it to eliminate $\nabla^2 P\left(\mathbf{r}, t\right)$ from Eq. (7.12). Note that we are switching from the variables $\beta\left(\mathbf{r}, t\right)$ and $\mu\left(\mathbf{r}, t\right)$ to $n\left(\mathbf{r}, t\right)$ and $P\left(\mathbf{r}, t\right)$. The solution to Eq. (7.13) is just

$$(P\left(\mathbf{r}, t\right) - P) = \frac{5}{3}\frac{P}{n}\left(n\left(\mathbf{r}, t\right) - n\right) \tag{7.14}$$

since $P\left(\mathbf{r}, t\right)$ is P and $n\left(\mathbf{r}, t\right)$ is n at the initial time $t = -\infty$. Then from Eq. (7.14)

$$\nabla^2 P\left(\mathbf{r}, t\right) = \frac{5}{3}\beta^{-1}\nabla^2 n\left(\mathbf{r}, t\right)$$

so that Eq. (7.12) becomes

$$\left[\frac{\partial^2}{\partial t^2} - \frac{5}{3}\frac{\beta^{-1}}{m}\nabla^2\right]n\left(\mathbf{r}, t\right) = \frac{n}{m}\nabla^2 U\left(\mathbf{r}, t\right) \tag{7.15}$$

This is the equation obeyed by forced, undamped sound waves. The velocity C of this sound wave is given by

$$C^2 = \frac{5}{3}\frac{\beta^{-1}}{m} = \frac{5}{3}\frac{k_B T}{m} \tag{7.16}$$

which is the adiabatic or Laplace sound velocity for a perfect gas. The restriction (7.13) is equivalent to the statement that the sound wave must propagate with constant entropy. In terms of the thermodynamic derivatives of a free gas, C^2 is given by

$$C^2 = \left[\frac{1}{m}\left(\frac{\partial P}{\partial n}\right)_{S/N}\right]_{\text{free gas}}$$

The analysis that we have just carried out is the lowest order in an expansion in powers of ω/Γ and $\left[\frac{(\mathbf{k}\cdot\mathbf{p})}{m}\right]/\Gamma$, where ω and $\mathbf{k}$ are the frequency and wavenumber of the disturbance U, and Γ is the typical collision rate.

The next-order terms in this expansion involve viscosity and thermal conductivity. These transport coefficients can, therefore, be calculated from the Boltzmann equation. They appear in the sound-wave damping.

It is interesting to note that these correction terms are of order ω/Γ relative to the terms we have just computed. Therefore, the analysis of transport is based on an expansion of $1/\Gamma$ or one over the square of the potential. Thus, in our Green's function analysis, we can hardly expect that any power-series expansion in the potential could describe transport.

This result for the sound velocity indicates both the strength and the weakness of the Boltzmann equation approach. The Boltzmann equation predicts the existence of sound waves, and it gives the correct sound velocity for low-density systems: the result $C^2 = 5k_B T/3m$ has been verified experimentally for dilute gases. However, the sound velocity $C^2 = \left(\frac{1}{m}\right)\left(\frac{\partial P}{\partial n}\right)_{S/N}$ is correct for a very wide range of fluids, even in situations in which thermodynamic derivative is very far from its free gas value. Yet the Boltzmann equation predicts the free gas value, which suggests that the Boltzmann equation approach cannot give a good description of any systems except those that are weakly interacting.

There is another hint that the Boltzmann equation is inherently limited to weakly interacting system. Look at the energy conservation law, Eq. (7.9b). This is actually a conservation law for kinetic energy,

$$\int \frac{d\mathbf{p}}{(2\pi)^3} \frac{p^2}{2m} f(\mathbf{p}, \mathbf{r}, t)$$

However, it is not merely the kinetic energy that is conserved but the total energy—kinetic plus potential. Any approximation that leads to an energy conservation law of the kinetic energy alone can be valid only when the average potential energy is much smaller than the kinetic energy, i.e., in the weak-interaction limit.

As we shall soon see, the Green's function approach overcomes this limitation of the Boltzmann equation and, in fact, is capable of going far beyond the Boltzmann equation in its range of applicability and its accuracy.

On the other hand, we must retain one very important feature of the Boltzmann equation in our Green's function approach: the conservation laws for the number of particles, the energy, and the momentum. In fact, we saw that the derivation of sound waves depends only on the assumption of local thermodynamic equilibrium and the use of these conservation laws. These conservation laws dominate the response of the system to slowly varying disturbances; they must be included to get a qualitatively correct description of this response.

7.2 Green's Function Description of Transport

The problem posed by transport theory, be it quantum or classical, is to calculate the space- and time-dependent responses induced in a system by external space- and time-dependent disturbances. In electrical transport, for example, one applies, starting at a certain time, an external disturbance in the form of an electric potential, like Eq. (7.1), and tries to find the current and charge distributions due to this potential.

To be specific, we shall consider only disturbances of the form Eq. (7.1),

$$H'(t) = \int d\mathbf{r}\, n(\mathbf{r}, t)\, U(\mathbf{r}, t) \tag{7.1}$$

We then want to calculate the expectation values of physical operators, as they develop in time when the system is influenced by U. In the Heisenberg representation, any operator, $\hat{X}(\mathbf{r}, t)$, develops in time according to the equation

$$i\frac{\partial}{\partial t}\hat{X}_U(\mathbf{r}, t) = \left[\hat{X}_U(\mathbf{r}, t),\, \hat{H}_U(t) + \int d\mathbf{r}'\, \hat{n}_U(\mathbf{r}', t)\, U(\mathbf{r}', t)\right] \tag{7.17}$$

Here $\hat{H}_U(t)$ is the Hamiltonian (2.2) of the system. It now depends on time because there is an external time-dependent perturbation. The subscript U on the operators indicates that their time development is given by Eq. (7.17) and, therefore, depends on U.

Let us suppose that at a very earlier time t_0, before U is turned on, the system is in a definite eigenstate, $|i, t_0\rangle$ of $\hat{H}$ and $\hat{N}$. The t_0 in the designation of the state means that

$$\hat{H}(t)\, |i, t_0\rangle = E_i\, |i, t_0\rangle \quad \text{when } t < t_0 \tag{7.18}$$

In the Heisenberg picture, the system will always remain in this state. Only the operators change in time. [The relation (7.18) will fail to hold as soon as t becomes later than the time when U is turned on.] The expectation value of the operator $\hat{X}$ at the time t and point $\mathbf{r}$ is

$$\left\langle \hat{X}\left(\mathbf{r}, t\right)\right\rangle_U = \langle i, t_0 | \hat{X}_U\left(\mathbf{r}, t\right) | i, t_0 \rangle \tag{7.19}$$

Now in an actual experiment, the system is not in a definite eigenstate of the Hamiltonian at time t_0 but is rather at a definite temperature β^{-1}. We start with a system in thermal equilibrium at a definite temperature (and chemical potential) when we begin the experiment, and then we observe how the system develops in time. We must, therefore, average Eq. (7.19) over a grand-canonical ensemble of eigenstates of the system, at time t_0. The expectation value becomes

$$\left\langle \hat{X}\left(\mathbf{r}, t\right)\right\rangle_U = \frac{\sum_i e^{-\beta(E_i - \mu N_i)} \langle i, t_0 | \hat{X}_U\left(\mathbf{r}, t\right) | i, t_0 \rangle}{\sum_i e^{-\beta(E_i - \mu N_i)}} \tag{7.20}$$

The ensemble can still be represented by a trace, but we must be careful to specify, by writing $\hat{H}\left(t_0\right)$, the time at which the ensemble was prepared. Actually, Eq. (7.20) is independent of t_0 as long as t_0 is before the time that U is turned on. The number operator is independent of time, since an external potential does not change the number of particles.

Next we notice that we can solve Eq. (7.17), at least formally, by going to the interaction representation. In this representation, the operators develop in time according to

$$i\frac{\partial \hat{X}\left(\mathbf{r}, t\right)}{\partial t} = \left[\hat{X}\left(\mathbf{r}, t\right), \hat{H}\left(t\right)\right] \tag{7.21}$$

The transformation between the interaction representation and the Heisenberg representation is given by[d]

$$\hat{X}_U\left(\mathbf{r}, t\right) = \hat{V}^{-1}(t)\hat{X}(t)\hat{V}(t) \tag{7.22}$$

---

[d] (*Original*) ‡One may check (7.22) by explicit differentiation with respect to t_0. Using

$$i\frac{\partial \hat{V}\left(t\right)}{\partial t} = \int d\mathbf{r}'\, n\left(\mathbf{r}', t\right) U\left(\mathbf{r}', t\right) \hat{V}\left(t\right)$$

where

$$\hat{V}(t) = \hat{T} \left\{ \exp \left[-i \int_{t_0}^t dt' \int d\mathbf{r}' \, \hat{n}\left(\mathbf{r}', t'\right) U\left(\mathbf{r}', t'\right) \right] \right\} \qquad (7.23)$$

is written in terms of the density operator in the interaction representation.

The problem of calculating the expectation value of an operator, developing in the presence of U, is then reduced to calculating

$$\left\langle \hat{X}_U\left(\mathbf{r}, t\right) \right\rangle = \left\langle \hat{X}\left(\mathbf{r}, t\right) \right\rangle_U$$

$$= \frac{\mathrm{tr}\left\{ e^{-\beta\left[\hat{H}(t_0) - \mu\hat{N}\right]} \hat{V}^{-1}(t) \, \hat{X}\left(\mathbf{r}, t\right) \hat{V}(t) \right\}}{\mathrm{tr}\left\{ e^{-\beta\left[\hat{H}(t_0) - \mu\hat{N}\right]} \right\}} \qquad (7.24a)$$

Since we are in the interacting representation, $\hat{H}(t_0)$ is independent of time so that we can drop the t_0 in $\hat{H}(t_0)$. Since t_0 can be any time before the disturbance is turned on, Eq. (7.24) does not depend on t_0. Then we can write

$$\left\langle \hat{X}\left(\mathbf{r}, t\right) \right\rangle_U = \left\langle \hat{V}^{-1}(t) \, \hat{X}\left(\mathbf{r}, t\right) \hat{V}(t) \right\rangle$$

$$\hat{V}(t) = \hat{T} \left\{ \exp \left[-i \int_{-\infty}^t dt' d\mathbf{r}' \, n\left(\mathbf{r}', t'\right) U\left(\mathbf{r}', t'\right) \right] \right\} \qquad (7.24b)$$

where the expectation value written without the U denotes the *equilibrium* expectation value. Equation (7.24b) is, in a certain sense, the solution to the problem of transport, since all the operators develop as they would in the equilibrium ensemble. All the dependence on the external field U is explicit in Eq. (7.24b).

Our program for determining quantities like Eq. (7.24b) will be to write equations of motion for generalized Green's functions in

---

one finds

$$i\frac{\hat{X}_U(t)}{\partial t} = \hat{V}^{-1}(t)\left(i\frac{\partial}{\partial t}\hat{X}(t)\right)\hat{V}(t)$$

$$+ \hat{V}^{-1}(t)\left[\hat{X}(t), \int d\mathbf{r}' \, \hat{n}\left(\mathbf{r}', t\right) U\left(\mathbf{r}, t\right)\right]\hat{V}(t)$$

$$= \hat{V}^{-1}(t)\left[\hat{X}(t), \hat{H}(t) + \int d\mathbf{r}' \, \hat{n}\left(\mathbf{r}', t\right) U\left(\mathbf{r}, t\right)\right]\hat{V}(t)$$

$$= \left[\hat{X}_U(t), \hat{H}_U(t) + \int d\mathbf{r}' \, \hat{n}_U\left(\mathbf{r}', t\right) U\left(\mathbf{r}', t\right)\right]$$

terms of which quantities like Eq. (7.24b) can be expressed. These equations of motion will bear a strong resemblance to Boltzmann equations.

We now use the Heisenberg representation creation and annihilation operators to define Green's functions

$$g\left(1, 1'; U\right) = \frac{1}{i} \left\langle \hat{T}\left(\hat{\psi}_U\left(1\right) \hat{\psi}_U^\dagger\left(1'\right)\right)\right\rangle$$

$$g^>\left(1, 1'; U\right) = \frac{1}{i} \left\langle \hat{\psi}_U\left(1\right) \hat{\psi}_U^\dagger\left(1'\right)\right\rangle$$

$$g^<\left(1, 1'; U\right) = \pm \frac{1}{i} \left\langle \hat{\psi}_U^\dagger\left(1'\right) \hat{\psi}_U\left(1\right)\right\rangle$$

$$g_2\left(12, 1'2'; U\right) = \left(\frac{1}{i}\right)^2 \left\langle \hat{T}\left(\hat{\psi}_U\left(1\right) \hat{\psi}_U\left(2\right) \hat{\psi}_U^\dagger\left(2'\right) \hat{\psi}_U^\dagger\left(1'\right)\right)\right\rangle$$

$$(7.25)$$

In terms of these Green's functions, we may describe the response of a system, initially in thermodynamic equilibrium, to the applied disturbance U. For example, the average density and current at the point $\mathbf{r}, t$ are given by

$$\langle \hat{n}\left(\mathbf{r}, t\right)\rangle_U = \left\langle \hat{\psi}_U^\dagger\left(\mathbf{r}, t\right) \hat{\psi}_U\left(\mathbf{r}, t\right)\right\rangle = \pm i g^<\left(\mathbf{r}t, \mathbf{r}t; U\right)$$

$$\left\langle \hat{\mathbf{j}}\left(\mathbf{r}, t\right)\right\rangle_U = \left\{\frac{\nabla - \nabla'}{2im}\left[\pm i g^<\left(\mathbf{r}t, \mathbf{r}'t; U\right)\right]\right\}_{\mathbf{r}'=\mathbf{r}} \qquad (7.26)$$

We use the "g" to distinguish these physical response functions, which are defined for real times, from their imaginary time counterparts $G(U)$, $G_2(U)$. We shall see later that there is a close connection between these two different sets of Green's functions. For the time being, we limit ourselves to discussing the real-time functions.

We now consider the equations of motion obeyed by $g(U)$. To derive these, we notice from Eq. (7.17) that[e]

$$i\frac{\partial}{\partial t}\hat{\psi}_U\left(\mathbf{r}, t\right) = -\left(\frac{\nabla^2}{2m}\right)\hat{\psi}_U\left(\mathbf{r}, t\right) + U\left(\mathbf{r}, t\right)\hat{\psi}_U\left(\mathbf{r}, t\right)$$

$$+ \int d\mathbf{r}' \, v\left(\mathbf{r} - \mathbf{r}'\right) \hat{\psi}_U^\dagger\left(\mathbf{r}', t\right) \hat{\psi}_U\left(\mathbf{r}', t\right) \hat{\psi}_U\left(\mathbf{r}, t\right)$$

$$(7.27)$$

---

[e]The typographic error of the omission of the superscript † at the first $\hat{\psi}$ in the second line in the original text is fixed.

It follows then that

$$\left[i\frac{\partial}{\partial t_1} + \frac{\nabla_1^2}{2m} - U(1)\right] g\left(1, 1'; U\right)$$

$$= \delta\left(1 - 1'\right) \pm i \int_{-\infty}^{\infty} dt_2 d\mathbf{r}_2 \, V\left(1 - 2\right) g_2\left(12, 1'2'; U\right)$$

$$\tag{7.28a}$$

Making use of the equation of motion of $\hat{\psi}_U^\dagger$, we can similarly derive

$$\left[-i\frac{\partial}{\partial t_{1'}} + \frac{\nabla_{1'}^2}{2m} - U(1)\right] g\left(1, 1'; U\right)$$

$$\tag{7.28b}$$

$$= \delta\left(1 - 1'\right) \pm i \int_{-\infty}^{\infty} d2 \, V\left(1' - 2\right) g_2\left(12^-, 1'2'; U\right)$$

Here $V(1 - 2) = v(\mathbf{r}_1 - \mathbf{r}_2)\delta(t_1 - t_2)$. As in the case of the equilibrium Green's functions, we shall construct approximations for $g(U)$ by substituting an approximation for $g_2(U)$ into these equations of motion.

7.3 Conservation Laws for $g(U)$

In our derivation of sound propagation from the Boltzmann equations, we saw that it was essentially to make use of the conservation laws for the number of particles, the energy, and the momentum. When a system is disturbed from equilibrium, the first thing that happens is that the collision forces the system to a situation that is close to local thermodynamic equilibrium. This happens in a comparatively short time, on the order of Γ^{-1}. After this rapid decay has occurred, there is much slower return to all-over equilibrium. During this latter stage, the behavior of the system is dominated by the conservation laws. These laws very strongly limit the ways in which the system can return to full equilibrium. For example, if there is an excess of energy in one portion of the system, this energy cannot just disappear; it must slowly spread itself out over the entire system. This slow spreading out is the transport process known as heat conduction. Therefore, in order to predict even the existence of transport phenomena—like heat conduction or sound propagation—it is absolutely essential that we include the effects of the conservation laws. The conservation laws must be woven into the very fabric of our Green's function approximation scheme.

For example, we must be sure that any approximate calculation leads to an $\langle \hat{n}(\mathbf{r}, t) \rangle_U$ and $\left\langle \hat{\mathbf{j}}(\mathbf{r}, t) \right\rangle_U$ which satisfy the differential number conservation law

$$\frac{\partial}{\partial T} \langle \hat{n}(\mathbf{r}, t) \rangle_U + \nabla \cdot \left\langle \hat{\mathbf{j}}(\mathbf{r}, t) \right\rangle_U = 0$$

This conservation law becomes a restriction on $g(U)$. Using Eq. (7.26), we can express this restriction as

$$\left[\left(i \frac{\partial}{\partial t_1} + i \frac{\partial}{\partial t_{1'}} \right) g(1, 1'; U) \right]_{1'=1^+} + \nabla \cdot \left[\frac{\nabla_1 - \nabla_{1'}}{2m} g(1, 1'; U) \right]_{1'=1^+} = 0 \tag{7.29}$$

where $1' = 1^+$ means $\mathbf{r}_{1'} = \mathbf{r}_1, t_{1'} = t_1^+$.

Fortunately, it is very simple to state criteria that will guarantee that an approximation for $g(U)$ is conserving, i.e., it satisfies the restrictions imposed by the number, momentum, and energy conservation laws. We get an approximation for $g(U)$ by substituting an approximation for $g_2(U)$ into Eqs. (7.28a) and (7.28b). This procedure really defines two different approximations for $g(U)$, one given by Eq. (7.28a) and the other by Eq. (7.28b). We shall show that the differential number conservation law is equivalent to the requirement on the approximation: [criterion A] $g(U)$ satisfies both Eqs. (7.28a) *and* (7.28b).

To derive the number conservation law from criterion A, it is only necessary to subtract Eq. (7.28b) from Eq. (7.28a) to find

$$\left[i \frac{\partial}{\partial t_1} + i \frac{\partial}{\partial t_{1'}} + (\nabla_1 + \nabla_{1'}) \cdot \frac{\nabla_1 - \nabla_{1'}}{2m} - U(1) + U(1') \right] g(1, 1'; U)$$

$$= \pm i \int d2 \left[V(1-2) - V(1'-2) \right] g_2(12^-, 1'2^+; U) \tag{7.30}$$

When we set $1' = 1^+$ in Eq. (7.30), we find Eq. (7.29), so that the approximation indeed satisfies the differential number conservation law exactly.

We shall not write differential momentum or energy conservation laws analogous to Eq. (7.30). Instead we shall only employ the integrated forms of these conservation laws. For example, the conservation law for the total momentum is

$$\frac{d}{dt}\left\langle\hat{\mathbf{P}}(t)\right\rangle_U = -\int d\mathbf{r}\,[\nabla U(\mathbf{r},t)]\,\langle\hat{n}(\mathbf{r},t)\rangle_U \tag{7.31}$$

This states that the time derivative of the total momentum is equal to the total force acting on the system.

In order to have an approximation that conserves the total momentum, we place one more restriction on the approximate $g_2(U)$ to be substituted into Eq. (7.28). This is: [criterion B] $g_2\left(12;1^+2^+;U\right) = g_2\left(21;2^+1^+;U\right)$.

In order to see that this additional restriction is sufficient to obtain a momentum-conserving approximation, we construct the time derivative of the total momentum in the system by applying $\frac{\nabla_1-\nabla_{1'}}{2i}$ to Eq. (7.27), setting $1' = 1^+$ and integrating over all $\mathbf{r}_1$. In this way, we find[f]

$$\frac{d}{dt_1}\left\{\int d\mathbf{r}_1\left[\frac{\nabla_1-\nabla_{1'}}{2i}ig^<\left(1,1';U\right)\right]_{1'=1^+}\right\}$$
$$+\int d\mathbf{r}_1\,\nabla\cdot\left[\frac{\nabla_1-\nabla_{1'}}{2i}\frac{\nabla_1-\nabla_{1'}}{2m}g^<\left(1,1';U\right)\right]_{1'=1}$$
$$=\pm\int d\mathbf{r}_1 d\mathbf{r}_2\left[\nabla_{\mathbf{r}_1}v\left(|\mathbf{r}_1-\mathbf{r}_2|\right)\right]g_2\left(\mathbf{r}_1 t_1,\mathbf{r}_2 t_1;\mathbf{r}_1 t_1^+,\mathbf{r}_2 t_1^+;U\right)$$
$$-i\int d\mathbf{r}_1\left[\nabla U(\mathbf{r}_1)\right]g^<\left(\mathbf{r}_1 t_1,\mathbf{r}_1 t_1;U\right) \tag{7.32}$$

The term proportional to a divergence on the left side of Eq. (7.32) vanishes after integration over all $\mathbf{r}_1$. The term proportional to g_2 vanishes in this equation because criterion B implies that this term changes sign when the labels $\mathbf{r}_1$ and $\mathbf{r}_2$ are interchanged. Therefore, this term must be zero. Equation (7.32) then becomes

$$i\frac{d}{dt_1}\left\{\int d\mathbf{r}_1\left[\frac{\nabla_1-\nabla_{1'}}{2i}g^<\left(1,1';U\right)\right]_{1'=1^+}\right\}$$
$$=-i\int d\mathbf{r}_1\,g^<\left(1,1\right)\nabla U(\mathbf{r}_1) \tag{7.33}$$

---

[f]The typographic error in the argument of the function $g_2\left(\mathbf{r}_1 t_1,\mathbf{r}_2 t_1;\mathbf{r}_1 t_1^+;\mathbf{r}_2 t_1^+;U\right)$ in the original text is fixed.

This is just the momentum conservation law that we wished to build into our approximations.

The discussion of the energy conservation law is no more complicated in principle, but it involves some algebraic complexities, so we shall only outline it here. By using the same device as we discussed in this section on equilibrium properties of Eq. (3.9), we can express the energy density in terms of U and of differential operators acting upon $\hat{\psi}_U^{\dagger}(1')\,\hat{\psi}_U(1)$. Then, with the aid of Eq. (7.30), we can construct the time derivative of the total energy. After a bit of algebraic manipulation, which employs only criteria A and B, we find

$$\frac{d}{dT}\left\langle \hat{H}(t)\right\rangle_U = -\int d\mathbf{r}\,[\nabla U(\mathbf{r},t)]\cdot\left\langle \hat{\mathbf{j}}(\mathbf{r},t)\right\rangle_U \qquad (7.34)$$

which says that the time derivative of the total energy in the system is equal to the total power fed into the system by the external disturbance.

To sum up: Any approximation that satisfies criteria A and B must automatically agree with the differential number conservation law and the integral conservation laws for energy and momentum. Therefore, we may expect that these conserving approximations for $g(U)$ lead to fitting descriptions of transport phenomena.

7.4 Relation of $g(U)$ to the Distribution Function $f(\mathbf{p}, \mathbf{R}, T)$[g]

The Green's function theory of transport is logically independent of the Boltzmann equation approach. However, it will be interesting for us to make contact between the two theories. We shall now indicate the connection between the distribution function $f(\mathbf{p}, \mathbf{R}, T)$ and the Green's function $g(U)$.

We have already noted that $f(\mathbf{p}, \mathbf{R}, T)$ has no well-defined quantum mechanical meaning. Therefore, the best that we can hope to do is to define an f (in terms of g) that has many properties analogous to those of the classical distribution function. To do this,

---

[g]Now, we introduce the coordinates $\mathbf{R}$ and T in terms of the Wigner distribution function. It was not necessary to introduce these coordinates before this section.

we write the real-time Green's function, $\pm i g^< (1, 1'; U)$, in terms of the variables

$$\begin{aligned} \mathbf{r} = \mathbf{r}_1 - \mathbf{r}_{1'} \quad & t = t_1 - t_{1'} \\ \mathbf{R} = \tfrac{\mathbf{r}_1 + \mathbf{r}_{1'}}{2} \quad & T = \tfrac{t_1 + t_{1'}}{2} \end{aligned} \tag{7.35}$$

Then we define

$$g^< (\mathbf{p}, \omega; \mathbf{R}, T; U) = \int d\mathbf{r} \int_{-\infty}^{\infty} dt \, e^{-i\mathbf{p}\cdot\mathbf{r} + i\omega t} \left[\pm i g^< (\mathbf{r}, t; \mathbf{R}, T; U) \right] \tag{7.36}$$

This function may be thought of as the density of particles with momentum $\mathbf{p}$ and energy ω at the space–time point $\mathbf{R}, T$ at least in the limit in which g varies slowly in $\mathbf{R}$ and T. Hence, $f(\mathbf{p}, \mathbf{R}, T)$ can be defined as

$$\begin{aligned} f(\mathbf{p}, \mathbf{R}, T) &= \int \frac{d\omega}{2\pi} g^< (\mathbf{p}, \omega; \mathbf{R}, T; U) \\ &= \int d\mathbf{r} \, e^{-i\mathbf{p}\cdot\mathbf{r}} \left\langle \hat{\psi}_U^\dagger \left(\mathbf{R} - \frac{\mathbf{r}}{2}, T \right) \hat{\psi}_U \left(\mathbf{R} + \frac{\mathbf{r}}{2}, T \right) \right\rangle \end{aligned} \tag{7.37}$$

This definition is originally due to Wigner.

The function f has many similarities to the classical distribution function. When it is integrated over all momenta, it gives the density at $\mathbf{R}, T$, that is,

$$\int \frac{d\mathbf{p}}{(2\pi)^3} f(\mathbf{p}, \mathbf{R}, T) = \left\langle \hat{\psi}_U^\dagger (\mathbf{R}, T) \hat{\psi}_U (\mathbf{R}, T) \right\rangle = \langle \hat{n} (\mathbf{R}, T) \rangle_U$$

When it is integrated over all $\mathbf{R}$, it gives the number of particles with momentum $\mathbf{p}$ at time T since

$$\begin{aligned} \int d\mathbf{R} f(\mathbf{p}, \mathbf{R}, T) &= \int d\mathbf{r}_1 d\mathbf{r}_{1'} \, e^{-i\mathbf{p}\cdot\mathbf{r}_1} e^{-i\mathbf{p}\cdot\mathbf{r}} \left\langle \hat{\psi}_U^\dagger (\mathbf{r}_{1'}, T) \hat{\psi}_U (\mathbf{r}_1, T) \right\rangle \\ &= \left\langle \hat{\psi}_U^\dagger (\mathbf{p}, T) \hat{\psi}_U (\mathbf{p}, T) \right\rangle \end{aligned}$$

Just as in the classical case, the particle current is

$$\left\langle \hat{\mathbf{j}} (\mathbf{R}, T) \right\rangle_U = \int \frac{d\mathbf{p}}{(2\pi)^3} \frac{\mathbf{p}}{m} f(\mathbf{p}, \mathbf{R}, T)$$

This identification of the distribution function f will enable us to see the relationship between Green's function transport equations and the Boltzmann equation.

Chapter 8

Hartree Approximation, Collision-Less Boltzmann Equation, and Random Phase Approximation

Our general procedure for describing transport phenomena will be based on approximations in which $g_2(U)$, which appears in the equation of motion for $g(U)$, is expanded in terms of $g(U)$. The simplest approximation of this nature is the Hartree approximation.

$$g_2\left(12; 1'2'; U\right) = g\left(1, 1'; U\right) g\left(2, 2'; U\right) \qquad (8.1)$$

The two particles added to the system are taken to propagate completely independently of each other. They do, however, feel the effects of the applied potential U as they propagate through the medium, and hence their propagation is described by $g(U)$.

When Eq. (8.1) is substituted in the equations of motion, Eq. (7.28), these become

$$\left[i\frac{\partial}{\partial t_1} + \frac{\nabla_1^2}{2m} - U_{\text{eff}}(1)\right] g\left(1, 1'; U\right) = \delta\left(1 - 1'\right) \qquad (8.2a)$$

$$\left[-i\frac{\partial}{\partial t_{1'}} + \frac{\nabla_{1'}^2}{2m} - U_{\text{eff}}\left(1'\right)\right] g\left(1, 1'; U\right) = \delta\left(1 - 1'\right) \qquad (8.2b)$$

where

$$U_{\text{eff}}\left(\mathbf{R}, T\right) = U\left(\mathbf{R}, T\right) \pm i \int d\mathbf{R}' \, v\left(\mathbf{R} - \mathbf{R}'\right) g^<\left(\mathbf{R}'T, \mathbf{R}'T\right) \qquad (8.3)$$

Annotations to Quantum Statistical Mechanics
In-Gee Kim
Copyright © 2018 Pan Stanford Publishing Pte. Ltd.
ISBN 978-981-4774-15-4 (Hardcover), 978-1-315-19659-6 (eBook)
www.panstanford.com

Equations (8.2) describe the propagation of free particles through the effective potential field $U_{\text{eff}}(\mathbf{R}, T)$. This potential is the sum of the applied potential U and the average potential produced by all the particles in the system. It is the potential that would be felt by a test charge added to the medium.

When the particles have internal degrees of freedom, such as spin, or there is more than one kind of particle in the system, we must sum the last term in Eq. (8.3) over the different degrees of freedom. If the internal degree of freedom is spin, and the interaction is spin-independent, then this summation just gives a factor[a] $2S + 1$, so that Eq. (8.3) becomes

$$U_{\text{eff}}(\mathbf{R}, T) = U(\mathbf{R}, T) \pm i(2S + 1) \int d\mathbf{R}'\, v(\mathbf{R} - \mathbf{R}')\, g^<(\mathbf{R}'T, \mathbf{R}'T)$$

$$(8.3a)$$

In general, we shall not explicitly write this summation in our formulas.

Before we go any further, we shall show that this approximation is conserving. From Eq. (8.1), we see directly that criterion B, the symmetry of $g_2(12, 1'2'; U)$ under the interchange $1 \leftrightarrow 2$, $1' \leftrightarrow 2'$ is trivially satisfied. Criterion A states that Eqs. (8.2a) and (8.2b) are consistent with one another. To check this, we construct

$$\Lambda = \left[i\frac{\partial}{\partial t_1} + \frac{\nabla_1^2}{2m} - U_{\text{eff}}(1) \right] \left[-i\frac{\partial}{\partial t_{1'}} + \frac{\nabla_{1'}^2}{2m} - U_{\text{eff}}(1') \right] g(1, 1'; U)$$

in two ways; first, by multiplying Eq. (8.2a) by

$$\left[-i\frac{\partial}{\partial t_{1'}} + \frac{\nabla_{1'}^2}{2m} - U_{\text{eff}}(1') \right]$$

and then multiplying Eq. (8.2b) by

$$\left[i\frac{\partial}{\partial t_1} + \frac{\nabla_1^2}{2m} - U_{\text{eff}}(1) \right]$$

---

[a] The variable S denotes the number of internal degrees of freedom. This statement was not mentioned in the original text.

These two operations imply, respectively, that

$$\Lambda = \left[-i\frac{\partial}{\partial t_{1'}} + \frac{\nabla_{1'}^2}{2m} - U_{\text{eff}}\left(1'\right) \right] \delta\left(1 - 1'\right)$$

and

$$\Lambda = \left[i\frac{\partial}{\partial t_1} + \frac{\nabla_1^2}{2m} - U_{\text{eff}}\left(1\right) \right] \delta\left(1 - 1'\right)$$
$$= \left[-i\frac{\partial}{\partial t_{1'}} + \frac{\nabla_{1'}^2}{2m} - U_{\text{eff}}\left(1'\right) \right] \delta\left(1 - 1'\right)$$

Therefore, we see that Eqs. (8.2a) and (8.2b) both lead to the same differential equation for g. When supplemented by suitable boundary conditions, they will both determine the same function g. Thus, the Hartree approximation is conserving.

If we take the difference of the two mutually consistent equations (8.2a) and (8.2b), we find

$$\left\{ i\left(\frac{\partial}{\partial t_1} + \frac{\partial}{\partial t_{1'}}\right) + \left(\nabla_1 + \nabla_{1'}\right)\cdot\frac{\nabla_1 - \nabla_{1'}}{2m} - \left[U_{\text{eff}}\left(1\right) - U_{\text{eff}}\left(1'\right)\right] \right\}$$
$$\times g\left(1,\ 1';U\right) = 0$$

We now set $t_{1'} = t_{1}^{+} = T$; thus

$$\left\{ i\left(\frac{\partial}{\partial t_1} + \frac{\partial}{\partial t_{1'}}\right) + \left(\nabla_1 + \nabla_{1'}\right)\cdot\frac{\nabla_1 - \nabla_{1'}}{2m} \right.$$
$$\left. - \left[U_{\text{eff}}\left(\mathbf{r}_1, T\right) - U_{\text{eff}}\left(\mathbf{r}_{1'}, T\right)\right] \right\} \times g^{<}\left(\mathbf{r}_1 T, \mathbf{r}_{1'} T; U\right) = 0$$

$$(8.4)$$

In the limit in which $U_{\text{eff}}\left(\mathbf{R}, T\right)$ varies slowly in space, Eq. (8.4) is equivalent to the collision-less Boltzmann equation. In order to show the relationship between the Green's function theory of transport and the Boltzmann equation approach, and to gain a deeper insight into the meaning of both, we shall now derive the collisions Boltzmann equation from Eq. (8.4).

8.1 Collision-Less Boltzmann Equation

When Eq. (8.4) is expressed in terms of the variables $\mathbf{r} = \mathbf{r}_1 - \mathbf{r}_{1'}$ and $\mathbf{R} = \frac{1}{2}(\mathbf{r}_1 - \mathbf{r}_{1'})$, it becomes[b]

$$\pm \left\{ \frac{\partial}{\partial T} + \frac{\nabla_{\mathbf{R}} \cdot \nabla_{\mathbf{r}}}{im} - \frac{1}{i} \left[U_{\text{eff}} \left(\mathbf{R} + \frac{\mathbf{r}}{2}, T \right) - U_{\text{eff}} \left(\mathbf{R} - \frac{\mathbf{r}}{2}, T \right) \right] \right\}$$

$$\times \int \frac{d\mathbf{p}'}{(2\pi)^3} e^{i\mathbf{p}' \cdot \mathbf{r}} f\left(\mathbf{p}', \mathbf{R}, T \right) = 0$$

(8.5)

where f, defined by Eq. (7.37), is the quantum analogue of the classical one-particle distribution function. We multiply Eq. (8.5) by $e^{-i\mathbf{p}\cdot\mathbf{r}}$ and integrate over all $\mathbf{r}$. Then Eq. (8.5) becomes

$$\left(\frac{\partial}{\partial T} + \frac{\mathbf{p} \cdot \nabla_{\mathbf{R}}}{m} \right) f\left(\mathbf{p}, \mathbf{R}, T \right) = \frac{1}{i} \int d\mathbf{r} \int \frac{d\mathbf{p}'}{(2\pi)^3} e^{i(\mathbf{p}'-\mathbf{p})\cdot\mathbf{r}}$$

$$\times \left[U_{\text{eff}} \left(\mathbf{R} + \frac{\mathbf{r}}{2}, T \right) - U_{\text{eff}} \left(\mathbf{R} - \frac{\mathbf{r}}{2}, T \right) \right] f(\mathbf{p}', \mathbf{R}, T)$$

(8.6)

So far this equation is an exact consequence of the Hartree approximation (8.1). Now let us suppose that $U_{\text{eff}}(\mathbf{R}, T)$ varies slowly in $\mathbf{R}$. In the integrand above, we may, therefore, expand $U_{\text{eff}}\left(\mathbf{R} \pm \frac{\mathbf{r}}{2}, T \right)$ as

$$U_{\text{eff}}\left(\mathbf{R} \pm \frac{\mathbf{r}}{2}, T \right) = U_{\text{eff}}(\mathbf{R}, T) \pm \left(\frac{\mathbf{r}}{2} \right) \cdot \nabla_{\mathbf{R}} U_{\text{eff}}(\mathbf{R}, T)$$

so that

$$\left(\frac{\partial}{\partial T} + \frac{\mathbf{p} \cdot \nabla_{\mathbf{R}}}{m} \right) f\left(\mathbf{p}, \mathbf{R}, T \right) = \nabla_{\mathbf{R}} U_{\text{eff}}(\mathbf{R}, T) \cdot \int d\mathbf{r}' \int \frac{d\mathbf{p}'}{(2\pi)^3}$$

$$\times f\left(\mathbf{p}', \mathbf{R}, T \right) \left[-\nabla_{\mathbf{p}'} e^{i(\mathbf{p}'-\mathbf{p})\cdot\mathbf{r}} \right]$$

On integrating by parts, we find precisely the collisions Boltzmann equation

$$\left[\frac{\partial}{\partial T} + \frac{\mathbf{p}}{m} \cdot \nabla_{\mathbf{R}} - \nabla_{\mathbf{R}} U_{\text{eff}}(\mathbf{R}, T) \cdot \nabla_{\mathbf{p}} \right] f(\mathbf{p}, \mathbf{R}, T) = 0 \qquad (8.7)$$

where, in terms of f,

$$U_{\text{eff}}(\mathbf{R}, T) = U(\mathbf{R}, T) + \int d\mathbf{R}' \, v\left(\mathbf{R} - \mathbf{R}' \right) \int \frac{d\mathbf{p}'}{(2\pi)^3} f\left(\mathbf{p}', \mathbf{R}', T \right)$$

(8.8)

---

[b]We also need the new time variables $t = t_1 - t_{1'}$ and $T = \frac{1}{2}(t_1 + t_{1'})$ for the completion. This statement was omitted in the original text.

8.2 Linearization of the Hartree Approximation: The Random Phase Approximation

We may solve Eq. (8.4), or equivalently Eq. (8.6), exactly in the limit in which the potential U $(\mathbf{R}, T)$ is small.

We consider only disturbances that vanishes as $T \to -\infty$. The boundary condition on Eq. (8.6) is an initial condition which states that at $T = -\infty$, the system is in equilibrium, i.e., that f $(\mathbf{p}, \mathbf{R}, T)$ is given by the equilibrium value of $\int \frac{d\omega}{2\pi} G^< (p, \omega)$, evaluated in the Hartree approximation. Thus,

$$\lim_{T \to -\infty} f \, (\mathbf{p}, \mathbf{R}, T) = \frac{1}{e^{\beta(E(p)-\mu)} \pm 1} = f \, (E(p)) \qquad (8.9)$$

where

$$E(p) = \frac{p^2}{2m} + n \int d\mathbf{r} \, v(r)$$

From the definition Eq. (7.25) of $g^<(U)$, we see that f $(\mathbf{p}, \mathbf{R}, T)$ depends on the values of U $(\mathbf{R}', T')$ only for times T' earlier than time T. We may, therefore, write, to first order in U, that

$$f \, (\mathbf{p}, \mathbf{R}, T) = f \, (E(p)) + \delta f \, (\mathbf{p}, \mathbf{R}, T)$$

where

$$\delta f \, (\mathbf{p}, \mathbf{R}, T) = \int_{-\infty}^{T} dT' \int d\mathbf{R}' \, \frac{\delta f}{\delta U} \, (\mathbf{R} - \mathbf{R}', T - T') \, U \, (\mathbf{R}', T')$$
$$(8.10)$$

This equation defines the linear response function, $\frac{\delta f}{\delta U}$, in the real time domain. It is closely related, as we shall soon see, to the functional derivative in the imaginary time domain, which was defined in Chapter 6.

Owing to the smallness of U, we may write Eqs. (8.6) and (8.8) in linearized form:

$$\left[\frac{\partial}{\partial T} + \frac{\mathbf{p} \cdot \nabla_{\mathbf{R}}}{m} \right] \delta f \, (\mathbf{p}, \mathbf{R}, T)$$

$$= \frac{1}{i} \int d\mathbf{r} \frac{d\mathbf{p}'}{(2\pi)^3} \, e^{i(\mathbf{p}'-\mathbf{p})\cdot\mathbf{r}} \left[\delta U_{\text{eff}} \left(\mathbf{R} + \frac{\mathbf{r}}{2}, T \right) - \delta U_{\text{eff}} \left(\mathbf{R} - \frac{\mathbf{r}}{2}, T \right) \right]$$

$$(8.11a)$$

and

$$\delta U_{\text{eff}} (\mathbf{R}, T) = U (\mathbf{R}, T) + \int d\mathbf{R}'\, v (\mathbf{R} - \mathbf{R}') \int \frac{d\mathbf{p}'}{(2\pi)^3}\, \delta f (\mathbf{p}', \mathbf{R}, T)$$

$$(8.11b)$$

The Hartree approximation, when linearized in the external field, is known as the "random phase approximation." Equation (8.11) is just one of many equivalent statements of this approximation.

To solve these equations, we consider the case in which $U (\mathbf{R}, T)$ is of the form

$$U (\mathbf{R}, T) = U (\mathbf{k}, \Omega)\, e^{i\mathbf{k}\cdot\mathbf{r} - i\Omega T} \qquad (8.12)$$

where Ω is a complex frequency such that $\Im\Omega > 0$. Then $U (\mathbf{R}, T)$ vanishes as $T \to -\infty$. We see from Eq. (8.10) that

$$\delta f (\mathbf{p}, \mathbf{R}, T) = e^{i\mathbf{k}\cdot\mathbf{r} - i\Omega T}\, \delta f (\mathbf{p}, \mathbf{k}, T) \qquad (8.13)$$

where

$$\delta f (\mathbf{p}, \mathbf{k}, \Omega) = \int_{-\infty}^{0} dT' \int d\mathbf{R}' e^{-i\Omega T' + i\mathbf{k}\cdot\mathbf{R}'} \frac{\delta f}{\delta U} (\mathbf{p}, -\mathbf{R}', -T')\, U (\mathbf{k}, \Omega)$$

$$(8.14)$$

Equation (8.11) then becomes

$$\left(\Omega - \frac{\mathbf{k} \cdot \mathbf{p}}{m}\right) \delta f (\mathbf{p}, \mathbf{k}, \Omega) = \left[f \left(E \left(\mathbf{p} - \frac{\mathbf{k}}{2} \right) \right) - f \left(E \left(\mathbf{p} + \frac{\mathbf{k}}{2} \right) \right) \right]$$

$$\times\, \delta U_{\text{eff}} (\mathbf{k}, \Omega) \qquad (8.15a)$$

where

$$\delta U_{\text{eff}} (\mathbf{k}, \Omega) = U (\mathbf{k}, \Omega) + v (k) \int \frac{d\mathbf{p}'}{(2\pi)^3}\, \delta f (\mathbf{p}', \mathbf{k}, \Omega) \qquad (8.15b)$$

Here

$$v (k) = \int d\mathbf{r}\, e^{-i\mathbf{k}\cdot\mathbf{r}}\, v(r)$$

We readily find

$$\delta f (\mathbf{p}, \mathbf{k}, \Omega) = \frac{f \left(E \left(\mathbf{p} - \frac{\mathbf{k}}{2} \right) \right) - f \left(E \left(\mathbf{p} + \frac{\mathbf{k}}{2} \right) \right)}{\Omega - \frac{\mathbf{k} \cdot \mathbf{p}}{m}}\, \delta U_{\text{eff}} (\mathbf{k}, \Omega)$$

$$(8.16a)$$

and

$$\delta U_{\text{eff}}\left(\mathbf{k}, \Omega\right) = U\left(\mathbf{k}, \Omega\right)$$

$$+ v(k) \int \frac{d\mathbf{p'}}{(2\pi)^3} \frac{f\left(E\left(\mathbf{p'} - \frac{\mathbf{k}}{2}\right)\right) - f\left(E\left(\mathbf{p'} + \frac{\mathbf{k}}{2}\right)\right)}{\Omega - \frac{\mathbf{k}\cdot\mathbf{p'}}{m}}$$

$$\times \delta U_{\text{eff}}\left(\mathbf{k}, \Omega\right)$$

$$= \frac{U\left(\mathbf{k}, \Omega\right)}{1 - v(k) \int \frac{d\mathbf{p'}}{(2\pi)^3} \frac{f\left(E\left(\mathbf{p'} - \frac{\mathbf{k}}{2}\right)\right) - f\left(E\left(\mathbf{p'} + \frac{\mathbf{k}}{2}\right)\right)}{\Omega - \frac{\mathbf{k}\cdot\mathbf{p'}}{m}}} \qquad (8.16\text{b})$$

There are two quantities of physical interest that we can determine from Eq. (8.16). The first is the change in the density

$$\delta n\left(\mathbf{k}, \Omega\right) = \int \frac{d\mathbf{p}}{(2\pi)^3} \, \delta f\left(\mathbf{p}, \mathbf{k}, \Omega\right)$$

Let

$$\left(\frac{\delta n}{\delta U}\right)_0 \left(\mathbf{k}, \Omega\right) = \int \frac{d\mathbf{p}}{(2\pi)^3} \frac{f\left(E\left(\mathbf{p'} - \frac{\mathbf{k}}{2}\right)\right) - f\left(E\left(\mathbf{p'} + \frac{\mathbf{k}}{2}\right)\right)}{\Omega - \frac{\mathbf{k}\cdot\mathbf{p'}}{m}} \qquad (8.17)$$

This is the density response of a system of free particles, with single-particle energies $E(p)$, to applied field. Then δn is given by

$$\delta n\left(\mathbf{k}, \Omega\right) = \frac{\left(\frac{\delta n}{\delta U}\right)_0 \left(\mathbf{k}, \Omega\right)}{1 - v\left(\mathbf{k}\right) \left(\frac{\delta n}{\delta U}\right)_0 \left(\mathbf{k}, \Omega\right)} U\left(\mathbf{k}, \Omega\right) \qquad (8.18)$$

The other function of direct interest is the dynamic dielectric response function K. This function, defined by

$$\delta U_{\text{eff}}\left(\mathbf{R}, T\right) = \int_{-\infty}^{T} dT' \int d\mathbf{R'} \, K\left(\mathbf{R} - \mathbf{R'}, T - T'\right) U\left(\mathbf{R'}, T'\right) \qquad (8.19\text{a})$$

or

$$K\left(\mathbf{R} - \mathbf{R'}, T - T'\right) = \frac{\delta U_{\text{eff}}\left(\mathbf{R}, T\right)}{\delta U\left(\mathbf{R'}, T'\right)} \qquad (8.19\text{b})$$

gives the change in the effective potential when one changes the externally applied potential. It is a generalization of the ordinary (inverse) dielectric constant to the case in which the external potential depends on space and time. When $U\left(\mathbf{R'}, T'\right)$ is one of the form (8.12), it follows that

$$\delta U_{\text{eff}}\left(\mathbf{R}, T\right) = e^{i\mathbf{k}\cdot\mathbf{R} - i\Omega T} K\left(\mathbf{k}, \Omega\right) U\left(\mathbf{k}, \Omega\right)$$

where

$$K(\mathbf{k}, \Omega) = \int_{-\infty}^{0} dT' \int d\mathbf{R}' \, e^{i\mathbf{k}\cdot\mathbf{R}' - i\Omega T'} K(-\mathbf{R}', -T')$$

In the $\Omega = 0$, $k \to 0$ limit, $K^{-1}(\mathbf{k}, \Omega)$ becomes the ordinary static dielectric constant ϵ. It is clear from Eq. (8.16b) that in the random phase approximation

$$K(\mathbf{k}, \Omega) = \frac{1}{1 - v(\mathbf{k}) \left(\frac{\delta n}{\delta U}\right)_0 (\mathbf{k}, \Omega)} \tag{8.20}$$

8.3 Coulomb Interaction

A particularly important application of the random phase approximation is to a system of charged particles. The interaction is through the Coulomb potential[c]

$$v(\mathbf{R}) = \frac{e^2}{R} \quad v(k) = \frac{4\pi e^2}{k^2} \tag{8.21}$$

If a system contains two kinds of oppositely charged particles, say electrons and ions, and the ions are much heavier than the electrons, then to a first approximation, we can think of the ions as producing a fixed uniform positive background potential, and consider only the dynamics of electrons. The positive background is a time-independent potential added to U_{eff}, whose only purpose is to guarantee overall electrical neutrality of the system.

In this case, $-e^{-1}U$ is the scalar potential for an externally applied electric field, and $-e^{-1}U_{\text{eff}}$ is the scalar potential for the total electric field seen by the particles—the external field plus the average field produced by the electrons plus the uniform background:

$$U_{\text{eff}}(\mathbf{R}, T) = U(\mathbf{R}, T) + \int d\mathbf{R}' \frac{e^2}{|\mathbf{R} - \mathbf{R}'|} \left(\langle \hat{n}(\mathbf{R}', T)\rangle_U - n\right) \tag{8.22}$$

n, representing the background, is the average density of particles.

---

[c]Instead of "coulomb" in the original text, we will use "Coulomb," which starts with the capital C, because it is the name of a person.

The random phase approximation is useful for calculating the dielectric response function of the system. From Eq. (8.17), we have

$$\left(\frac{\delta n}{\delta U}\right)_0 (\mathbf{k}, \Omega) = \int \frac{d\mathbf{p}}{(2\pi)^3} \frac{f\left(\frac{\left(\mathbf{p}-\frac{\mathbf{k}}{2}\right)^2}{2m}\right) - f\left(\frac{\left(\mathbf{p}+\frac{\mathbf{k}}{2}\right)^2}{2m}\right)}{\Omega - \frac{\mathbf{k}\cdot\mathbf{p}'}{m}} \qquad (8.23)$$

Let us consider first the limit in which the disturbance varies so slowly in space that $\left(\frac{\mathbf{k}\cdot\mathbf{p}}{m}\right)^2 \ll \Omega^2$ for all momenta $\mathbf{p}$ that are appreciably represented in the system. Then,

$$\frac{1}{\Omega - \frac{\mathbf{k}\cdot\mathbf{p}}{m}} \approx \frac{1}{\Omega}\left[1 + \frac{\mathbf{k}\cdot\mathbf{p}}{m\Omega} + \left(\frac{\mathbf{k}\cdot\mathbf{p}}{m\Omega}\right)^2 + \left(\frac{\mathbf{k}\cdot\mathbf{p}}{m\Omega}\right)^3 + \cdots\right]$$

By symmetry, the terms even in $\mathbf{p}$ here do not contribute to the integral in Eq. (8.23). Thus

$$\left(\frac{\delta n}{\delta U}\right)_0 (\mathbf{k}, \Omega) = \frac{1}{\Omega^2}\int \frac{d\mathbf{p}}{(2\pi)^3}\left[f\left(\frac{\left(\mathbf{p}-\frac{\mathbf{k}}{2}\right)^2}{2m}\right) - f\left(\frac{\left(\mathbf{p}+\frac{\mathbf{k}}{2}\right)^2}{2m}\right)\right]$$
$$\times \left[\frac{\mathbf{k}\cdot\mathbf{p}}{m} + \frac{\left(\frac{\mathbf{k}\cdot\mathbf{p}}{m}\right)^3}{\Omega^2}\right]$$

Shifting the origin of the $\mathbf{p}$ integrations and keeping only terms up to order k^4, we find

$$\left(\frac{\delta n}{\delta U}\right)_0 (\mathbf{k}, \Omega) = \frac{k^2}{m\Omega^2}\int \frac{d\mathbf{p}}{(2\pi)^3} f\left(\frac{p^2}{2m}\right)\left[1 + 3\left(\frac{\mathbf{k}\cdot\mathbf{p}}{m\Omega}\right)^2\right]$$
$$= \frac{nk^2}{m\Omega^2}\left[1 + \frac{k^2}{\Omega^2}\langle v^2\rangle\right] \qquad (8.24)$$

where

$$\langle v^2\rangle = \frac{1}{n}\int \frac{d\mathbf{p}}{(2\pi)^3} f\left(\frac{p^2}{2m}\right)\frac{p^2}{m^2} \qquad (8.25)$$

In the classical limit

$$\langle v^2\rangle = \frac{3}{m\beta} \qquad (8.26a)$$

and for zero temperature fermions

$$\langle v^2 \rangle = \left(\frac{3}{5} \right) \left(\frac{p_F}{m} \right)^2 \tag{8.26b}$$

where $p_F = (2m\mu)^{1/2}$ is the Fermi momentum.

Upon substituting Eq. (8.24) into Eq. (8.20) for the dielectric function, we find

$$K(k, \Omega) = \frac{1}{1 - \frac{4\pi e^2}{k^2} \frac{nk^2}{m\Omega^2} \left(1 + \frac{k^2}{\Omega^2} \langle v^2 \rangle \right)}$$

$$= \frac{\Omega^2}{\Omega^2 - \frac{4\pi ne^2}{m} - \frac{4\pi ne^2}{m\Omega^2} \langle v^2 \rangle k^2} \tag{8.27}$$

We notice at once that there are poles in this response function at

$$\Omega^2 = \left(\frac{4\pi ne^2}{m} \right) + \langle v^2 \rangle k^2 = \omega_p^2 + \langle v^2 \rangle k^2 \tag{8.28}$$

Exactly as a pole in the one-particle Green's function $G(z)$ indicated a single-particle excited state, so does a pole in K indicate a possible excitation, or resonant response, of the system. This resonance occurs also in $\delta n (\mathbf{k}, \Omega)$, as we see from Eq. (8.18). It, therefore, corresponds to a possible density oscillation of the system with frequency $(\omega_p^2 + \langle v^2 \rangle k^2)^{1/2}$. This resonance is called a plasma oscillation, and the frequency ω_p is called the plasma frequency. Plasma oscillations have been observed experimentally in systems as diverse as the upper atmosphere and metals. The upper atmosphere is partially ionized; a metal, to the first approximation, can be described as an electron gas.

We may see the physical significance of the plasma oscillation quite clearly if we examine the density change, $\delta n (\mathbf{R}, T)$, caused by an external field $U (\mathbf{R}) = e^{i\mathbf{k}\cdot\mathbf{R}} U_{\mathbf{k}}$, which is switched on at time T_0 and switched off at a later time T_1, i.e.,

$$U (\mathbf{R}, T) = \begin{cases} e^{i\mathbf{k}\cdot\mathbf{R}} U_{\mathbf{k}}, & T_0 < T < T_1 \\ 0, & \text{otherwise} \end{cases}$$

This U may be written in terms of its Fourier transform as

$$U (\mathbf{R}, T) = e^{i\mathbf{k}\cdot\mathbf{R}} U_{\mathbf{k}} \int_{-\infty}^{\infty} \frac{d\Omega}{2\pi i} e^{-i\Omega T} \left(\frac{e^{i\Omega T_1} - e^{i\Omega T_0}}{\Omega} \right)$$

Hence, U may be regarded as a superposition of potentials of the form $U(\mathbf{R}, \Omega) = U_{\mathbf{k}} \frac{e^{i\Omega T_1} - e^{i\Omega T_0}}{i\Omega}$. Thus, $\delta n(\mathbf{R}, T)$ is found from Eq. (8.18) by the formula

$$\delta n(\mathbf{R}, T)$$

$$= e^{i\mathbf{k}\cdot\mathbf{R}} U_{\mathbf{k}} \int_{-\infty}^{\infty} \frac{d\Omega}{2\pi i} \frac{e^{-i\Omega(T-T_1)} - e^{-i\Omega(T-T_0)}}{\Omega} \frac{\left(\frac{\delta n}{\delta U}\right)_0 (k, \Omega)}{1 - \left(\frac{4\pi e^2}{k^2}\right) \left(\frac{\delta n}{\delta U}\right)_0 (k, \Omega)}$$

Using $\left(\frac{\delta n}{\delta U}\right)_0$ from Eq. (8.24), we find

$$\delta n(\mathbf{R}, T) = e^{i\mathbf{k}\cdot\mathbf{R}} U_{\mathbf{k}} \int \frac{d\Omega}{2\pi i} \frac{nk^2}{m\Omega} \frac{\Omega^2 + k^2 \langle v^2 \rangle}{\Omega^4 - \Omega^2 \omega_p^2 - \omega_p^2 k^2 \langle v^2 \rangle}$$

$$\times \left(\frac{e^{-i\Omega(T-T_1)} - e^{-i\Omega(T-T_0)}}{\Omega} \right)$$

$$= e^{i\mathbf{k}\cdot\mathbf{R}} U_{\mathbf{k}} \int \frac{d\Omega}{2\pi i} \frac{\frac{nk^2}{m\Omega}}{\Omega^2 - \omega_p^2 - k^2 \langle v^2 \rangle}$$

$$\times \left(\frac{e^{-i\Omega(T-T_1)} - e^{-i\Omega(T-T_0)}}{\Omega} \right)$$

since to order k^2, we may make the replacement

$$\Omega^4 - \Omega^2 \omega_p^2 - \omega_p^2 k^2 \langle v^2 \rangle \to \left(\Omega^2 - \omega_p^2 - k^2 \langle v^2 \rangle\right)\left(\Omega^2 + k^2 \langle v^2 \rangle\right)$$

The integrand has plasma oscillation poles at

$$\Omega = \pm \left(\omega_p^2 + k^2 \langle v^2 \rangle\right)^{1/2}$$

and hence the integral will be well defined only when we specify the integration contour near these poles. The contour is determined from the fact that the response to the external potential is causal, i.e., $\delta n(\mathbf{R}, T) = 0$ if T is earlier than T_0. This implies that the contour must be chosen to pass above the poles, since when $T < T_0$, we may close the path of integration in the upper-half Ω plane and the integral vanishes.

When $T > T_1$, we may close the integration contour in the lower-half Ω plane and find, from the sum of the residues,

$$\delta n(\mathbf{R}, T) = e^{i\mathbf{k}\cdot\mathbf{R}} U_{\mathbf{k}} \frac{nk^2}{m\Omega_p^2(k)}$$

$$\times \left\{ \cos\left[\Omega_p(k)(T - T_0)\right] - \cos\left[\Omega_p(k)(T - T_1)\right] \right\}$$

$$(8.29)$$

where

$$\Omega_p(k) = + \left(\omega_p^2 + k^2 \langle v^2 \rangle\right)^2$$

It is clear from Eq. (8.29) that the effect of the external potential is to set the density in oscillation with frequency $\Omega_p(k)$. The spatial dependence is the same as the spatial dependence of the external field.

In a zero-temperature fermion system, the plasma oscillations are undamped. However, a more careful calculation of Eq. (8.23) at finite temperature would show that the plasma oscillations decay in time.

From the evaluation (8.27) of the dielectric response function, we see that in the limit of very high frequencies, $K \approx 1$; therefore, the total field is almost exactly the same as the applied field. This is because at very high frequencies, the particles in the system hardly have time to move in response to the applied field. The first correction to this result is

$$K(k, \Omega) = 1 + \frac{\omega_p^2}{\Omega^2} \tag{8.30}$$

On the other hand, when the external field is very slowly varying, the particles have time to respond, and they move to as to practically cancel the applied field. We may see this very clearly in the limit in which the frequency goes to zero, and the wavenumber is small but nonzero. Then, from Eq. (8.27)

$$
\begin{aligned}
\left(\frac{\delta n}{\delta U}\right)_0 (\mathbf{k}, 0) &= -\int \frac{d\mathbf{p}}{(2\pi)^3} \frac{f\left(\frac{(\mathbf{p}-\frac{\mathbf{k}}{2})^2}{2m}\right) - f\left(\frac{(\mathbf{p}+\frac{\mathbf{k}}{2})^2}{2m}\right)}{\frac{\mathbf{k}\cdot\mathbf{p}'}{m}} \\
&= \int \frac{d\mathbf{p}}{(2\pi)^3} \frac{\frac{\mathbf{k}\cdot\mathbf{p}}{m}\frac{\partial}{\partial\left(\frac{p^2}{2m}\right)} f\left(\frac{p^2}{2m}\right)}{\frac{\mathbf{k}\cdot\mathbf{p}}{m}} \\
&= -\int \frac{d\mathbf{p}}{(2\pi)^3} \frac{\partial}{\partial\mu} f\left(\frac{p^2}{2m}\right) \\
&= -\left(\frac{\partial n}{\partial\mu}\right)_\beta,
\end{aligned} \tag{8.31}
$$

Hence

$$K(k, 0) = \frac{k^2}{k^2 + r_{\mathrm{D}}^{-2}} \tag{8.32}$$

where the Debye shielding distance r_D is given by

$$r_D^{-2} = 4\pi e^2 \left(\frac{\partial n}{\partial \mu}\right)_\beta \tag{8.33}$$

In the classical limit, $\left(\frac{\partial n}{\partial \mu}\right)_\beta = n\beta$, so

$$r_D^2 = \frac{1}{4\pi e^2 n\beta} \tag{8.34}$$

For a zero-temperature fermion system, $\frac{\partial n}{\partial \mu} = \frac{3n}{mv_F^2}$, where v_F is the velocity of particles at the edge of the Fermi sea.[d] Thus,

$$r_D^2 = \frac{\pi \hbar a_0}{4mv_F} \tag{8.35}$$

where a_0 is the Bohr radius, $\frac{\hbar}{me^2}$.

The particles in the system, therefore, move in such a way as to reduce the total field by the factor $k^2 / \left(k^2 + r_D^{-2}\right)$. In particular, if the external field is a static Coulomb potential,

$$U(\mathbf{R}, T) = \frac{C}{|\mathbf{R}|} \quad U(\mathbf{k}, \Omega) = 2\pi\delta(\Omega)\frac{4\pi C}{k^2}$$

then in the long-wavelength limit,

$$U_{\text{eff}}(\mathbf{k}, \Omega) = \frac{4\pi C}{k^2 + r_D^{-2}} 2\pi\delta(\Omega)$$

The long-ranged applied field is shielded by the particles in the system, and the effective field is short-ranged.

In the classical limit, Eq. (8.32) is valid for all wavelengths, so that

$$U_{\text{eff}}(\mathbf{R}, T) = C \frac{e^{-R/r_D}}{R} \tag{8.36}$$

Thus, the total field produced by a point charge drops off with exponential rapidity, with a range equal to the shielding radius r_D. This screening effect is a very fundamental property of a Coulomb gas.

---

[d]We note that v_F, which was written as v_f in the original text, is nothing more than the Fermi velocity.

8.4 Low-Temperature Fermion System and Zero Sound

In a low-temperature, highly degenerate fermion system, the evaluation of $\left(\frac{\delta n}{\delta U}\right)_0$ in Eq. (8.17) is particularly simple. In the long-wavelength limit, as $\beta\mu \to \infty$,

$$f\left(E\left(\mathbf{p} - \frac{\mathbf{k}}{2}\right)\right) - f\left(E\left(\mathbf{p} + \frac{\mathbf{k}}{2}\right)\right)$$

$$= -\frac{\mathbf{k}\cdot\mathbf{p}}{m}\frac{\partial}{\partial\left(\frac{p^2}{2m}\right)}\frac{1}{e^{\beta\left(\frac{p^2}{2m} + nv - \mu\right)} + 1}$$

$$= \frac{\mathbf{k}\cdot\mathbf{p}}{m}\delta\left(\frac{p^2}{2m} + nv - \mu\right)$$

Therefore, Eq. (8.17) becomes

$$\left(\frac{\delta n}{\delta U}\right)_0(k,\Omega) = \frac{1}{4\pi}\int_0^\infty p^2 dp \int_{-1}^1 dz \frac{\frac{kpz}{m}}{\Omega - \frac{kpz}{m}}\delta\left(\frac{p^2}{2m} - \frac{p_F^2}{2m}\right)$$

where p_F, the Fermi momentum, is defined by

$$\frac{p_F^2}{2m} = \mu - n\int d\mathbf{r}\, v(r)$$

and z is the direction cosine between $\mathbf{k}$ and $\mathbf{p}$, then

$$\left(\frac{\delta n}{\delta U}\right)_0(k,\Omega) = \rho_E\int_{-1}^1 \frac{dz}{2}\frac{\frac{kp_F z}{m}}{\Omega - \frac{kp_F z}{m}} \tag{8.37}$$

where $\rho_E = \frac{mp_F}{2\pi^2}$ is the density of energy states at the top of the Fermi sea; i.e., $\frac{d\mathbf{p}}{(2\pi)^3} = \rho_E dE\frac{dz}{2}$.

The inverse of the response function K is thus given by

$$K^{-1}(k,\Omega) = 1 + v(k)\rho_E\left(1 + \frac{m\Omega}{2kp_F}\int_{-\frac{kp_F}{m}}^{\frac{kp_F}{m}}\frac{dx}{x - \Omega}\right)$$

Let $\Omega = \omega + i\epsilon$, where ω be a real positive frequency and ϵ be an infinitesimal positive number. Then we find explicitly,

$$K^{-1}(k,\Omega) = 1 + v(k)\rho_E$$

$$\times\left\{1 + \frac{m\omega}{2kp_F}\left[\ln\left|\frac{\frac{kp_F}{m} - \omega}{\frac{kp_F}{m} + \omega}\right| + \pi i\eta_+\left(\frac{kp_F}{m} - \omega\right)\right]\right\} \tag{8.38}$$

where $\eta_+(x) = 1$ if x is positive, and 0 otherwise. If the interaction is sufficiently weak, so that the dimensionless parameter $v(k)\rho_E$ is much less than unity, then K will be very close to unity except when the logarithm is very large, and this happens when $\omega \approx \frac{kp_F}{m}$. In this case

$$K^{-1}(k, \omega + i\epsilon) = 1 + \frac{v(k)\rho_E}{2}$$

$$\times \left\{ 1 + \log \frac{1}{2} \left| 1 - \frac{m\omega}{kp_F} \right| + \pi i \eta_+ \left(\frac{kp_F}{m} - \omega \right) \right\}$$

$$(8.39)$$

When the interaction is attractive, $v(k) < 0$, a very special condensation occurs in a low-temperature fermion system—the transition to the superconducting state, as described by Bardeen, Cooper, and Schrieffer. This condensation leads to new physical effects that completely invalidate the Hartree approximation. In Eq. (8.39), all we see is that K becomes very small in the neighborhood of $\omega = \frac{kp_F}{m}$, indicating that a disturbance with this frequency and wavenumber would be screened out.

On the other hand, when the interaction is repulsive, $v(k)$ is positive, and K has the form indicated in Fig. 8.1.[e] (We have drawn the $v < 0$ case in dashed lines for comparison.) Notice the sharp resonance at

$$\omega = \frac{kp_F}{m} \left\{ 1 + 2 \exp\left[-\left(\frac{2 + 2v\rho_E}{v\rho_E} \right) \right] \right\}$$

This corresponds to a resonant phenomenon in the system, which is called zero sound. It is characterized by the sound velocity

$$C_0 \approx \frac{p_F}{m} = v_F \qquad (8.40)$$

An analogue of zero sound is observed in the giant dipole resonance of nuclei.

It is interesting to compare the phenomenon of zero sound with ordinary sound in a highly degenerate fermion system. The relation $C^2 = \frac{1}{m} \left(\frac{\partial P}{\partial n} \right)_S$ implies that $C = v_F/\sqrt{3}$. We see that the dispersion

---

[e]The figure differs from the original one significantly. The figure has been tried to be reproduced from Eq. (8.38), but the low-frequency region results are completely different from the original one.

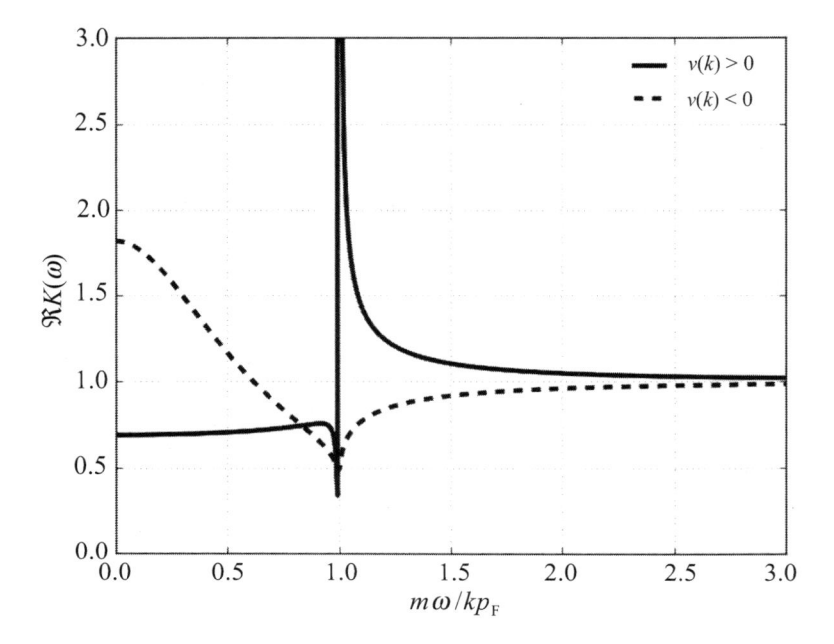

Figure 8.1 $\Re K(\omega)$ for a weak repulsive interaction. The dashed line is $\Re K(\omega)$ for a weak attractive interaction.

relation for zero sound, $\omega = v_F k$, is different from that for ordinary sound, $\omega = \left(v_F/\sqrt{3}\right) k$. For ordinary sound, the system is in local thermodynamic equilibrium, so that

$$\delta f(\mathbf{p}, \mathbf{k}, \Omega) = [\mathbf{p} \cdot \delta \mathbf{v}(\mathbf{k}, \Omega) - \delta \mu(\mathbf{k}, \Omega)] \frac{\partial}{\partial \left(\frac{p^2}{2m}\right)} f(E(p))$$

(8.41a)

(At low temperatures, $\frac{\delta \beta}{\beta}$ is negligible in a sound wave.) On the other hand, Eq. (8.16a) implies that for zero sound

$$\delta f(\mathbf{p}, \mathbf{k}, \Omega) = \frac{\frac{\mathbf{k} \cdot \mathbf{p}}{m}}{\frac{\mathbf{k} \cdot \mathbf{p}}{m} - \Omega} \frac{\partial}{\partial \left(\frac{p^2}{2m}\right)} f(E(p))$$

(8.41b)

This is clearly not a form for a local equilibrium phenomenon. Ordinary sound is just an oscillating translation and an oscillating expansion of the Fermi surface, but its shape remains spherical. Zero sound is a complex oscillation of the surface of the Fermi sphere. Atkins[f] describes this oscillation as follows: "At a particular instant the Fermi surface is considerably elongated in the forward direction of propagation and slightly shortened in the backward direction (like an egg), but half a cycle later it is slightly elongated in the backward direction and considerably shortened in the forward direction, the amplitude of oscillation being greater at the forward pole than at the backward pole."

Finally, the change in density for zero sound is

$$\delta n\,(\mathbf{k},\,\Omega) = \rho_E v_F^2 k^2 \int_{-1}^{1} \frac{dz}{2} \frac{z^2}{\Omega^2 - v_F^2 k^2 z^2} \delta U_{\text{eff}}\,(\mathbf{k},\,\Omega)$$

whereas, in ordinary sound it is

$$\delta n\,(\mathbf{k},\,\Omega) = \frac{\rho_E C^2 k^2}{\Omega^2 - C^2 k^2} U_{\text{eff}}\,(\mathbf{k},\,\Omega)$$

Zero sound is certainly a more complex phenomena.

Ordinary sound was derived from a better Boltzmann equation than was zero sound—a Boltzmann equation that included not only the effect of the average fields, but also the effect of collisions. It was just the collision terms that determined that the distribution function in the low-frequency, low-wavenumber limit be a local equilibrium one. In fact, when we examine this problem more carefully, we find that the quasi-equilibrium result must hold whenever the disturbance is so slowly varying that even the longest-lived single-particle excited states have ample time to decay. Since these states are at the edge of the Fermi sea, the criterion for the correctness of the ordinary sound solution is

$$\Omega \ll \Gamma\,(p_F,\,\mu) \quad \frac{\mathbf{k}\cdot\mathbf{p}}{m} \ll \Gamma\,(p_F,\,\mu) \tag{8.42}$$

In the opposite limit, the fields are oscillating too rapidly for the collisions to exert a damping effect. Therefore, the zero sound calculation, which neglected collisions, may be expected to be valid in the limit of high-frequency, short-wavelength disturbances:

$$\Omega \gg \Gamma\,(p_F,\,\mu) \quad \frac{\mathbf{k}\cdot\mathbf{p}}{m} \gg \Gamma\,(p_F,\,\mu) \tag{8.43}$$

---

[f]*(Original)* ‡K. R. Atkins, "Liquid Helium," Cambridge University Press, New York, 1959, p. 249.

At zero temperature, the single-particle excited states at the edge of the Fermi sea are infinitely long lived; $\Gamma(p, \mu) = 0$. Thus, the domain of existence of ordinary sound essentially disappears, but one can have zero sound at very low frequencies. In particular, nuclei in their ground states are zero-temperature system; therefore, they may be expected to exhibit an analogue of zero sound.

8.5 Breakdown of the Random Phase Approximation

The Hartree approximation and the random phase approximation do not always lead to sensible results. In particular, the pressure derived from the Hartree approximation does not always obey the basic statistical mechanical inequality[g]

$$\left(\frac{\partial P}{\partial n}\right)_\beta \geq 0 \tag{8.44}$$

We recall that the classical limit of the Hartree approximation gives

$$P = \left(\frac{1}{2}\right) n^2 v\,(k = 0) + \beta^{-1} n$$

and, therefore,

$$\left(\frac{\partial P}{\partial n}\right)_\beta = nv(0) + \beta^{-1} \tag{8.45}$$

---

[g] *(Original)* ‡To derive this inequality we write

$$\left(\frac{\partial P}{\partial n}\right)_\beta = \left(\frac{\partial P}{\partial \mu}\right)_\beta \Big/ \left(\frac{\partial n}{\partial \mu}\right)_\beta$$

Now

$$\left(\frac{\partial P}{\partial \mu}\right)_\beta = \frac{1}{\beta\Omega}\frac{\partial}{\partial \mu}\log \Xi = n \geq 0 \quad (\Omega = \text{volume of system}) \tag{8.44a}$$

and

$$\left(\frac{\partial n}{\partial \mu}\right)_\beta = \frac{1}{\Omega}\frac{\partial}{\partial \mu}\frac{\text{tr}\left[e^{-\beta(\hat{H}-\mu\hat{N})}\hat{N}\right]}{\text{tr}\,e^{-\beta(\hat{H}-\mu\hat{N})}} \tag{8.44b}$$

$$= \frac{\beta}{\Omega}\left\langle(\hat{N} - \langle\hat{N}\rangle)^2\right\rangle$$

so that $\left(\frac{\partial P}{\partial n}\right)_\beta$ is the ratio of two nonnegative quantities.

Suppose that the interaction is attractive, so that $v(0)$ is negative. Thus, if we keep n fixed, we can make $\left(\frac{\partial P}{\partial n}\right)_\beta$ negative by choosing

$$\beta^{-1} = k_\mathrm{B} T \leq -nv(0)$$

Thus, for temperatures too low, the Hartree approximation violates $\left(\frac{\partial P}{\partial n}\right)_\beta \geq 0$

As the temperature is lowered and $\left(\frac{\partial P}{\partial n}\right)_\beta = n / \left(\frac{\partial n}{\partial \mu}\right)_\beta$ approaches zero, it is clear that $\left\langle \left(\hat{N} - \langle \hat{N} \rangle \right)^2 \right\rangle$ becomes arbitrarily large. Such a tremendous fluctuation in the number of particles can be a signal that the system is about to undergo a phase transition.

This thermodynamic instability in the Hartree approximation is reflected as a dynamic instability in the response of the system to external fields, as calculated in the random phase approximation. To see this, we calculate K in the classical and long-wavelength limit.

In this limit, Eq. (8.17) becomes

$$
\left(\frac{\partial n}{\partial U}\right)_0 (\mathbf{k}, \Omega)
$$

$$
= -\int \frac{d\mathbf{p}}{(2\pi)^3} \frac{\frac{\mathbf{k}\cdot\mathbf{p}}{m}}{\Omega - \frac{\mathbf{k}\cdot\mathbf{p}}{m}} \frac{\partial}{\partial \left(\frac{p^2}{2m}\right)} \exp\left[-\beta \left(\frac{p^2}{2m} + nv(0) - \mu \right) \right]
$$

$$
= \beta \int \frac{d\mathbf{p}}{(2\pi)^3} \frac{\frac{\mathbf{k}\cdot\mathbf{p}}{m} - \Omega + \Omega}{\Omega - \frac{\mathbf{k}\cdot\mathbf{p}}{m}} \exp\left[-\beta \left(\frac{p^2}{2m} + nv(0) - \mu \right) \right]
$$

$$
= -\beta n + \beta \Omega e^{\beta(\mu - nv(0))} \frac{1}{4\pi} \int_{-1}^{1} dz \int_0^\infty dp \frac{p^2 e^{-\beta \left(\frac{p^2}{2m}\right)}}{\Omega - \frac{kpz}{m}}
$$

Now let Ω be very small and in the upper half plane. Then to lowest order in Ω, we may replace the Ω in the denominator of the integral by $i\epsilon$. Then Ω integral becomes

$$
\int_{-1}^{1} dz \frac{1}{i\epsilon - \frac{kpz}{m}} = -\frac{\pi i m}{kp}
$$

in the $\epsilon \to 0$ limit. Thus,

$$
\left(\frac{\delta n}{\delta U}\right)_0 (k, \Omega) = -\beta n \left(1 + i \frac{\Omega}{k} \sqrt{\frac{\beta m}{2}} \right) \tag{8.46}
$$

We see then that

$$K(k, \Omega) = \left[1 + \beta n v(k) - i \frac{\Omega}{k} \sqrt{\frac{\beta m}{2}} \beta v(k) \right]^{-1} \qquad (8.47)$$

has a pole at

$$\Omega = \Omega_c = ik \sqrt{\frac{2}{\beta m} \frac{\beta^{-1} + n v(k)}{n v(k)}} \qquad (8.48)$$

As long as Ω_c is in the lower half-plane, there is no difficulty, since we have assumed Ω to be in the upper half-plane in deriving the form (8.47) for K. However, when

$$\frac{\beta^{-1} + n v(k)}{n v(k)} \geq 0 \qquad (8.49)$$

the pole is in the upper half-plane. To produce such a pole with an attractive interaction, $v(k) < 0$, we need only increase β, i.e., lower the temperature until $1 + \beta n v(k)$ is negative. If $v(0) \geq v(k)$, the temperature at which poles in the upper half-plane begin to appear in K is the same temperature at which $\left(\frac{\partial P}{\partial n} \right)_\beta$ becomes negative.

It is very easy to see how a pole in K in the upper half-plane represents a dynamic instability. Consider, for example, an external disturbance of the form

$$U(\mathbf{R}, T) = \begin{cases} e^{i\mathbf{k} \cdot \mathbf{R} + \zeta T} U_{\mathbf{k}}, & T < 0 \\ 0, & T > 0 \end{cases}$$

where $\zeta > -i\Omega_c$. This U may be written in terms of its Fourier transform as

$$U(\mathbf{R}, T) = e^{i\mathbf{k} \cdot \mathbf{R} + \zeta T} U_{\mathbf{k}} \int_{-\infty}^{\infty} \frac{d\omega}{2\pi i} \frac{e^{-i\omega T}}{\omega - i\epsilon}$$

Then the density fluctuations induced by this U are given by

$$\delta n(\mathbf{R}, T) = e^{i\mathbf{k} \cdot \mathbf{R}} U_{\mathbf{k}} \int_{-\infty}^{T} dT' \frac{\delta n}{\delta U}(k, T - T') e^{\zeta T'} \int_{-\infty}^{\infty} \frac{d\omega}{2\pi i} \frac{e^{-i\omega T'}}{\omega - i\epsilon}$$

$$= e^{i\mathbf{k} \cdot \mathbf{R}} U_{\mathbf{k}} \int_{-\infty}^{\infty} \frac{d\omega}{2\pi i} \frac{e^{-i(\omega + i\zeta)T'}}{\omega - i\epsilon} \int_{-\infty}^{0} dT' e^{-i(\omega + i\zeta)T'} \frac{\delta n}{\delta U}(k, -T')$$

$$= e^{i\mathbf{k} \cdot \mathbf{R}} U_{\mathbf{k}} \int_{-\infty + i\zeta}^{\infty + i\zeta} \frac{d\Omega}{2\pi i} \frac{e^{-i\Omega T}}{\Omega - i(\zeta + \epsilon)} \frac{\delta n}{\delta U}(k, \Omega)$$

$$= e^{i\mathbf{k} \cdot \mathbf{R}} U_{\mathbf{k}} \int_{-\infty + i\zeta}^{\infty + i\zeta} \frac{d\Omega}{2\pi i} \frac{e^{-i\Omega T}}{\Omega - i(\zeta + \epsilon)} \left(\frac{\delta n}{\delta U} \right)_0 (k, \Omega) K(k, \Omega)$$

Suppose that K has the pole in the upper half-plane at $\Omega = \Omega_c$, but is otherwise analytic in the upper half-plane. Then, since $\zeta > -i\Omega_c$, we can write the Ω integral as a loop around the pole and an integral from $-\infty$ to ∞ just above the real axis. The contribution to δn from the pole is, therefore,

$$e^{i\mathbf{k}\cdot\mathbf{R}} e^{-i\Omega_c T} \, (-2\pi i) \, (\text{residue at } \Omega_c).$$

This term increases exponentially in time, which would seem to indicate that the potential U has excited an unstable density fluctuation. It really implies that the random phase approximation is unable to describe the system (except for very short time), and that there are physical processes occurring in the system that call for a better mathematical approximation. The appearance of the pole in the upper half-plane has been suggested as a way of seeing dynamically that the collection of particles with attractive interactions has undergone a transition from a gas to liquid.[h]

Later we shall see a similar instability occurring in fermion systems with an attractive short-range interaction. The onset of this instability represents the transition to a "superconducting" phase.

---

[h] (*Original*) ‡N. D. Mermin, doctoral thesis, Harvard University, 1961.

Chapter 9

Relation between Real and Imaginary Time Response Functions

In the last chapter, we used the Hartree approximation to describe nonequilibrium phenomena. Unfortunately, we cannot directly write more complicated approximations in the real-time domain because we have no simple boundary conditions that can act as a guide in determining $g_2(U)$. Therefore, we have, at this stage, no complete theory for determining the physical response function $g(U)$. [As we saw in Chapter 4, simple physical arguments do not suffice to determine approximations for the two-particle Green's function; it is necessary to use the boundary conditions to determine the range of the time integrations in, e.g., Eqs. (5.6) and (5.7).]

In Chapter 6, we developed a theory for approximating Σ and, therefore, $G_2(U)$ in the imaginary time domain. Now we shall discuss the relationship between $g(U)$, the physical response function, and $G(U)$, the imaginary time response function, and show how the theory already developed suffices to determine $g(U)$.

9.1 Linear Response

There is a particularly simple relation between the linear responses of the density in the two time domains. In the imaginary time

Annotations to Quantum Statistical Mechanics
In-Gee Kim
Copyright © 2018 Pan Stanford Publishing Pte. Ltd.
ISBN 978-981-4774-15-4 (Hardcover), 978-1-315-19659-6 (eBook)
www.panstanford.com

domain,

$$\frac{\delta G\left(1, 1'; U\right)}{\delta U\left(2\right)} = \pm \left[G_2\left(12, 1'2^+\right) - G\left(1, 1'\right) G\left(2, 2^+\right)\right]$$

Hence, the response of the density can be written

$$\pm i \frac{\delta G\left(1, 1^+; U\right)}{\delta U\left(2\right)} = i \left[G_2\left(12, 1^+2^+\right) - G\left(1, 1^+\right) G\left(2, 2^+\right)\right]$$

$$= \frac{1}{i} \left[\langle \hat{T}\left(\hat{n}\left(1\right) \hat{n}\left(2\right)\right)\rangle - \langle \hat{n}\rangle \langle \hat{n}\rangle\right] \tag{9.1}$$

In discussing this response, it is convenient to define

$$L\left(1 - 2\right) = \pm i \left[\frac{\delta G\left(1, 1^+; U\right)}{\delta U\left(2\right)}\right]$$

$$= \frac{1}{i} \left\langle \hat{T}\left[\left(\hat{n}\left(1\right) - \langle \hat{n}\rangle\right)\left(\hat{n}\left(2\right) - \langle \hat{n}\rangle\right)\right]\right\rangle \tag{9.2}$$

We should notice that $L\left(1 - 2\right)$ is quite analogous in structure to the one-particle Green's function. Just as $G\left(1 - 1'\right)$ is composed of the two analytic functions of time $G^>\left(1 - 1'\right)$ and $G^<\left(1 - 1'\right)$, so

$$L\left(1 - 2\right) = \begin{cases} L^>\left(1 - 2\right) & \text{for } t_1 > t_2 \\ L^<\left(1 - 2\right) & \text{for } t_1 < t_2 \end{cases} \tag{9.2a}$$

where

$$L^>\left(1 - 2\right) = \frac{1}{i} \left\langle \left(\hat{n}\left(1\right) - \langle \hat{n}\rangle\right)\left(\hat{n}\left(2\right) - \langle \hat{n}\rangle\right)\right\rangle$$

$$L^<\left(1 - 2\right) = \frac{1}{i} \left\langle \left(\hat{n}\left(2\right) - \langle \hat{n}\rangle\right)\left(\hat{n}\left(1\right) - \langle \hat{n}\rangle\right)\right\rangle \tag{9.2b}$$

As G satisfies the boundary condition,

$$G\left(1 - 1'\right)\big|_{t_1=0} = \pm e^{\beta\mu}\, G\left(1 - 1'\right)\big|_{t_1=-i\beta}$$

so $L\left(1 - 2\right)$ the boundary condition

$$L\left(1 - 2\right)\big|_{t_1=0} = L\left(1 - 2\right)\big|_{t_1=-i\beta} \tag{9.3}$$

Therefore, L can also be written in terms of a Fourier series as[a]

$$L\left(1 - 2\right) = \int \frac{d\mathbf{k}}{\left(2\pi\right)^3} \sum_{\nu} L\left(k,\, \Omega_{\nu}\right) e^{i\mathbf{k}\cdot\left(\mathbf{r}_1 - \mathbf{r}_2\right) - i\Omega_{\nu}\left(t_1 - t_2\right)} \tag{9.4a}$$

---

[a]The function symbol L just after the summation symbol $\sum_{\nu}$ was omitted in the original text.

where

$$\Omega_\nu = \frac{\pi \nu}{-i\beta} \qquad \nu = \text{even integer} \qquad (9.4b)$$

In exactly the same way as we establish that the Fourier coefficient for $G(1 - 1')$ is

$$G(\mathbf{p}, z) = \int \frac{d\omega'}{2\pi} \frac{A(p, \omega')}{z - \omega'}$$
$$= \int \frac{d\omega'}{2\pi} \frac{G^>(p, \omega') \mp G^<(p, \omega')}{z - \omega'}$$

we find that

$$L(k, \Omega) = \int \frac{d\omega'}{2\pi} \frac{L^>(k, \omega') - L^<(k, \omega')}{\Omega - \omega'} \qquad (9.5a)$$

where

$$L^\gtrless(k, \omega) = \int d\mathbf{r}_1 \int_{-\infty}^{\infty} dt_1 \, e^{-i\mathbf{k}\cdot(\mathbf{r}_1 - \mathbf{r}_2) + i\omega(t_1 - t_2)} i L^\gtrless(\mathbf{r}_1 - \mathbf{r}_2, t_1 - t_2)$$
$$(9.5b)$$

The function $L(k, \Omega)$ is the quantity that is most directly evaluated by a Green's function analysis in the imaginary time domain. The linear response of the density to a physical disturbance can be easily expressed in terms of $L(k, \Omega)$. The physical response is given by

$$\langle \hat{n}(1) \rangle_U = \left\langle \hat{U}^\dagger(t_1) \hat{n}(1) \hat{U}(t_1) \right\rangle$$

where

$$\hat{U} = \hat{T} \left\{ \exp\left[-i \int_{-\infty}^{t_1} d2 \, U(2) \hat{n}(2) \right] \right\}$$

and all the times are real. Hence, the linear response of $\langle \hat{n}(1) \rangle_U$ to U is

$$\delta \left[\pm i g\left(1, 1^+; U\right) \right] = \delta \langle \hat{n}(1) \rangle$$
$$= \frac{1}{i} \int_{-\infty}^{t_1} d2 \, \langle [\hat{n}(1), \hat{n}(2)] \rangle U(2)$$
$$= \int_{-\infty}^{t_1} d2 \, [L^>(1 - 2) - L^<(1 - 2)] U(2)$$
$$(9.6)$$

These functions $L^>$ and $L^<$ are exactly the same analytic functions that appear in the coefficient of Eq. (9.2) of the linear term in the expansion of $G(U)$. This is the fundamental connection between the two linear responses.

If

$$U(\mathbf{R}, T) = U_0 e^{i\mathbf{k}\cdot\mathbf{R} - i\Omega T} \tag{9.7}$$

then

$$\delta \langle \hat{n}(1) \rangle_U = \left(\frac{\delta n}{\delta U}\right)(k, \Omega) U(\mathbf{R}, T)$$

where

$$\left(\frac{\delta n}{\delta U}\right)(k, \Omega) = \int_{-\infty}^{t_1} dt_2 \int d\mathbf{r}_2 \, e^{-i\mathbf{k}\cdot(\mathbf{r}_1 - \mathbf{r}_2) + i\Omega(t_1 - t_2)}$$
$$\times [L^>(\mathbf{r}_1 - \mathbf{r}_2, t_1 - t_2) - L^<(\mathbf{r}_1 - \mathbf{r}_2, t_1 - t_2)]$$
$$= \frac{1}{i} \int_{-\infty}^{t_1} \int_{-\infty}^{\infty} \frac{d\omega'}{2\pi} [L^>(k, \omega') - L^<(k, \omega')] e^{i(\Omega - \omega')(t_1 - t_2)}$$
$$= \int \frac{d\omega'}{2\pi} \frac{L^>(k, \omega') - L^<(k, \omega')}{\Omega - \omega'}$$

However, we can recognize this last expression as just $L(k, \Omega)$, so that

$$\left(\frac{\delta n}{\delta U}\right)(k, \Omega) = L(k, \Omega) \tag{9.8}$$

Therefore, the Fourier coefficient function $L(k, \Omega)$ is exactly the linear response of $\langle \hat{n}(1) \rangle_U$ to a disturbance with wavenumber $\mathbf{k}$ and frequency Ω in the upper half-plane.

Let us determine this Fourier coefficient by using the Hartree approximation in the complex time domain. We certainly expect that this approximation has the same physical content as the real-time Hartree approximation. Therefore, we anticipate that the linear response $L(\mathbf{k}, \Omega)$ computed from this approximation for $G(U)$ should be identical to the $\left(\frac{\delta n}{\delta U}\right)(\mathbf{k}, \Omega)$ that we computed in the last chapter by means of the random phase approximation.

In the imaginary time domain, the Hartree approximation is

$$G^{-1}(1, 1'; U) = \left[i\frac{\partial}{\partial t_1} + \frac{\nabla_1^2}{2m} - U_{\text{eff}}(1)\right] \delta(1 - 1')$$
$$= \left[i\frac{\partial}{\partial t_1} + \frac{\nabla_1^2}{2m} - U(1)\right.$$
$$\left. \mp i \int_0^{-i\beta} d2 \, V(1 - 2) G^<(2, 2; U)\right] \delta(1 - 1')$$

We can compute

$$\frac{\delta G\left(1, 1'; U\right)}{\delta U\left(2\right)} = - \int_0^{-i\beta} d3 \int_0^{-i\beta} d3'\, G\left(1, 3; U\right) G\left(3, 1'; U\right)$$

$$\times \frac{\delta G^{-1}\left(3, 3'; U\right)}{\delta U\left(2\right)}$$

$$= \int_0^{-i\beta} d3\, G\left(1, 3; U\right) G\left(3, 1'; U\right) \frac{\delta U_{\text{eff}}(3)}{\delta U\left(2\right)}$$

$$= G\left(1, 2; U\right) G\left(2, 1'; U\right) \pm i \int_0^{-i\beta} d3 \int_0^{-i\beta} d4$$

$$\times G\left(1, 3; U\right) G\left(3, 1'; U\right) V\left(3 - 4\right) \frac{\delta G\left(4, 4^+; U\right)}{\delta U\left(2\right)} \tag{9.9}$$

Therefore, in the Hartree approximation,

$$L(1 - 2) = \pm i \left[\frac{\delta G\left(1, 1^+; U\right)}{\delta U\left(2\right)}\right]_{U=0}$$

$$= \pm i G\left(1 - 2\right) G\left(2 - 1\right) + \int_0^{-i\beta} d3 \int_0^{-i\beta} d4$$

$$\times \left[\pm i G\left(1 - 3\right) G\left(3 - 1\right)\right] V\left(3 - 4\right) L\left(4 - 2\right)$$

If we define

$$L_0\left(1 - 2\right) = \pm i G\left(1 - 2\right) G\left(2 - 1\right) \tag{9.10}$$

we can write this approximation as

$$L(1 - 2) = L_0(1 - 2) + \int_0^{-i\beta} d3 \int_0^{-i\beta} d4 L_0(1 - 3) V(3 - 4) L(4 - 2) \tag{9.11}$$

By employing the boundary conditions on G

$$G\left(1 - 2\right)|_{t_1=0} = \pm e^{\beta\mu}\, G\left(1 - 2\right)|_{t_1=-i\beta}$$

$$G\left(1 - 2\right)|_{t_2=0} = \pm e^{\beta\mu}\, G\left(1 - 2\right)|_{t_2=-i\beta}$$

we can see that L_0 satisfies the same boundary condition (9.3) as L. Thus, L_0 may also be expanded in a Fourier series of the form Eq. (9.4), with a Fourier coefficient $L_0\left(k, \Omega_\nu\right)$. From Eq. (9.10), it follows that

$$L_0^>\left(1 - 2\right) = \pm i G^>\left(1 - 2\right) G^<\left(2 - 1\right)$$

$$L_0^<\left(1 - 2\right) = \pm i G^<\left(1 - 2\right) G^>\left(2 - 1\right)$$

and hence

$$L_0^{\gtrless}(k, \omega) = \int \frac{d\mathbf{p}'\, d\omega'}{(2\pi)^3\, 2\pi}\, G^{\gtrless}\left(\mathbf{p}' + \frac{\mathbf{k}}{2}, \omega' + \frac{\omega}{2}\right) G^{\lessgtr}\left(\mathbf{p}' - \frac{\mathbf{k}}{2}, \omega' - \frac{\omega}{2}\right)$$

so that

$$L_0^{>}(k, \omega) - L_0^{<}(k, \omega)$$

$$= \int \frac{d\mathbf{p}'\, d\omega'}{(2\pi)^3\, 2\pi}\, A\left(\mathbf{p}' + \frac{\mathbf{k}}{2}, \omega' + \frac{\omega}{2}\right) A\left(\mathbf{p}' - \frac{\mathbf{k}}{2}, \omega' - \frac{\omega}{2}\right)$$

$$\times \left\{ \left[1 \pm f\left(\omega' + \frac{\omega}{2}\right)\right] f\left(\omega' - \frac{\omega}{2}\right) \right.$$

$$\left. - f\left(\omega' + \frac{\omega}{2}\right)\left[1 \pm f\left(\omega' - \frac{\omega}{2}\right)\right] \right\}$$

Because Eq. (9.11) is derived by differentiating the Hartree approximation, the G's that appear in Eq. (9.10) must be the Hartree Green's functions, and for these

$$A(p, \omega) = 2\pi\delta\left(\omega - E(p)\right)$$

$$= 2\pi\delta\left(\omega - \frac{p^2}{2m} - nv\right)$$

Therefore, $L_0^{>} - L_0^{<}$ takes the simple form

$$L_0^{>}(k, \omega) - L_0^{<}(k, \omega) = \int \frac{d\mathbf{p}}{(2\pi)^3}\, 2\pi\delta\left(\omega - E\left(\mathbf{p} + \frac{\mathbf{k}}{2}\right) + E\left(\mathbf{p} - \frac{\mathbf{k}}{2}\right)\right)$$

$$\times \left[f\left(E\left(\mathbf{p} + \frac{\mathbf{k}}{2}\right)\right) - f\left(E\left(\mathbf{p} - \frac{\mathbf{k}}{2}\right)\right) \right]$$

It follows then that the Fourier coefficient $L_0(k, \Omega)$ is

$$L_0(k, \Omega) = \int \frac{d\omega'}{2\pi}\, \frac{L_0^{>}(k, \omega') - L_0^{<}(k, \omega')}{\Omega - \omega'}$$

$$= \int \frac{d\mathbf{k}}{(2\pi)^3}\, \frac{f\left(E\left(\mathbf{p} + \frac{\mathbf{k}}{2}\right)\right) - f\left(E\left(\mathbf{p} - \frac{\mathbf{k}}{2}\right)\right)}{\Omega - \frac{\mathbf{k}\cdot\mathbf{p}}{m}} \qquad (9.12)$$

If we compare Eq. (9.12) with Eq. (8.23), we see that

$$L_0(k, \Omega) = \left(\frac{\delta n}{\delta U}\right)_0 (k, \Omega) \qquad (9.12a)$$

The latter function is the quantity that appears in the solution to the real-time Hartree approximation.

Now it is trivial to solve Eq. (9.11). We multiply it by $e^{-i\mathbf{k}\cdot(\mathbf{r}_1 - \mathbf{r}_2) + i\Omega_\mu(t_1 - t_2)}$ and integrate over all $\mathbf{r}_1$ and all t_1 between 0 and

$-i\beta$. In this way, we pick out the Fourier coefficients on both sides of the equation and find

$$L(k, \Omega_\nu) = L_0(k, \Omega_\nu)[1 + v(k)L(k, \Omega_\nu)]$$

and therefore

$$L(k, \Omega) = L_0(k, \Omega)[1 + v(k)L(k, \Omega)]$$

Thus

$$L(k, \Omega) = \frac{L_0(k, \Omega)}{1 - v(k)L_0(k, \Omega)}$$

or

$$L(k, \Omega) = \frac{\left(\frac{\delta n}{\delta U}\right)_0(k, \Omega)}{1 - v(k)\left(\frac{\delta n}{\delta U}\right)_0(k, \Omega)} \tag{9.13}$$

We recognize this expression for $L(k, \Omega)$ as exactly that derived for $\left(\frac{\delta n}{\delta U}\right)(k, \Omega)$ in the random phase approximation [cf. Eq. (8.18)]. Therefore, we see that $\left(\frac{\delta n}{\delta U}\right)(k, \Omega)$ can be determined equally well from the imaginary time theory. One just has to solve for $L(k, \Omega)$, using an approximation for $G(U)$, to find the physical response $\left(\frac{\delta n}{\delta U}\right)(k, \Omega)$.

Unfortunately, this procedure for determining the physical response from the imaginary time response is very difficult to employ for approximations fancier than the Hartree approximation. It is only for this approximation that we can solve exactly for the response and hence obtain an exact solution for the Fourier coefficient. In other situations, we cannot obtain an explicit form for $L(k, \Omega)$ from the imaginary time Green's function approximation, and hence we cannot employ the simple analysis that we have developed here.

9.2 Continuation of Imaginary Time Response to Real Times

We should really like to have approximate equations of motion for $g(U)$. However, these are hard to obtain directly, because $g_2(U)$ satisfies a somewhat complicated boundary condition. Instead of working with $g_2(U)$ directly, we shall show how, $g_2(U)$ in terms of $G_2(U)$, we obtain a theory of the physical response function.

We begin this analysis by introducing an essentially trivial generalization of $G(U)$ and $G_2(U)$. These functions were originally defined for pure imaginary times in the interval $0 < it$, $it' < i\beta$. However, there is nothing very special about the time zero. We could just as well define Green's functions in the interval $[t_0, t_0 - i\beta]$, i.e.,

$$0 < i(t - t_0) < \beta \quad (t_0 \text{ real}) \tag{9.14}$$

For times in this interval, we write

$$G(1, 1'; U; t_0) = \frac{1}{i} \frac{\langle \hat{T} [\hat{S}\hat{\psi}(1) \hat{\psi}^\dagger(1')] \rangle}{\langle \hat{T} [\hat{S}] \rangle} \tag{9.15a}$$

where

$$\hat{S} = \exp\left[-i \int_{t_0}^{t_0 - i\beta} d2\, U(2)\hat{n}(2)\right] \tag{9.15b}$$

Here $\hat{T}$ orders according to the size of $i(t - t_0)$; operators with larger value of $i(t - t_0)$ appear on the left. When $t_0 = 0$, the $G(U; t_0)$ defined by Eq. (9.15) reduces to the $G(U)$ discussed in Chapter 6.

The theory of $G(U; t_0)$ is identical to the theory of $G(U)$. This generalized response function satisfies the boundary condition

$$G(1, 1'; U, t_0)\big|_{t_1=t_0} = \pm e^{\beta\mu}\, G(1, 1'; U, t_0)\big|_{t_1=t_0-i\beta}$$

instead of

$$G(1, 1'; U)\big|_{t_1=0} = \pm e^{\beta\mu}\, G(1, 1'; U)\big|_{t_1=-i\beta}$$

Therefore, the only change that has to be made in the formulas of Chapter 6 to make them apply to $G(U; t_0)$ is to replace all time integrals over the time interval $[0, -i\beta]$ by integrals over $[t_0, t_0 - i\beta]$. In particular, $G(U; t_0)$ satisfies the equations of motion:[b]

$$\left[i\frac{\partial}{\partial t_1} + \frac{\nabla_1^2}{2m} - U(1)\right] G(1, 1'; U; t_0)$$
$$+ \int_{t_0}^{t_0 - i\beta} d\bar{1}\, \Sigma(1, \bar{1}; U; t_0)\, G(\bar{1}, 1'; U; t_0) = \delta(1 - 1') \tag{9.16a}$$

---

[b]The $+$ symbol in front of the integral sign $\int$ in Eq. (9.16a) in the original text was omitted.

and

$$\left[-i\frac{\partial}{\partial t_{1'}} + \frac{\nabla_{1'}^2}{2m} - U(1)\right] G\left(1, 1'; U; t_0\right)$$
$$-\int_{t_0}^{t_0-i\beta} d\bar{1}\, G\left(\bar{1}, \bar{1}; U; t_0\right) \Sigma\left(\bar{1}, 1'; U; t_0\right) = \delta\left(1-1'\right) \tag{9.16b}$$

We shall now establish a relationship between $G\left(U; t_0\right)$ and $g(U)$ in order that we may convert Eq. (9.16) into equations of motion for $g(U)$. To do this, we consider the case $i\left(t_1 - t_0\right) < i\left(t_{1'} - t_0\right)$. Then

$$G\left(1, 1'; U; t_0\right) = G^{<}\left(1, 1'; U; t_0\right)$$
$$= \pm\frac{1}{i}\frac{\left\langle \hat{T}\left[\hat{S}\hat{\psi}^{\dagger}\left(1'\right)\hat{\psi}\left(1\right)\right]\right\rangle}{\left\langle \hat{T}\left[\hat{S}\right]\right\rangle}$$
$$= \pm\left(\frac{1}{i}\right)\left\langle \hat{\mathcal{U}}\left(t_0, t_0 - i\beta\right)\left[\hat{\mathcal{U}}^{\dagger}\left(t_0, t_{1'}\right)\hat{\psi}^{\dagger}\left(1'\right)\hat{\mathcal{U}}\left(t_0, t_{1'}\right)\right]\right.$$
$$\left.\times\left[\hat{\mathcal{U}}^{\dagger}\left(t_0, t_1\right)\hat{\psi}\left(1\right)\hat{\mathcal{U}}\left(t_0, t_1\right)\right]\right\rangle \Big/ \left\langle \hat{U}\left(t_0, t_0 - i\beta\right)\right\rangle \tag{9.17}$$

where

$$\hat{\mathcal{U}}\left(t_0, t_1\right) = \hat{T}\left\{\exp\left[-i\int_{t_0}^{t_1} d2\, U\left(2\right)\hat{n}(2)\right]\right\} \tag{9.17a}$$

For comparison, we write the physical response function, which is defined for real times. For example,

$$g^{<}\left(1, 1'; U\right) = \pm\left(\frac{1}{i}\right)\left\langle \hat{\psi}_U^{\dagger}\left(1'\right)\hat{\psi}_U\left(1\right)\right\rangle$$
$$= \pm\left(\frac{1}{i}\right)\left\langle \left[\hat{\mathcal{U}}^{\dagger}\left(t_{1'}\right)\hat{\psi}^{\dagger}\left(1'\right)\hat{\mathcal{U}}\left(t_{1'}\right)\right]\left[\hat{\mathcal{U}}^{\dagger}\left(t_1\right)\hat{\psi}\left(1\right)\hat{\mathcal{U}}\left(t_1\right)\right]\right\rangle \tag{9.18}$$

where

$$\hat{\mathcal{U}}\left(t_1\right) = \hat{T}\left\{\exp\left[-i\int_{-\infty}^{t_1} d2\, U\left(2\right)\hat{n}(2)\right]\right\} \tag{9.18a}$$

Let us consider the case in which $U(1)$ is an analytic function of $t_{1'}$ for $0 > \Im t_1 > -\beta$, which satisfies

$$\lim_{\Re t_1 \to -\infty} U\left(t_1\right) = 0 \tag{9.19}$$

For example, $U\left(\mathbf{R}, T\right)$ might be $U_0 e^{i\mathbf{k}\cdot\mathbf{r}-i\Omega T}$ where $\Im\Omega > 0$. If $U\left(\mathbf{R}, T\right)$ is an analytic function of the time, then $\hat{\mathcal{U}}\left(t_0, t_1\right)$ and

$\hat{U}(t_0)$ are analytic functions of their time variables in the sense that every matrix element of each term in their power-series expansion is analytic. If all sums converge uniformly, as we shall assume, $G^<(1, 1'; U; t_0)$ and $g^<(1, 1'; U)$ are then each analytic functions of their time arguments. The analytic functions $\hat{U}(t_0, t_1)$ and $\hat{U}(t_0)$ can also be defined by

$$i\left(\frac{\partial}{\partial t_1}\right)\hat{U}(t_1) = \int d\mathbf{r}_1 \, \hat{n}(1) \, U(1) \hat{U}(1)$$

$$\hat{U}(-\infty) = 1 \qquad\qquad (9.20)$$

and

$$i\left(\frac{\partial}{\partial t_1}\right)\hat{U}(t_0, t_1) = \int d\mathbf{r}_1 \, \hat{n}(1) \, U(1) \hat{U}(t_0, t_1)$$

$$\hat{U}(t_0, t_0) = 1$$

Because of this analyticity, it follows that

$$\lim_{t_0 \to -\infty} \hat{U}(t_0, t_1) = \hat{U}(t_0)$$

and, because of (9.19)

$$\lim_{t_0 \to -\infty} \hat{U}(t_0, t_0 - i\beta) = 1$$

Therefore, the analytic functions $G^<(1, 1'; U; t_0)$ and $g(1, 1'; U)$ are connected by

$$\lim_{t_0 \to -\infty} G^<(1, 1'; U; t_0) = g^<(1, 1'; U) \qquad (9.21a)$$

and, similarly,

$$\lim_{t_0 \to -\infty} G^>(1, 1'; U; t_0) = g^>(1, 1'; U) \qquad (9.21b)$$

In order to have a simple confirmation of the result that we have just obtained, let us compute $\pm i G^<(1, 1'; U; t_0)$ and $\pm i g^<(1, 1; U)$ to first order in U. These are

$$\pm i G^<(1, 1; U; t_0)$$

$$= \langle \hat{n} \rangle + \int_{t_0}^{t_0 - i\beta} d2 \left(\frac{1}{i}\right) \left\langle \hat{T} \left\{[\hat{n}(1) - \langle \hat{n} \rangle][\hat{n}(2) - \langle \hat{n} \rangle]\right\}\right\rangle U(2)$$

$$= \langle \hat{n} \rangle + \int_{t_0}^{t_1} d2 \left(\frac{1}{i}\right) \langle [\hat{n}(2) - \langle \hat{n} \rangle][\hat{n}(1) - \langle \hat{n} \rangle]\rangle U(2)$$

$$- \int_{t_0 - i\beta}^{t_1} d2 \left(\frac{1}{i}\right) \langle [\hat{n}(1) - \langle \hat{n} \rangle][\hat{n}(2) - \langle \hat{n} \rangle]\rangle U(2)$$

$$= \langle \hat{n} \rangle + \int_{t_0}^{t_1} d2 \, L^>(1 - 2) U(2) - \int_{t_0 - i\beta}^{t_1} d2 \, L^<(1 - 2) U(2)$$

$$(9.22)$$

Since $L^>$ and $L^<$ are analytic function of their time variable, when U is also analytic, the right side of Eq. (9.22) is clearly an analytic function of t_1 and t_0. If we take the limit $t_0 \to -\infty$, Eq. (9.22) becomes

$$\lim_{t_0 \to -\infty} [\pm i G^< (1, 1; U ; t_0)] = \langle \hat{n} \rangle + \int_{-\infty}^{t_1} d2 \ [L^> (1-2) - L^< (1-2)] \, U (2)$$

(9.22a)

This should be compared with Eq. (9.6), which indicates the physical response is

$$\langle \hat{n}(1) \rangle_U = \pm i g^< (1, 1; U)$$

$$= \langle \hat{n} \rangle + \int_{-\infty}^{t_1} d2 \ [L^> (1-2) - L^< (1-2)] \, U (2) \quad (9.22b)$$

This is, of course, the same as Eq. (9.22a).

9.3 Equations of Motion in the Real-Time Domain

We now describe how approximate equations of motion for $G\,(U ; t_0)$ may be continued into equations of motion for the physical response function $g(U)$.

Let us begin with the very simple example, the Hartree approximation. In this approximation, Eq. (9.16a) is

$$\left[i \frac{\partial}{\partial t_1} + \frac{\nabla_1^2}{2m} - U_{\text{eff}} (1; t_0) \right] G \left(1, 1'; U ; t_0 \right) = \delta \left(1 - 1' \right) \quad (9.23a)$$

where

$$U_{\text{eff}} (\mathbf{R}, T ; t_0) = U (\mathbf{R}, T) \pm i \int d\mathbf{R}' \, v \left(\mathbf{R} - \mathbf{R}' \right) G^< \left(\mathbf{R}'T ; \mathbf{R}'T ; U ; t_0 \right)$$

(9.23b)

We consider the case in which $i \, (t_1 - t_0) < i \, (t_{1'} - t_0)$. Then

$$\left[i \frac{\partial}{\partial t_1} + \frac{\nabla_1^2}{2m} - U_{\text{eff}} (1; t_0) \right] G^< \left(1, 1'; U ; t_0 \right) = 0$$

Using the analyticity of $U\,(\mathbf{R}, T)$, we take the limit $t_0 \to -\infty$ to find

$$U_{\text{eff}} (\mathbf{R}, T ; -\infty) = U \left(\mathbf{R}, T \right) \pm i \int d\mathbf{R}' \, v \left(\mathbf{R} - \mathbf{R}' \right) g^< \left(\mathbf{R}'T ; \mathbf{R}'T ; U \right)$$

(9.24a)

and

$$\left[i\frac{\partial}{\partial t_1} + \frac{\nabla_1^2}{2m} - U_{\text{eff}}(1; -\infty)\right] g^<(1, 1'; U) = 0 \qquad (9.24b)$$

These equations hold for all complex values of t_1 and $t_{1'}$ such that $\beta > \Im(t_1 - t_{1'}) > 0$. When they are specialized to the case of real values of the time variables, they become just the familiar statement of the real-time Hartree approximation.

Our original derivation of the Hartree approximation depended in no way on the analytic properties of $U(\mathbf{R}, T)$. In fact, the validity of the equations for $g(U)$ that we shall derive does not depend on the analyticity of U at all. The analytic continuation device is just a convenient way of handling the boundary conditions on the real-time response functions. It also gives a particularly simple way of seeing the connection between the imaginary time $G(U)$ and the physical response function $g(U)$.

This same continuation device can be applied in a much more general discussion of the equations of motion for $g(U)$. The self-energy $\Sigma(1, 1'; U; t_0)$ can be split into two parts as

$$\Sigma(1, 1'; U; t_0) = \Sigma_{\text{HF}}(1, 1'; U; t_0) + \Sigma_c(1, 1'; U; t_0) \qquad (9.25)$$

where the Hartree–Fock part of Σ is

$$\Sigma_{\text{HF}}(1, 1'; U; t_0) = \delta(t_1 - t_{1'})\left\{\pm i\delta(\mathbf{r}_1 - \mathbf{r}_{1'})\int d\mathbf{r}_2\, v(\mathbf{r}_1 - \mathbf{r}_2)\right.$$
$$\left. \times G^<(\mathbf{r}_2 t_1; \mathbf{r}_2 t_1; U; t_0) + iv(\mathbf{r}_1 - \mathbf{r}_{1'}) G^<(1, 1'; U; t_0)\right\}$$
$$(9.25a)$$

and the collisional part of Σ is composed of two analytic functions of the time variables $\Sigma^>$ and $\Sigma^<$:

$$\Sigma_c(1, 1'; U; t_0) = \begin{cases} \Sigma^>(1, 1'; U; t_0) & \text{for } i(t_1 - t_{1'}) > 0 \\ \Sigma^<(1, 1'; U; t_0) & \text{for } i(t_1 - t_{1'}) < 0 \end{cases} \qquad (9.25b)$$

For example, in the Born collision approximation

$$\Sigma_c(1, 1'; U; t_0) = \pm i^2 \int d\mathbf{r}_2 d\mathbf{r}_{2'}\, v(\mathbf{r}_1 - \mathbf{r}_2) v(\mathbf{r}_{1'} - \mathbf{r}_{2'})$$
$$\times \left\{ G(1, 1'; U; t_0) G(2, 2'; U; t_0) G(2', 2; U; t_0) \right.$$
$$\left. - G(1, 2'; U; t_0) G(2, 1'; U; t_0) G(2', 2; U; t_0)\right\}_{\substack{t_2 = t_1 \\ t_{2'} = t_{1'}}}$$
$$(9.26a)$$

so that $\Sigma^>$ and $\Sigma^<$ are

$$\Sigma^{\gtrless}\left(1, 1'; U; t_0\right)$$

$$= \pm i^2 \int d\mathbf{r}_2 d\mathbf{r}_{2'} \, v\left(\mathbf{r}_1 - \mathbf{r}_2\right) v\left(\mathbf{r}_{1'} - \mathbf{r}_{2'}\right)$$

$$\times \left\{ G^{\gtrless}\left(1, 1'; U; t_0\right) G^{\gtrless}\left(2, 2'; U; t_0\right) G^{\lessgtr}\left(2', 2; U; t_0\right) \right.$$

$$\left. \pm G^{\gtrless}\left(1, 2'; U; t_0\right) G^{\lessgtr}\left(2, 1'; U; t_0\right) G^{\lessgtr}\left(2', 2; U; t_0\right) \right\}_{\substack{2=t_1 \\ 2'=t_{1'}}}$$

$$(9.26\mathrm{b})$$

Since $G^>$ and $G^<$ are analytic functions of their time variables, so is $\Sigma^{\gtrless}$.

For the sake of simplicity in writing, let us for the moment drop the exchange term in Σ_{HF}, i.e., the term proportional to $v\left(\mathbf{r}_1 - \mathbf{r}_{1'}\right)$ in Eq. (9.25a). Then Eq. (9.16a) becomes

$$\left[i\frac{\partial}{\partial t_1} + \frac{\nabla_1^2}{2m} - U_{\mathrm{eff}}\left(1; t_0\right)\right] G\left(1, 1'; U; t_0\right)$$

$$= \delta\left(1 - 1'\right) + \int_{t_0}^{t_0 - i\beta} d\bar{1} \, \Sigma_{\mathrm{c}}\left(1, \bar{1}; U; t_0\right) G\left(\bar{1}, 1'; U; t_0\right)$$

For the case $i\left(t_1 - t_0\right) < i\left(t_{1'} - t_0\right)$, this gives

$$\left[i\frac{\partial}{\partial t_1} + \frac{\nabla_1^2}{2m} - U_{\mathrm{eff}}\left(1; t_0\right)\right] G^<\left(1, 1'; U; t_0\right)$$

$$= \int_{t_0}^{t_1} d\bar{1} \, \Sigma^>\left(1, \bar{1}; U; t_0\right) G^<\left(\bar{1}, 1'; U; t_0\right)$$

$$+ \int_{t_1}^{t_{1'}} d\bar{1} \, \Sigma^<\left(1, \bar{1}; U; t_0\right) G^<\left(\bar{1}, 1'; U; t_0\right)$$

$$+ \int_{t_{1'}}^{t_0 - i\beta} d\bar{1} \, \Sigma^<\left(1, \bar{1}; U; t_0\right) G^<\left(\bar{1}, 1'; U; t_0\right)$$

If we now take the limit $t_0 \to -\infty$, we find that $g^<(U)$ obeys

$$\left[i\frac{\partial}{\partial t_1} + \frac{\nabla_1^2}{2m} - U_{\mathrm{eff}}\left(1\right)\right] g^<\left(1, 1'; U\right)$$

$$= \int_{-\infty}^{t_1} d\bar{1} \left[\Sigma^>\left(1, \bar{1}; U\right) - \Sigma^<\left(1, \bar{1}; U\right)\right] g^<\left(\bar{1}, 1'; U\right)$$

$$- \int_{-\infty}^{t_{1'}} d\bar{1} \, \Sigma^<\left(1, \bar{1}; U\right) \left[g^>\left(\bar{1}, 1'; U\right) - g^<\left(\bar{1}, 1'; U\right)\right]$$

$$(9.27\mathrm{a})$$

where

$$U_{\text{eff}}(1) = U_{\text{eff}}(1; -\infty)$$
$$\Sigma^{\gtrless}(1, 1'; U) = \Sigma^{\gtrless}(1, 1'; U; -\infty)$$

Applying the same arguments (9.16a) in the case $i(t_1 - t_0) > i(t_{1'} - t_0)$, we find[c]

$$\left[i\frac{\partial}{\partial t_1} + \frac{\nabla_1^2}{2m} - U_{\text{eff}}(1)\right] g^>(1, 1'; U)$$
$$= \int_{-\infty}^{t_1} d\bar{1}\,\left[\Sigma^>(1, \bar{1}; U) - \Sigma^<(1, \bar{1}; U)\right] g^>(\bar{1}, 1'; U)$$
$$- \int_{-\infty}^{t_{1'}} d\bar{1}\,\Sigma^>(1, \bar{1}; U)\left[g^>(\bar{1}, 1'; U) - g^<(\bar{1}, 1'; U)\right] \tag{9.27b}$$

Similarly, Eq. (9.16b) implies

$$\left[-i\frac{\partial}{\partial t_{1'}} + \frac{\nabla_{1'}^2}{2m} - U_{\text{eff}}(1')\right] g^<(1, 1'; U)$$
$$= \int_{-\infty}^{t_1} d\bar{1}\,\left[g^>(1, \bar{1}; U) - g^<(1, \bar{1}; U)\right] \Sigma^<(\bar{1}, 1'; U)$$
$$- \int_{-\infty}^{t_{1'}} d\bar{1}\,g^<(1, \bar{1}; U)\left[\Sigma^>(\bar{1}, 1'; U) - \Sigma^<(\bar{1}, 1'; U)\right] \tag{9.28a}$$

and

$$\left[-i\frac{\partial}{\partial t_{1'}} + \frac{\nabla_{1'}^2}{2m} - U_{\text{eff}}(1')\right] g^>(1, 1'; U)$$
$$= \int_{-\infty}^{t_1} d\bar{1}\,\left[g^>(1, \bar{1}; U) - g^<(1, \bar{1}; U)\right] \Sigma^>(\bar{1}, 1'; U)$$
$$- \int_{-\infty}^{t_{1'}} d\bar{1}\,g^>(1, \bar{1}; U)\left[\Sigma^>(\bar{1}, 1'; U) - \Sigma^<(\bar{1}, 1'; U)\right] \tag{9.28b}$$

When $\Sigma^>(U; t_0 = -\infty)$ and $\Sigma^<(U; t_0 = -\infty)$ are expressed in terms of $g^>(U)$ and $g^<(U)$, Eqs. (9.27) and (9.28) can be used to determine the real-time response functions $g^>(U)$ and $g^<(U)$. For

---

[c]The integration variable symbol $d\bar{1}$ in Eqs. (9.27b) and (9.28a) was omitted in the original text.

example, the Born collision approximation for $g(U)$ is derived by using Eq. (9.26b) to find[d]

$$\Sigma^{\gtrless}\left(1, 1'; U\right)$$

$$\equiv \Sigma^{\gtrless}\left(1, 1'; U; t_0 = -\infty\right)$$

$$= \pm i^2 \int d\mathbf{r}_2 d\mathbf{r}_{2'} \; v\left(\mathbf{r}_1 - \mathbf{r}_2\right) v\left(\mathbf{r}_{1'} - \mathbf{r}_{2'}\right)$$

$$\times \left[g^{\gtrless}\left(1, 1'; U\right) g^{\gtrless}\left(2, 2'; U\right) g^{\lessgtr}\left(2', 2; U\right)\right.$$

$$\left.\pm g^{\gtrless}(1, 2'; U) g^{\gtrless}(2, 1'; U) g^{\gtrless}(2', 2; U)\right]_{t_2 = t_{1'},\, t_{2'} = t_{1'}}$$

$$(9.29)$$

Equations (9.27) and (9.28) are exact, except for the trivial omission of the exchange term in Σ_{HF}. In Chapter 10, we shall discuss how these equations may be used to describe transport. In particular, we shall use the approximation (9.29) to derive a generalization of the Boltzmann equation. We shall also use these equations to discuss sound propagation in many-particle systems.

---

[d]The typographic errors of the wrong parenthesis $[\cdots]$ in the last line and of the missing comma at the subscript in the original text are corrected.

Chapter 10

Slowly Varying Disturbances and the Boltzmann Equation

Equations (9.27)–(9.29) are, in general, exceedingly complicated. Fortunately, they become much simpler in the limit in which $U(\mathbf{R}, T)$ varies slowly in space and time. This is exactly the situation in which simple transport processes occur.

When U varies slowly, $g^{>}(1, 1'; U)$ and $g^{<}(1, 1'; U)$ are slowly varying functions of the coordinates

$$\mathbf{R} = \frac{\mathbf{r}_1 + \mathbf{r}_{1'}}{2} \quad T = \frac{t_1 + t_{1'}}{2} \tag{10.1a}$$

but sharply peak about zero values of

$$\mathbf{r} = \mathbf{r}_1 - \mathbf{r}_{1'} \quad t = t_1 - t_{1'} \tag{10.1b}$$

The equilibrium Green's functions are sharply peaked about $\mathbf{r} = 0$ and $t = 0$, as can be seen, for example, from $G_0^{<}(r, t)$ in the low-density limit:

$$G_0^{<}(r, t) = \int \frac{d\mathbf{p}}{(2\pi)^3} e^{\beta\left(\frac{p^2}{2m} - \mu\right) - i\left(\frac{p^2}{2m}\right)t + i\mathbf{p}\cdot\mathbf{r}}$$

$$= \frac{1}{i} \left(\frac{m}{2\pi(\beta + it)}\right)^{3/2} \exp\left[\beta\mu - \frac{mr^2(\beta - it)}{2(\beta^2 + t^2)}\right]$$

Annotations to Quantum Statistical Mechanics
In-Gee Kim
Copyright © 2018 Pan Stanford Publishing Pte. Ltd.
ISBN 978-981-4774-15-4 (Hardcover), 978-1-315-19659-6 (eBook)
www.panstanford.com

This function has a spatial range on the order of a thermal wavelength, $\lambda_{th} = \frac{\hbar\beta}{2m}$, and in time, it decreases with a $t^{-3/2}$ dependence. Actually, if one includes a lifetime, then $G^<$ would decay exponentially in time. We may expect then that external disturbances with wavelengths much longer than the thermal wavelength and frequencies much smaller than the single-particle collision rates will not change this sharp r, t dependence of g.

It is, therefore, convenient to consider $g^\gtrless (1, 1'; U)$ as functions of the variables (10.1). We, therefore, write $g^\gtrless (1, 1'; U)$ as $g^\gtrless (\mathbf{r}, t; \mathbf{R}, T)$. We recall that

$$g^< (\mathbf{p}, \omega; \mathbf{R}, T) = \int d\mathbf{r}dt \, e^{-i\mathbf{p}\cdot\mathbf{r}+i\omega t} \left[\pm i g^< (\mathbf{r}, t; \mathbf{R}, T)\right] \quad (10.2a)$$

can be interpreted as the density of particles with momentum $\mathbf{p}$ and energy ω at the space–time point $\mathbf{R}, T$. Also

$$g^> (\mathbf{p}, \omega; \mathbf{R}, T) = \int d\mathbf{r}dt \, e^{-i\mathbf{p}\cdot\mathbf{r}+i\omega t} \, i g^> (\mathbf{r}, t; \mathbf{R}, T) \quad (10.2b)$$

is essentially the density of states available to a particle that is added to the system at $\mathbf{R}, T$ with momentum $\mathbf{p}$ and energy ω.

10.1 Derivation of the Boltzmann Equation

We may derive an equation of motion for $g^< (\mathbf{p}, \omega; \mathbf{R}, T)$ by subtracting Eq. (9.27a) from Eq. (9.28a). We find

$$\left[i\frac{\partial}{\partial t_1} + i\frac{\partial}{\partial t_{1'}} + \frac{\nabla_1^2}{2m} - \frac{\nabla_{1'}^2}{2m} - U_{\text{eff}}(1) + U_{\text{eff}}(1') \right] g^< (1, 1'; U)$$

$$= \int_{-\infty}^{t_1} d\bar{1} \left[\Sigma^> (1, \bar{1}; U) - \Sigma^< (1, \bar{1}; U) \right] g^< (\bar{1}, 1'; U)$$

$$+ \int_{-\infty}^{t_{1'}} d\bar{1} \, g^< (1, \bar{1}; U) \left[\Sigma^> (\bar{1}, 1'; U) - \Sigma^< (\bar{1}, 1'; U) \right]$$

$$- \int_{-\infty}^{t_{1'}} d\bar{1} \, \Sigma^< (1, \bar{1}; U) \left[g^> (\bar{1}, 1'; U) - g^< (\bar{1}, 1'; U) \right]$$

$$- \int_{-\infty}^{t_1} d\bar{1} \left[g^> (1, \bar{1}; U) - g^< (1, \bar{1}; U) \right] \Sigma^< (\bar{1}, 1'; U)$$

$$(10.3)$$

We now rewrite Eq. (10.3) in terms of the variable $\mathbf{r}, t; \mathbf{R}, T$ by expressing the g's that appear in this equation in terms of these variables and also writing Σ as[a]

$$\Sigma^{\gtrless} (1, 1'; U) = \Sigma^{\gtrless} (\mathbf{r}, t; \mathbf{R}, T)$$

Then after this change of variables, the left side of Eq. (10.3) becomes

$$\left[i \frac{\partial}{\partial T} + \frac{\nabla_{\mathbf{R}} \cdot \nabla_{\mathbf{r}}}{m} - U_{\text{eff}} \left(\mathbf{R} + \frac{\mathbf{r}}{2}, T + \frac{t}{2} \right) + U_{\text{eff}} \left(\mathbf{R} - \frac{\mathbf{r}}{2}, T - \frac{t}{2} \right) \right]$$
$$\times g^{<} (\mathbf{r}, t; \mathbf{R}, T)$$

$$(10.3a)$$

Because $g^{<} (\mathbf{r}, t; \mathbf{R}, T)$ is very sharply peaked about $\mathbf{r} = 0, t = 0$, we can consider $\mathbf{r}$ and t to be small in Eq. (10.3a). Then we can expand the difference of U_{eff}'s in the powers of $\mathbf{r}$ and t, retaining only the lowest-order terms. In this way, we see that Eq. (10.3a) may be approximately replaced by

$$\left\{ i \frac{\partial}{\partial T} + \frac{\nabla_{\mathbf{R}} \cdot \nabla_{\mathbf{r}}}{m} - \left[\left(\mathbf{r} \cdot \nabla_{\mathbf{R}} + t \frac{\partial}{\partial T} \right) U_{\text{eff}} (\mathbf{R}, T) \right] \right\} g^{<} (\mathbf{r}, t; \mathbf{R}, T)$$

$$(10.4a)$$

In terms of the variables $\mathbf{r}, t; \mathbf{R}, T$, the first term on the right side of Eq. (10.3) may be written as[b]

$$\int_{-\infty}^{t} d\bar{t} \int d\bar{\mathbf{r}} \left[\Sigma^{>} \left(\mathbf{r} - \bar{\mathbf{r}}, t - \bar{t}; \mathbf{R} + \frac{\bar{\mathbf{r}}}{2}, T + \frac{\bar{t}}{2} \right) \right.$$
$$\left. - \Sigma^{<} \left(\mathbf{r} - \bar{\mathbf{r}}, t - \bar{t}; \mathbf{R} + \frac{\bar{\mathbf{r}}}{2}, T + \frac{\bar{t}}{2} \right) \right] \quad (10.3b)$$
$$\times g^{<} \left(\bar{\mathbf{r}}, \bar{t}; \mathbf{R} - \left(\mathbf{r} - \frac{\bar{\mathbf{r}}}{2} \right), T - \left(t - \frac{\bar{t}}{2} \right) \right)$$

where we have made the change of integration variables

$$\bar{\mathbf{r}} = \bar{\mathbf{r}}_1 - \mathbf{r}_{1'} = \bar{\mathbf{r}}_1 - \left(\mathbf{R} - \frac{\mathbf{r}}{2} \right)$$
$$\bar{t} = \bar{t}_1 - t_{1'} = \bar{t}_1 - \left(T - \frac{t}{2} \right)$$

---

[a] The typographic error, ...(10.3) terms of the variable ..., of this sentence in the original text is corrected to ...(10.3) in terms of the variable ...

[b] The notation for the integration symbol $d\bar{r}$ in the original text is corrected to be a vector symbol $d\bar{\mathbf{r}}$.

Because $\Sigma^{>}(\mathbf{r}, t; \mathbf{R}, T)$ and $g^{<}(\mathbf{r}, t; \mathbf{R}, T)$ are each sharply peaked about $\mathbf{r} = 0, t = 0$, and slowly varying in $\mathbf{R}, T$, we can neglect the necessarily small quantities added to $\mathbf{R}$ and T in Eq. (10.3b). Then Eq. (10.3b) becomes[c]

$$\int_{-\infty}^{t} d\bar{t} \int d\bar{\mathbf{r}} \; [\Sigma^{>}(\mathbf{r} - \bar{\mathbf{r}}, t - \bar{t}; \mathbf{R}, T) \\ - \Sigma^{<}(\mathbf{r} - \bar{\mathbf{r}}, t - \bar{t}; \mathbf{R}, T)] \, g^{<}(\bar{\mathbf{r}}, \bar{t}; \mathbf{R}, T) \tag{10.4b}$$

The second term on the right side of Eq. (10.3) can be written in terms of the variable $\mathbf{r}, T; \mathbf{R}, T$ as[d]

$$\int_{t}^{\infty} d\bar{t} \int d\bar{\mathbf{r}} \; g^{<}\left(\mathbf{r}, \bar{t}; \mathbf{R} + \frac{(\mathbf{r} - \bar{\mathbf{r}})}{2}, T + \frac{(t - \bar{t})}{2}\right) \\ \times \left[\Sigma^{>}\left(\mathbf{r} - \bar{\mathbf{r}}, t - \bar{t}; \mathbf{R} - \frac{\bar{\mathbf{r}}}{2}, T - \frac{\bar{t}}{2}\right) \right. \tag{10.5a} \\ \left. - \Sigma^{<}\left(\mathbf{r} - \bar{\mathbf{r}}, t - \bar{t}; \mathbf{R} - \frac{\bar{\mathbf{r}}}{2}, T - \frac{\bar{t}}{2}\right)\right]$$

after the change in integration variable

$$\bar{t} = t_1 - \bar{t}_1 \quad \bar{\mathbf{r}} = \mathbf{r}_1 - \bar{\mathbf{r}}_1$$

We again realize that only small values of $\mathbf{r}$ and $\bar{\mathbf{r}}$, t and $\bar{t}$ are important, so that this term becomes[e]

$$\int_{t}^{\infty} d\bar{t} \int d\bar{\mathbf{r}} g(\bar{\mathbf{r}}, \bar{t}; \mathbf{R}, T)[\Sigma^{>}(\mathbf{r} - \bar{\mathbf{r}}, t - \bar{t}; \mathbf{R}, T) - \Sigma^{<}(\mathbf{r} - \bar{\mathbf{r}}, t - \bar{t}; \mathbf{R}, T)] \tag{10.5b}$$

When Eqs. (10.4b) and (10.5b) are added together, we see that the first two terms on the right side of Eq. (10.3) can be approximated by

$$\int_{-\infty}^{\infty} d\bar{t} \int d\bar{\mathbf{r}} g^{<}(\bar{\mathbf{r}}, \bar{t}; \mathbf{R}, T)[\Sigma^{>}(\mathbf{r} - \bar{\mathbf{r}}, t - \bar{t}; \mathbf{R}, T) - \Sigma^{<}(\mathbf{r} - \bar{\mathbf{r}}, t - \bar{t}; \mathbf{R}, T)]$$

Similarly, the remaining two terms in Eq. (10.3) can be evaluated as

$$-\int_{-\infty}^{\infty} d\bar{t} \int d\bar{\mathbf{r}} [g^{>}(\bar{\mathbf{r}}, \bar{t}; \mathbf{R}, T) - g^{<}(\bar{t}, \bar{t}; \mathbf{R}, T)]\Sigma^{<}(\mathbf{r} - \bar{\mathbf{r}}, t - \bar{t}; \mathbf{R}, T)$$

---

[c]The integral symbol $\int$ is added in front of the integration variable symbol $d\bar{\mathbf{r}}$, which is also vectorized.
[d]The integral symbol $\int$ is added in front of the integral variable symbol $d\bar{\mathbf{r}}$.
[e]The integral symbol $\int$ is added in front of the integration variable symbol $d\bar{\mathbf{r}}$, which is also vectorized.

Therefore, Eq. (10.3) may be approximately replaced by

$$\left[i\frac{\partial}{\partial T} + \frac{\nabla_\mathbf{R} \cdot \nabla_\mathbf{r}}{m} - (\mathbf{r} \cdot \nabla_\mathbf{R} U_{\text{eff}}(\mathbf{R}, T)) - t\frac{\partial}{\partial T}U_{\text{eff}}(\mathbf{R}, T) \right] g^<(\mathbf{r}, t; \mathbf{R}, T)$$

$$= \int d\bar{\mathbf{r}}d\bar{t} \, \{g^<(\bar{\mathbf{r}}, \bar{t}; \mathbf{R}, T)[\Sigma^>(\mathbf{r}-\bar{\mathbf{r}}, t-\bar{t}; \mathbf{R}, T) - \Sigma^<(\mathbf{r}-\bar{\mathbf{r}}, t-\bar{t}; \mathbf{R}, T)]$$

$$- [g^>(\bar{\mathbf{r}}, \bar{t}; \mathbf{R}, T) - g^<(\bar{\mathbf{r}}, \bar{t}; \mathbf{R}, T)]\Sigma^<(\mathbf{r}-\bar{\mathbf{r}}, t-\bar{t}; \mathbf{R}, T)\}$$

$$= \int d\bar{\mathbf{r}}d\bar{t} \, \{g^<(\bar{\mathbf{r}}, \bar{t}; \mathbf{R}, T)\Sigma^>(\mathbf{r}-\bar{\mathbf{r}}; \mathbf{R}, T)$$

$$-g^>(\bar{\mathbf{r}}, \bar{t}; \mathbf{R}, T)\Sigma^<(\mathbf{r}-\bar{\mathbf{r}}, t-\bar{t}; \mathbf{R}, T)\} \tag{10.6}$$

To convert this equation into a more useful form, we multiply by $\pm e^{-i\mathbf{p}\cdot\mathbf{r}-i\omega t}$ and integrate over all $\mathbf{r}$ and t. Then we find

$$\left[\frac{\partial}{\partial T} + \frac{\mathbf{p} \cdot \nabla_\mathbf{R}}{m} - \nabla_\mathbf{R} U_{\text{eff}}(\mathbf{R}, T) \cdot \nabla_\mathbf{p} + \frac{\partial U_{\text{eff}}(\mathbf{R}, T)}{\partial T}\frac{\partial}{\partial \omega} \right] g^<(\mathbf{p}, \omega; \mathbf{R}, T)$$

$$= -g^<(\mathbf{p}, \omega; \mathbf{R}, T)\Sigma^>(\mathbf{p}, \omega; \mathbf{R}, T) + g^>(\mathbf{p}, \omega; \mathbf{R}, T)\Sigma^<(\mathbf{p}, \omega; \mathbf{R}, T) \tag{10.7a}$$

where

$$\Sigma^>(\mathbf{p}, \omega; \mathbf{R}, T) = \int d\mathbf{r}dt \, e^{-i\mathbf{p}\cdot\mathbf{r}+i\omega t}i\Sigma^>(\mathbf{r}, t; \mathbf{R}, T)$$

$$\Sigma^<(\mathbf{p}, \omega; \mathbf{R}, T) = \int d\mathbf{r}dt \, e^{-i\mathbf{p}\cdot\mathbf{r}+i\omega t}[\pm i\Sigma^<(\mathbf{r}, t; \mathbf{R}, T)] \tag{10.8}$$

Exactly the same analysis applied to Eqs. (9.27b)[f] and (9.28b) yields the equation of motion for $g^>(\mathbf{p}, \omega; \mathbf{R}, T)$:[g]

$$\pm\left[\frac{\partial}{\partial T} + \frac{\mathbf{p} \cdot \nabla_\mathbf{R}}{m} - \nabla_\mathbf{R} U_{\text{eff}}(\mathbf{R}, T) \cdot \nabla_\mathbf{p} + \frac{\partial U_{\text{eff}}(\mathbf{R}, T)}{\partial T}\frac{\partial}{\partial \omega} \right] g^>(\mathbf{p}, \omega; \mathbf{R}, T)$$

$$= -g^<(\mathbf{p}, \omega; \mathbf{R}, T)\Sigma^>(\mathbf{p}, \omega; \mathbf{R}, T) + g^>(\mathbf{p}, \omega; \mathbf{R}, T)\Sigma^<(\mathbf{p}, \omega; \mathbf{R}, T) \tag{10.7b}$$

In order to gain some insight into the result we have just obtained, we consider the Born collision approximation in which $\Sigma^{\gtrless}$ are given by Eq. (9.29),[h] where this equation is written in terms of

---

[f]The wrong equation number (9-27b) in the original text is corrected.
[g]The wrong superscript < of g at the left-hand side in the original text is corrected to >.
[h]The wrong equation number (8-2) in the original text is corrected.

the variables $\mathbf{r}, t; \mathbf{R}, T$:

$$\Sigma^{\gtrless} (\mathbf{r}, t; \mathbf{R}, T) = i^2 \int d\bar{\mathbf{R}} d\bar{\mathbf{r}} \; v \left(\mathbf{R} + \frac{\mathbf{r}}{2} - \bar{\mathbf{R}} - \frac{\bar{\mathbf{r}}}{2} \right) v \left(\mathbf{R} - \frac{\mathbf{r}}{2} - \bar{\mathbf{R}} + \frac{\bar{\mathbf{r}}}{2} \right)$$

$$\times g^{\lessgtr} (-\bar{\mathbf{r}}, -t; \mathbf{R}, T) \left[g^{\gtrless} (\mathbf{r}, t; \mathbf{R}, T) g^{\gtrless} (\bar{\mathbf{r}}, t; \bar{\mathbf{R}}, T) \right.$$

$$\pm g^{\gtrless} \left(\bar{\mathbf{R}} + \frac{\bar{\mathbf{r}}}{2} - \mathbf{R} + \frac{\mathbf{r}}{2}, t; \frac{(\mathbf{R} + \bar{\mathbf{R}})}{2} + \frac{(\bar{\mathbf{r}} - \mathbf{r})}{4}, T \right)$$

$$\left. \times g^{\gtrless} \left(\mathbf{R} + \frac{\mathbf{r}}{2} - \bar{\mathbf{R}} + \frac{\bar{\mathbf{r}}}{2}, t; \frac{(\mathbf{R} + \mathbf{R})}{2} + \frac{(\bar{\mathbf{r}} - \bar{\mathbf{r}})}{4}, T \right) \right]$$

If the disturbance varies very little within a distance on the order of the potential range, the second spatial argument of all the g's may be taken to be $\mathbf{R}$, i.e.,

$$\Sigma^{\gtrless} (\mathbf{r}, t; \mathbf{R}, T) \approx i^2 \int d\bar{\mathbf{R}} d\bar{\mathbf{r}} \; v \left(\mathbf{R} + \frac{\mathbf{r}}{2} - \bar{\mathbf{R}} - \frac{\bar{\mathbf{r}}}{2} \right) v \left(\mathbf{R} - \frac{\mathbf{r}}{2} - \bar{\mathbf{R}} + \frac{\bar{\mathbf{r}}}{2} \right)$$

$$\times g^{\lessgtr} (-\bar{\mathbf{r}}, -t; \mathbf{R}, T) \left[g^{\gtrless} (\mathbf{r}, t; \mathbf{R}, T) g^{\gtrless} (\bar{\mathbf{r}}, t; \mathbf{R}, T) \right.$$

$$\pm g^{\gtrless} \left(\bar{\mathbf{R}} + \frac{\bar{\mathbf{r}}}{2} - \mathbf{R} + \frac{\mathbf{r}}{2}, t; \mathbf{R}, T \right)$$

$$\left. \times g^{\gtrless} \left(\mathbf{R} + \frac{\bar{\mathbf{r}}}{2} - \bar{\mathbf{R}} + \frac{\bar{\mathbf{r}}}{2}, t; \mathbf{R}, T \right) \right]$$

This may now be Fourier transformed in $\mathbf{r}, t$ to give[i]

$$\Sigma^{\gtrless} (\mathbf{p}, \omega; \mathbf{R}, T) = \int \frac{d\mathbf{p}'}{(2\pi)^3} \frac{d\omega'}{2\pi} \frac{d\bar{\mathbf{p}}}{(2\pi)^3} \frac{d\bar{\omega}}{2\pi} \frac{d\bar{\mathbf{p}}'}{(2\pi)^3} \frac{d\bar{\omega}'}{2\pi}$$

$$\times (2\pi)^4 \, \delta (\mathbf{p} + \mathbf{p}' - \bar{\mathbf{p}} - \bar{\mathbf{p}}') \, \delta (\omega + \omega' - \bar{\omega} - \bar{\omega}')$$

$$\times \left(\frac{1}{2} \right) \left[v (\mathbf{p} - \bar{\mathbf{p}}) \pm v (\mathbf{p} - \bar{\mathbf{p}}') \right]^2$$

$$\times g^{\lessgtr} (\mathbf{p}', \omega'; \mathbf{R}, T) g^{\gtrless} (\bar{\mathbf{p}}, \bar{\omega}; \mathbf{R}, T) g^{\gtrless} (\bar{\mathbf{p}}', \bar{\omega}'; \mathbf{R}, T)$$

$$\text{(10.9)}$$

In interpreting Eq. (10.7a), we should notice that $\Sigma^{>} (\mathbf{p}, \omega; \mathbf{R}, T)$ is the collision rate for a particle with momentum $\mathbf{p}$ and energy ω at $\mathbf{R}, T$, while $\Sigma^{<} (\mathbf{p}, \omega; \mathbf{R}, T)$ is the rate of scattering into $\mathbf{p}, \omega$ at the space–time point $\mathbf{R}, T$, assuming that the state is initially

---

[i]The typographic error at the last frequency integration variable symbol $\frac{d\omega'}{2\pi}$ in the original text is corrected to $\frac{d\bar{\omega}'}{2\pi}$.

unoccupied. Therefore, the right-hand side of Eq. (10.7a) is the net rate of change of the density of particles with momentum **p** and energy ω at **R**, T. This right side has then exactly the same interpretation as the right side of the Boltzmann equation (7.3). The contributions $-\mathbf{p} \cdot \nabla_\mathbf{R} g^<$ and $+\nabla_\mathbf{R} U_{\text{eff}} \cdot \nabla_\mathbf{p} g^<$ to the rate of change of $g^<$ can also be recognized in the Boltzmann equation. They are, respectively, the result of the drift of particles into the volume element about **R** and the change in the momentum due to the average force acting on the particles at **R**.[j] The eastern on the left-hand side of Eq. (10.7a), $\left(\frac{\partial U_{\text{eff}}}{\partial T}\right) \left(\frac{\partial}{\partial \omega}\right) g^<$, results from the change in the average energy of a particle at **R**, T caused by the time variation of the potential field through which it moves. This term does not appear in the usual Boltzmann equation because this equation does not include the particle energy as an independent variable.

Therefore, Eq. (10.7a) has the same physical content as the usual Boltzmann equation. To see whether these equations are mathematically identical, we subtract Eq. (10.7a) from Eq. (10.7b). The result is

$$\pm \left[\frac{\partial}{\partial T} + \frac{\mathbf{p} \cdot \nabla_\mathbf{R}}{m} - \nabla_\mathbf{R} U_{\text{eff}}(\mathbf{R}, T) \cdot \nabla_\mathbf{p} + \frac{\partial U_{\text{eff}}(\mathbf{R}, T)}{\partial T} \frac{\partial}{\partial \omega} \right]$$
$$\times [g^>(\mathbf{p}, \omega; \mathbf{R}, T) \mp g^<(\mathbf{p}, \omega; \mathbf{R}, T)] = 0$$

Just as in the equilibrium case, we define a spectral function a by

$$a(\mathbf{p}, \omega; \mathbf{R}, T) = g^>(\mathbf{p}, \omega; \mathbf{R}, T) \mp g^<(\mathbf{p}, \omega; \mathbf{R}, T) \qquad (10.10)$$

Thus, we may write

$$\left[\frac{\partial}{\partial T} + \frac{\mathbf{p} \cdot \nabla_\mathbf{R}}{m} - \nabla_\mathbf{R} U_{\text{eff}}(\mathbf{R}, T) \cdot \nabla_\mathbf{p} + \frac{\partial U_{\text{eff}}(\mathbf{R}, T)}{\partial T} \frac{\partial}{\partial \omega} \right] \times a(\mathbf{p}, \omega; \mathbf{R}, T) = 0$$
$$(10.10a)$$

This has the solution

$$a(\mathbf{p}, \omega; \mathbf{R}, T) = y \left(\omega - \frac{p^2}{2m} - U_{\text{eff}}(\mathbf{R}, T) \right) \qquad (10.10b)$$

where y is an arbitrary function.

We are now faced with a rather embarrassing situation. Because we claim that Eqs. (10.7) and (10.9) are just extensions of the equilibrium Born collision approximation to a nonequilibrium

---

[j]The notation R in the original text is replaced by the vector notation **R**.

situation, we must demand that as $T \to -\infty$, $a\,(\mathbf{p},\,\omega;\,\mathbf{R},\,T)$ reduce to the equilibrium $A\,(p,\,\omega)$, which emerges from the Born collision approximation. However, this equilibrium $A\,(p,\,\omega)$, which was determined in Chapter 5, is not a function only of $\omega - \frac{p^2}{2m}$. Therefore, the $a\,(\mathbf{p},\,\omega;\,\mathbf{R},\,T)$ determined as a solution to Eq. (10.10a) cannot possibly reduce to this $A\,(p,\,\omega)$ as $T \to -\infty$. Therefore, we must have made a mistake in our analysis.

Later, we shall look back and find the mistake. Now let us proceed as if no mistake had been made. We do know one very simple $A\,(p,\,\omega)$, which is of the form of Eq. (10.10b), namely, the Hartree result:

$$A\,(p,\,\omega) = 2\pi\delta\,(\omega - E(p))$$

$$E\,(p) = \frac{p^2}{2m} + nv$$

If we take this to be the initial value of $a\,(\mathbf{p},\,\omega;\,\mathbf{R},\,T)$, we find from Eq. (10.10b) that

$$a\,(\mathbf{p},\,\omega;\,\mathbf{R},\,T) = 2\pi\delta\,(\omega - E\,(\mathbf{p},\,\mathbf{R},\,T)) \tag{10.11}$$

where

$$E\,(\mathbf{p},\,\mathbf{R},\,T) = \frac{p^2}{2m} + U_{\text{eff}}\,(\mathbf{R},\,T)$$

We can now simplify the equation of motion (10.7a) for $g^<\,(\mathbf{p},\,\omega;\,\mathbf{R},\,T)$ considerably. We assume that $g^<$ is of the form

$$\begin{aligned}
g^<\,(\mathbf{p},\,\omega;\,\mathbf{R},\,T) &= a\,(\mathbf{p},\,\omega;\,\mathbf{R},\,T)\,f\,(\mathbf{p},\,\mathbf{R},\,T) \\
&= 2\pi\delta\,(\omega - E\,(\mathbf{p},\,\mathbf{R},\,T))\,f\,(\mathbf{p},\,\mathbf{R},\,T) \tag{10.12}
\end{aligned}$$

and, therefore,

$$g^>\,(\mathbf{p},\,\omega;\,\mathbf{R},\,T) = a\,(\mathbf{p},\,\omega;\,\mathbf{R},\,T)\,[1 \pm f\,(\mathbf{p},\,\mathbf{R},\,T)]$$

Here, $f\,(\mathbf{p},\,\mathbf{R},\,T)$ is the distribution function that appears in the Boltzmann equation, i.e., the density of particles with momentum $\mathbf{p}$ at $\mathbf{R}$, T. The left side of Eq. (10.7a) can be written as

$$\begin{aligned}
\left[\frac{\partial}{\partial T} + \frac{\mathbf{p} \cdot \nabla_{\mathbf{R}}}{m} - \nabla_{\mathbf{R}} U_{\text{eff}}\,(\mathbf{R},\,T) \cdot \nabla_{\mathbf{p}} + \frac{\partial U_{\text{eff}}\,(\mathbf{R},\,T)}{\partial T}\frac{\partial}{\partial \omega}\right] \\
\times a\,(\mathbf{p},\,\omega;\,\mathbf{R},\,T)\,f\,(\mathbf{p},\,\mathbf{R},\,T) \tag{10.13}
\end{aligned}$$

We have explicitly constructed $a\,(\mathbf{p}, \omega; \mathbf{R}, T)$ to commute with the differential operator appearing in Eq. (10.13). Therefore, Eq. (10.13) can just as well be written as

$$a\,(\mathbf{p}, \omega; \mathbf{R}, T)\left[\frac{\partial}{\partial T} + \frac{\mathbf{p}\cdot\nabla_{\mathbf{R}}}{m} - \nabla_{\mathbf{R}}U_{\text{eff}}\,(\mathbf{R}, T)\cdot\nabla_{\mathbf{p}}\right] f\,(\mathbf{p}, \mathbf{R}, T)$$

(10.13a)

The right side of Eq. (10.7a) is

$$a(\mathbf{p}, \omega; \mathbf{R}, T)[-f(\mathbf{p}, \mathbf{R}, T)\Sigma^{>}(\mathbf{p}, \omega; \mathbf{R}, T)$$
$$+ (1 \pm f(\mathbf{p}, \mathbf{R}, T))\Sigma^{<}(\mathbf{p}, \omega; \mathbf{R}, T)]$$

(10.14)

Therefore, when we integrate Eq. (10.7a) over all ω, it reduces to

$$\left[\frac{\partial}{\partial T} + \frac{\mathbf{p}\cdot\nabla_{\mathbf{R}}}{m} - \nabla_{\mathbf{R}}U_{\text{eff}}\,(\mathbf{R}, T)\cdot\nabla_{\mathbf{p}}\right] f\,(\mathbf{p}, \mathbf{R}, T)$$
$$= -f\,(\mathbf{p}, \mathbf{R}, T)\,\Sigma^{>}\,(\mathbf{p}, \omega = E\,(\mathbf{p}, \mathbf{R}, T)\,; \mathbf{R}, T)$$
$$+ [1 \pm f\,(\mathbf{p}, \mathbf{R}, T)]\,\Sigma^{<}\,(\mathbf{p}, \omega = E\,(\mathbf{p}, \mathbf{R}, T)\,; \mathbf{R}, T)$$

(10.15)

By using the expressions (10.9) for $\Sigma^{\lessgtr}$, we find

$$\left[\frac{\partial}{\partial T} + \frac{\mathbf{p}\cdot\nabla_{\mathbf{R}}}{m} - \nabla_{\mathbf{R}}U_{\text{eff}}\,(\mathbf{R}, T)\cdot\nabla_{\mathbf{p}}\right] f\,(\mathbf{p}, \mathbf{R}, T)$$
$$= -\int \frac{d\mathbf{p}'}{(2\pi)^3}\frac{d\omega'}{2\pi}\frac{d\bar{\mathbf{p}}}{(2\pi)^3}\frac{d\bar{\omega}}{2\pi}\frac{d\bar{\mathbf{p}}'}{(2\pi)^3}\frac{d\bar{\omega}'}{2\pi}$$
$$\times (2\pi)^4\,\delta\,(\mathbf{p}+\mathbf{p}'-\bar{\mathbf{p}}-\bar{\mathbf{p}}')\,\delta\,(\omega+\omega'-\bar{\omega}-\bar{\omega}')$$
$$\times \left(\frac{1}{2}\right)[v\,(\mathbf{p}-\bar{\mathbf{p}})\pm v\,(\mathbf{p}-\bar{\mathbf{p}}')]^2$$
$$\times \left[ff'\,(1\pm\bar{f})\,(1\pm\bar{f}') - (1\pm f)\,(1\pm f')\,\bar{f}\bar{f}'\right]$$

(10.16)

where $f = f\,(\mathbf{p}, \mathbf{R}, T)$, $f' = f\,(\mathbf{p}', \mathbf{R}, T)$, etc. Except for the trivial substitution of U_{eff} for U, this is exactly the ordinary Boltzmann equation with Born approximation collision cross section.

10.2 Generalization of the Boltzmann Equation

We have to go back and remove the inconsistency from our analysis of the previous section. We derived a value for $a\,(\mathbf{p}, \omega; \mathbf{R}, T)$ that did not agree with the Born collision approximation from which we began. Since our Boltzmann equation purports to be nothing more than the extension of the Born collision approximation to the case in which there is a slowly varying external disturbance, this lack of agreement with the equilibrium analysis is indeed a serious defect.

When we look back at our derivation, we can see our error at once. We were trying to find an expansion, Eq. (10.3), that is valid in the limit in which all the functions involved vary slowly in the variables $\mathbf{R}$, T. One the left side of Eq. (10.3), we held on to all terms of order $\frac{\partial}{\partial T}$ or $\nabla_{\mathbf{R}}$. However, in evaluating the right side of Eq. (10.3), we only considered terms that involved no space and time derivatives; we left out terms of order $\frac{\partial}{\partial T}$ and $\nabla_{\mathbf{R}}$. This procedure is clearly inconsistent. The correct analysis would include all terms of order $\nabla_{\mathbf{R}}$ and $\frac{\partial}{\partial T}$ on both sides of Eq. (10.3).

We shall now go back and find the terms that should not have been neglected. For example, let us re-examine the first two terms on the right side of Eq. (10.3). By employing exactly the same change of variables as we used earlier, we can write these terms as[k]

$$
\int_{-\infty}^{t} d\bar{t}d\bar{\mathbf{r}} \; (\Sigma^{>} - \Sigma^{<}) \left(\mathbf{r} - \bar{\mathbf{r}}, t - \bar{t}; \mathbf{R} + \frac{\bar{\mathbf{r}}}{2}, T + \frac{\bar{t}}{2} \right)
$$

$$
\times g^{<} \left(\bar{\mathbf{r}}, \bar{t}; \mathbf{R} - \frac{(\mathbf{r} - \bar{\mathbf{r}})}{2}, T - \frac{(t - \bar{t})}{2} \right)
$$

$$
+ \int_{t}^{\infty} d\bar{t}d\bar{\mathbf{r}} \; (\Sigma^{>} - \Sigma^{<}) \left(\mathbf{r} - \bar{\mathbf{r}}, t - \bar{t}; \mathbf{R} - \frac{\bar{\mathbf{r}}}{2}, T - \frac{\bar{t}}{2} \right)
$$

$$
\times g^{<} \left(\bar{\mathbf{r}}, \bar{t}; \mathbf{R} + \frac{(\mathbf{r} - \bar{\mathbf{r}})}{2}, T + \frac{(t - \bar{t})}{2} \right) \tag{10.17}
$$

where

$$
(\Sigma^{>} - \Sigma^{<}) (\mathbf{r}, t; \mathbf{R}, T) = \Sigma^{>} (\mathbf{r}, t; \mathbf{R}, T) - \Sigma^{<} (\mathbf{r}, t; \mathbf{R}, T)
$$

Because $\mathbf{r}, \bar{\mathbf{r}}$, and $t, \bar{t}$ are small, compared to the characteristic distances and times over which $g^{<} (\mathbf{p}, \omega; \mathbf{R}, T)$ and $\Sigma^{<} (\mathbf{p}, \omega; \mathbf{R}, T)$ vary, we can expand the various quantities that appear in the expression (10.17) as, for example,

$$
g^{<} \left(\bar{\mathbf{r}}, \bar{t}; \mathbf{R} - \frac{(\mathbf{r} - \bar{\mathbf{r}})}{2}, T - \frac{(t - \bar{t})}{2} \right)
$$

$$
= g^{<} (\bar{\mathbf{r}}, \bar{t}; \mathbf{R}, T) - \left[\frac{\mathbf{r} - \bar{\mathbf{r}}}{2} \cdot \nabla_{\mathbf{R}} + \frac{t - \bar{t}}{2} \frac{\partial}{\partial T} \right] g^{<} (\bar{\mathbf{r}}, \bar{t}; \mathbf{R}, T)
$$

---

[k]The argument variable $\bar{r}$ of $g^{<}$ function at the first term in the original text is corrected to be a vector $\bar{\mathbf{r}}$.

We can now see that to order ∇_R and $\frac{\partial}{\partial T}$ the expression in Eq. (10.17) is

$$\int_{-\infty}^{\infty} d\bar{t}d\bar{r}\ (\Sigma^> - \Sigma^<)\,(r - \bar{r}, t - \bar{t}; R, T)\,g^<\,(\bar{r}, t; R, T)$$

$$+ \int_{-\infty}^{\infty} d\bar{t}d\bar{r}\ \left\{ \left[\bar{r}\cdot\nabla_R + \bar{t}\frac{\partial}{\partial T} - (r - \bar{r})\cdot\nabla_{R'} - (t - \bar{t})\frac{\partial}{\partial T'} \right] \right.$$

$$\left. \times\ \sigma\,(r - \bar{r}, t - \bar{t}; R, T)\,g^<\,(\bar{r}, \bar{t}; R', T') \right\}_{R=R', T=T'}$$

$$(10.18)$$

where

$$\sigma\,(r, t; R, T) = \frac{1}{2}\frac{t}{|t|}\,(\Sigma^> - \Sigma^<)\,(r, t; R, T)$$

The first integral in Eq. (10.18) was included in our earlier discussion; it appears on the right side of Eq. (10.6). The second integral was not included, and it should be added to this right side. The last two terms in Eq. (10.3) also give an extra term:

$$-\int_{-\infty}^{\infty} d\bar{t}d\bar{r}\ \left\{ \left[\bar{r}\cdot\nabla_R + \bar{t}\frac{\partial}{\partial T} - (r - \bar{r})\cdot\nabla_{R'} - (t - \bar{t})\frac{\partial}{\partial T'} \right] \right.$$

$$\left. \times\ b\,(r - \bar{r}, t - \bar{t}; R, T)\,\Sigma^<\,(\bar{r}, \bar{t}; R', T') \right\}_{R=R', T=T'}$$

$$(10.19)$$

In this equation,

$$b\,(r, t; R, T) = \frac{1}{2}\frac{t}{|t|}\,(g^> - g^<)\,(r, t; R, T)$$

which also should be added to the right-hand side of Eq. (10.6). When these extra terms are included, this equation is correct to order ∇_R and $\frac{\partial}{\partial T}$.

We derived the ordinary Boltzmann equation by taking the Fourier transform of Eq. (10.6) and hence finding Eq. (10.7a). To obtain a generalized Boltzmann equation, we must add the Fourier transforms of these two extra terms to the right-hand side of Eq. (10.7a). If we define $b\,(p, \omega; R, T)$ as the Fourier transform of $b\,(r, t; R, T)$ in the r, t variables, we can write the transform of the term (10.19) as[1]

$$\pm i \int d\mathbf{r} dt d\bar{\mathbf{r}} d\bar{t}\, e^{-i\mathbf{p}\cdot\mathbf{r}+i\omega t} \left[\bar{\mathbf{r}}\cdot\nabla_{\mathbf{R}} + \bar{t}\frac{\partial}{\partial T} - (\mathbf{r}-\bar{\mathbf{r}})\cdot\nabla_{\mathbf{R}'} - (t-\bar{t})\frac{\partial}{\partial T'} \right]$$

$$\times \int \frac{d\mathbf{p}''}{(2\pi)^3}\frac{d\omega''}{2\pi}\frac{d\mathbf{p}'}{(2\pi)^3}\frac{d\omega'}{2\pi}\, e^{i\mathbf{p}''\cdot(\mathbf{r}-\bar{\mathbf{r}})+i\mathbf{p}'\cdot\bar{\mathbf{r}}-i\omega''(t-\bar{t})-i\omega'\bar{t}}$$

$$\times b\left(\mathbf{p}'',\omega'';\mathbf{R},T\right)\Sigma^<\left(\mathbf{p}',\omega';\mathbf{R}',T'\right)$$

$$= \pm \int d\mathbf{r} dt d\bar{\mathbf{r}} d\bar{t}\frac{d\mathbf{p}'d\omega'}{(2\pi)^4}\frac{d\mathbf{p}''d\omega''}{(2\pi)^4}\, e^{-i(\mathbf{p}-\mathbf{p}'')\cdot\mathbf{r}+i(\omega-\omega'')t+i(\mathbf{p}'-\mathbf{p}'')\cdot\bar{\mathbf{r}}-i(\omega'-\omega'')\bar{t}}$$

$$\times \left[-\nabla_{\mathbf{p}'}\cdot\nabla_{\mathbf{R}} + \frac{\partial}{\partial\omega'}\frac{\partial}{\partial T} + \nabla_{\mathbf{p}''}\cdot\nabla_{\mathbf{R}'} - \frac{\partial}{\partial\omega''}\frac{\partial}{\partial T'} \right]$$

$$\times b\left(\mathbf{p}'',\omega'';\mathbf{R},T\right)\Sigma^<\left(\mathbf{p}',\omega';\mathbf{R}',T'\right)\Big|_{\substack{\mathbf{R}'=\mathbf{R}\\T'=T}}$$

$$= \pm \left[-\nabla_{\mathbf{p}'}\cdot\nabla_{\mathbf{R}} + \frac{\partial}{\partial\omega'}\frac{\partial}{\partial T} + \nabla_{\mathbf{p}}\cdot\nabla_{\mathbf{R}'} - \frac{\partial}{\partial\omega}\frac{\partial}{\partial T'} \right]$$

$$\times b\left(\mathbf{p},\omega;\mathbf{R},T\right)\Sigma^<\left(\mathbf{p}',\omega';\mathbf{R}',T'\right)\Big|_{\substack{\mathbf{R}'=\mathbf{R},T'=T\\\mathbf{p}'=\mathbf{p},\omega'=\omega}}$$

In order to write expressions like this in a compact form, we define a generalization of the Poisson bracket

$$[X, Y] = \frac{\partial X}{\partial\omega}\left(\mathbf{p},\omega;\mathbf{R},T\right)\frac{\partial Y}{\partial T}\left(\mathbf{p},\omega;\mathbf{R},T\right)$$

$$- \frac{\partial X}{\partial T}\left(\mathbf{p},\omega;\mathbf{R},T\right)\frac{\partial Y}{\partial\omega}\left(\mathbf{p},\omega;\mathbf{R},T\right) \qquad (10.20)$$

$$- \nabla_{\mathbf{p}}X\left(\mathbf{p},\omega;\mathbf{R},T\right)\cdot\nabla_{\mathbf{R}}Y\left(\mathbf{p},\omega;\mathbf{R},T\right)$$

$$+ \nabla_{\mathbf{R}}X\left(\mathbf{p},\omega;\mathbf{R},T\right)\cdot\nabla_{\mathbf{p}}Y\left(\mathbf{p},\omega;\mathbf{R},T\right)$$

Using this Poisson bracket notation, we can write the Fourier transform of Eq. (10.19) as

$$\mp [b, \Sigma^<]$$

Similarly, the Fourier transform of the previously neglected term in Eq. (10.18) is

$$\pm [\sigma, g^<]$$

By adding these extra two terms, we can correct Eq. (10.7) so that it includes *all* terms of order $\nabla_{\mathbf{R}}$ and $\frac{\partial}{\partial T}$. This corrected version

---

[1]The wrong time argument T of $\Sigma^<$ at the second equality in the original text is corrected to be T'.

of Eq. (10.7) is

$$\left[\frac{\partial}{\partial T} + \frac{\mathbf{p} \cdot \nabla_{\mathbf{R}}}{m} - \nabla_{\mathbf{R}} U_{\text{eff}}(\mathbf{R}, T) \cdot \nabla_{\mathbf{p}} + \frac{\partial U_{\text{eff}}}{\partial T}(\mathbf{R}, T) \frac{\partial}{\partial \omega} \right] g^<(\mathbf{p}, \omega; \mathbf{R}, T)$$

$$- [\sigma, g^<] + [b, \Sigma^<]$$

$$= -\Sigma^<(\mathbf{p}, \omega; \mathbf{R}, T) g^>(\mathbf{p}, \omega; \mathbf{R}, T) + \Sigma^>(\mathbf{p}, \omega; \mathbf{R}, T) g^<(\mathbf{p}, \omega; \mathbf{R}, T)$$

$$(10.21)$$

Now we have to evaluate the Fourier transforms $\sigma(\mathbf{p}, \omega; \mathbf{R}, T)$ and $b(\mathbf{p}, \omega; \mathbf{R}, T)$. The latter is given by

$$b(\mathbf{p}, \omega; \mathbf{R}, T) = \int d\mathbf{r} dt \, e^{-i\mathbf{p} \cdot \mathbf{r} + i\omega t} \frac{t}{|t|} [g^>(\mathbf{r}, t; \mathbf{R}, T) - g^<(\mathbf{r}, t; \mathbf{R}, T)]$$

Since the Fourier transform of $i[g^> - g^<]$ is $a(\mathbf{p}, \omega; \mathbf{R}, T)$, we can write

$$b(\mathbf{p}, \omega; \mathbf{R}, T) = \int dt \, e^{i\omega t} \frac{t}{|t|} \int \frac{d\omega'}{2\pi i} e^{-i\omega' t} a(\mathbf{p}, \omega'; \mathbf{R}, T)$$

$$= \int \frac{d\omega'}{2\pi i} a(\mathbf{p}, \omega'; \mathbf{R}, T)$$

$$\times \left[\int_0^\infty dt \, e^{i(\omega - \omega')t} - \int_{-\infty}^0 dt \, e^{i(\omega - \omega')t} \right]$$

$$= \wp \int \frac{d\omega'}{2\pi} \frac{a(\mathbf{p}, \omega'; \mathbf{R}, T)}{\omega - \omega'}$$

where $\wp$ denotes the principal value integral.

In our discussion of the equilibrium Green's functions, we introduced the function

$$G(p, z) = \int \frac{d\omega'}{2\pi} \frac{A(p, \omega')}{z - \omega'}$$

As z approaches the real axis from above or below, $z \to \omega \pm i\epsilon$,

$$G(p, z) \to \int \frac{d\omega'}{2\pi} \frac{A(p, \omega')}{\omega - \omega'} \mp \pi i A(p, \omega)$$

In either case, we can write

$$\Re G(p, \omega) = \wp \int \frac{d\omega'}{2\pi} \frac{A(p, \omega')}{\omega - \omega'}$$

Similarly, for the nonequilibrium case, we define

$$g(\mathbf{p}, z; \mathbf{R}, T) = \int \frac{d\omega'}{2\pi} \frac{a(\mathbf{p}, \omega'; \mathbf{R}, T)}{z - \omega'} \qquad (10.22a)$$

and we write

$$b\left(\mathbf{p}, \omega; \mathbf{R}, T\right) = \wp \int \frac{d\omega'}{2\pi} \frac{a\left(\mathbf{p}, \omega'; \mathbf{R}, T\right)}{\omega - \omega'}$$

as

$$b\left(\mathbf{p}, \omega; \mathbf{R}, T\right) = \Re g\left(\mathbf{p}, \omega; \mathbf{R}, T\right) \tag{10.22b}$$

Moreover, in the equilibrium case, we define a collisional self-energy as

$$\Sigma_c\left(p, z\right) = \int \frac{d\omega'}{2\pi} \frac{\Sigma^>\left(p, \omega'\right) - \Sigma^<\left(p, \omega'\right)}{z - \omega'}$$
$$= \int \frac{d\omega'}{2\pi} \frac{\Gamma\left(p, \omega'\right)}{z - \omega'}$$

We now define the analogous nonequilibrium quantities:

$$\Gamma\left(\mathbf{p}, \omega; \mathbf{R}, T\right) = \Sigma^>\left(\mathbf{p}, \omega; \mathbf{R}, T\right) \mp \Sigma^<\left(\mathbf{p}, \omega; \mathbf{R}, T\right) \tag{10.23a}$$

and

$$\Sigma_c\left(\mathbf{p}, z; \mathbf{R}, T\right) = \int \frac{d\omega'}{2\pi} \frac{\Gamma\left(\mathbf{p}, \omega'; \mathbf{R}, T\right)}{z - \omega'} \tag{10.23b}$$

By just the same arguments as we used to derive Eq. (10.22b), we can see that $\sigma\left(\mathbf{p}, \omega; \mathbf{R}, T\right)$, the Fourier transform of $(t/|t|)\left[\Sigma^>\left(\mathbf{r}, t; \mathbf{R}, T\right) - \Sigma^<\left(\mathbf{r}, t; \mathbf{R}, T\right)\right]$, is

$$\sigma\left(\mathbf{p}, \omega; \mathbf{R}, T\right) = \Re\Sigma_c\left(\mathbf{p}, \omega; \mathbf{R}, T\right) \tag{10.23c}$$

Now we can rewrite Eq. (10.21) in the form

$$\left[\frac{\partial}{\partial T} + \frac{\mathbf{p} \cdot \nabla_{\mathbf{R}}}{m} - \nabla_{\mathbf{R}} U_{\text{eff}} \cdot \nabla_{\mathbf{p}} + \frac{\partial U_{\text{eff}}}{\partial T} \frac{\partial}{\partial \omega}\right] g^< - \left[\Re\Sigma_c, g^<\right] + \left[\Re g, \Sigma^<\right]$$
$$= -\Sigma^> g^< + \Sigma^> g^< \tag{10.24}$$

The last two terms on the left side of Eq. (10.24) are written in terms of the generalized Poisson bracket (10.20). This equation can be simplified in form a bit if we notice that the other terms on the left also form a Poisson bracket, i.e.,

$$\left[\frac{\partial}{\partial T} + \frac{\mathbf{p} \cdot \nabla_{\mathbf{R}}}{m} - \nabla_{\mathbf{R}} U_{\text{eff}} \cdot \nabla_{\mathbf{p}} + \frac{\partial U_{\text{eff}}}{\partial T} \frac{\partial}{\partial \omega}\right] g^< = \left[\omega - \left(\frac{p^2}{2m}\right) - U_{\text{eff}}, g^<\right]$$

Therefore, Eq. (10.24) becomes

$$\left[\omega - \left(\frac{p^2}{2m}\right) - U_{\text{eff}} - \Re\Sigma_c, g^<\right] + \left[\Re g, \Sigma^<\right] = -\Sigma^> g^< + \Sigma^< g^> \tag{10.25a}$$

By exactly the same procedure, we can derive the following equation of motion for $g^>$:

$$\pm \left[\omega - \left(\frac{p^2}{2m} \right) - U_{\text{eff}} - \Re\Sigma_c, g^> \right] + [\Re g, \Sigma^>] = -\Sigma^> g^< + \Sigma^< g^>$$

(10.25b)

Equations (10.25a) and (10.25b) are coupled integro-differential equations for the unknown functions $g^>$ ($\mathbf{p}, \omega; \mathbf{R}, T$) and $g^>$ ($\mathbf{p}, \omega; \mathbf{R}, T$). The self-energies $\Sigma^>$ and $\Sigma^<$ are expressed in terms of $g^>$ and $g^<$ by the particular Green's function approximation being considered. For example, in the Born collision approximation, $\Sigma^>$ and $\Sigma^<$ are given by Eq. (10.9). The auxiliary quantities $\Re g$ and $\Re\Sigma_c$ are expressed, respectively, in terms of $g^>$ and $g^<$ and $\Sigma^>$ and $\Sigma^<$ by Eqs. (10.22) and (10.23).

Equations (10.25) are generally correct except for one rather trivial omission: So far, we have left the exchange term in Σ_{HF} out of our discussion. The direct (Hartree) term is included; it appears in U_{eff} ($\mathbf{R}, T$). When $g^<$ ($\mathbf{p}, \omega; \mathbf{R}, T$) varies little within distances of the order of the potential range, we can approximately evaluate

$$U_{\text{eff}} (\mathbf{R}, T) = U (\mathbf{R}, T) + \int \frac{d\mathbf{p}'}{(2\pi)^3} \frac{d\omega'}{2\pi} \int d\mathbf{R}' \, v (\mathbf{R} - \mathbf{R}') \, g^< (\mathbf{p}', \omega'; \mathbf{R}', T)$$

as

$$U_{\text{eff}} (\mathbf{R}, T) = U (\mathbf{R}, T) + \int \frac{d\mathbf{p}'}{(2\pi)^3} \frac{d\omega'}{2\pi} \int d\mathbf{R}' \, v (\mathbf{R} - \mathbf{R}') \, g^< (\mathbf{p}', \omega; \mathbf{R}, T)$$

$$= U (\mathbf{R}, T) + \Sigma_{\text{Hartree}} (\mathbf{R}, T)$$

With the inclusion of the exchange term in Eq. (10.25),

$$U_{\text{eff}} (\mathbf{R}, T) + \Re\Sigma_c (\mathbf{p}, \omega; \mathbf{R}, T) \rightarrow U (\mathbf{R}, T) + \Re\Sigma (\mathbf{p}, \omega; \mathbf{R}, T)$$

(10.26a)

where, just as in the equilibrium case, the total self-energy is a sum of the Hartree–Fock and the collisional contribution

$$\Re\Sigma (\mathbf{p}, \omega; \mathbf{R}, T) = \Sigma_{\text{HF}} (\mathbf{p}, \mathbf{R}, T) + \Re\Sigma_c (\mathbf{p}, \omega; \mathbf{R}, T)$$

$$\Sigma_{\text{HF}} (\mathbf{p}, \mathbf{R}, T) = \int \frac{d\mathbf{p}'}{(2\pi)^3} d\omega' 2\pi \, [v \pm v (\mathbf{p} - \mathbf{p}')] \, g^< (\mathbf{p}', \omega'; \mathbf{R}, T)$$

(10.26b)

where

$$v = \int d\mathbf{r} \, v(r)$$

(10.26c)

When Eqs. (10.25) are modified using Eq. (10.26), they are exact for slowly varying disturbances.

These generalized Boltzmann equations can be integrated partially. We notice that the collision term on the right side of Eq. (10.25b) is exactly the same as the collision term in Eq. (10.25a). Therefore, when we subtract these two equations, the collision terms cancel and we find

$$\left[\omega - \left(\frac{p^2}{2m}\right) - U\left(\mathbf{R}, T\right) - \Re\Sigma\left(\mathbf{p}, \omega; \mathbf{R}, T\right), a\left(\mathbf{p}, \omega; \mathbf{R}, T\right)\right]$$

$$+ \left[\Re g\left(\mathbf{p}, \omega; \mathbf{R}, T\right), \Gamma\left(\mathbf{p}, \omega; \mathbf{R}, T\right)\right] = 0$$

$$(10.27)$$

where

$$a = g^> \mp g^< \qquad \Gamma = \Sigma^> \mp \Sigma^<$$

Equation (10.27) may be integrated simply. In fact, the solution to Eq. (10.27) gives almost exactly the same evaluation of a as in the equilibrium case. In equilibrium,

$$G\left(p, z\right) = \frac{1}{z - \frac{p^2}{2m} - \Sigma\left(p, z\right)}$$

Therefore,

$$G\left(p, \omega - i\epsilon\right) = \Re G\left(p, \omega\right) + \left(\frac{i}{2}\right) A\left(p, \omega\right)$$

$$= \left[\Re G^{-1}\left(p, \omega\right) - \left(\frac{i}{2}\right)\Gamma\left(p, \omega\right)\right]^{-1}$$

where $\Re G^{-1}$ is an abbreviation for $\omega - \left(\frac{p^2}{2m}\right) - \Re\Sigma\left(p, \omega\right)$. Also

$$G\left(p, \omega + i\epsilon\right) = \Re G\left(p, \omega\right) - \left(\frac{i}{2}\right) A\left(p, \omega\right)$$

$$= \left[\Re G^{-1}\left(p, \omega\right) + \left(\frac{i}{2}\right)\Gamma\left(p, \omega\right)\right]^{-1}$$

Thus

$$\Re G\left(p, \omega\right) = \frac{\Re G^{-1}\left(p, \omega\right)}{\left[\Re G^{-1}\left(p, \omega\right)\right]^2 + \left[\frac{\Gamma(p, \omega)}{2}\right]^2}$$

$$A\left(p, \omega\right) = \frac{\Gamma\left(p, \omega\right)}{\left[\Re G^{-1}\left(p, \omega\right)\right]^2 + \left[\frac{\Gamma(p, \omega)}{2}\right]^2}$$

Let us see whether there is a similar solution to Eq. (10.27). We try

$$g\left(\mathbf{p}, z; \mathbf{R}, T\right) = \frac{1}{z - \left(\frac{p^2}{2m}\right) - U\left(\mathbf{R}, T\right) - \Sigma\left(\mathbf{p}, z; \mathbf{R}, T\right)} \tag{10.28a}$$

Then

$$a\left(\mathbf{p}, \omega; \mathbf{R}, T\right) = \frac{1}{i}\left[\frac{1}{\Re g^{-1}\left(\mathbf{p}, \omega; \mathbf{R}, T\right) - \left(\frac{i}{2}\right)\Gamma\left(\mathbf{p}, \omega; \mathbf{R}, T\right)}\right.$$

$$\left. - \frac{1}{\Re g^{-1}\left(\mathbf{p}, \omega; \mathbf{R}, T\right) + \left(\frac{i}{2}\right)\Gamma\left(\mathbf{p}, \omega; \mathbf{R}, T\right)}\right]$$

$$= \frac{\Gamma\left(\mathbf{p}, \omega; \mathbf{R}, T\right)}{\left[\Re g^{-1}\left(\mathbf{p}, \omega; \mathbf{R}, T\right)\right]^2 + \left[\frac{\Gamma\left(\mathbf{p}, \omega; \mathbf{R}, T\right)}{2}\right]^2} \tag{10.28b}$$

and

$$\Re g\left(\mathbf{p}, \omega; \mathbf{R}, T\right) = \frac{1}{2}\left[\frac{1}{\Re g^{-1}\left(\mathbf{p}, \omega; \mathbf{R}, T\right) - \left(\frac{i}{2}\right)\Gamma\left(\mathbf{p}, \omega; \mathbf{R}, T\right)}\right.$$

$$\left. - \frac{1}{\Re g^{-1}\left(\mathbf{p}, \omega; \mathbf{R}, T\right) + \left(\frac{i}{2}\right)\Gamma\left(\mathbf{p}, \omega; \mathbf{R}, T\right)}\right]$$

$$= \frac{\Re g^{-1}\left(\mathbf{p}, \omega; \mathbf{R}, T\right)}{\left[\Re g^{-1}\left(\mathbf{p}, \omega; \mathbf{R}, T\right)\right]^2 + \left[\frac{\Gamma\left(\mathbf{p}, \omega; \mathbf{R}, T\right)}{2}\right]^2} \tag{10.28c}$$

where

$$\Re g^{-1}\left(\mathbf{p}, \omega; \mathbf{R}, T\right) = \omega - \left(\frac{p^2}{2m}\right) - U\left(\mathbf{R}, T\right) - \Re\Sigma\left(\mathbf{p}, \omega; \mathbf{R}, T\right) \tag{10.28d}$$

Then, the left side of Eq. (10.27) becomes

$$\left[\Re g^{-1}, a\right] + \left[\Re g^{-1}, \Gamma\right] = \frac{1}{i}\left[\Re g^{-1}, \frac{1}{\Re g^{-1} - i\frac{\Gamma}{2}}\right]$$

$$- \frac{1}{i}\left[\Re g^{-1}, \frac{1}{\Re g^{-1} + i\frac{\Gamma}{2}}\right]$$

$$+ \frac{1}{2}\left[\frac{1}{\Re g^{-1} - i\frac{\Gamma}{2}} \cdot \Gamma\right] + \frac{1}{2}\left[\frac{1}{\Re g^{-1} + i\frac{\Gamma}{2}} \cdot \Gamma\right] \tag{10.29}$$

Like the commutator, our Poisson bracket has the property $[A, B] = -[B, A]$. Hence, expression (10.29) may be rearranged in the form

$$\frac{1}{i}\left[\Re g^{-1} - \frac{i}{2}\Gamma, \frac{1}{\Re g^{-1} + i\frac{\Gamma}{2}}\right] - \frac{1}{i}\left[\Re g^{-1} + \frac{i}{2}\Gamma, \frac{1}{\Re g^{-1} - i\frac{\Gamma}{2}}\right]$$

$$(10.29a)$$

However, the Poisson bracket of any quantity A with any function of A is zero, since

$$[A, f(A)] = \frac{\partial A}{\partial \omega}\frac{\partial f(A)}{\partial T} - \frac{\partial A}{\partial T}\frac{\partial f(A)}{\partial \omega} - \nabla_{\mathbf{p}}A \cdot \nabla_{\mathbf{R}}f(A) + \nabla_{\mathbf{R}}A \cdot \nabla_{\mathbf{p}}f(A)$$

$$= \frac{\partial f}{\partial A}\left[\frac{\partial A}{\partial \omega}\frac{\partial A}{\partial T} - \frac{\partial A}{\partial T}\frac{\partial A}{\partial \omega} - \nabla_{\mathbf{p}}A \cdot \nabla_{\mathbf{R}}A + \nabla_{\mathbf{R}}A \cdot \nabla_{\mathbf{p}}A\right] = 0$$

Therefore, expression (10.29) is, in fact, zero, providing that Eq. (10.28) is a solution to Eq. (10.29). Since the solution (10.28a) is of exactly the same form as the equilibrium solution, it must reduce to the equilibrium solution as $T \to -\infty$. Therefore, it satisfies the initial condition on the equation of motion.

To sum up, the equation of motion

$$\left[\omega - \left(\frac{p^2}{2m}\right) - U(\mathbf{R}, T) - \Re\Sigma(\mathbf{p}, \omega; \mathbf{R}, T), g^<(\mathbf{p}, \omega; \mathbf{R}, T)\right]$$
$$+ [\Re g(\mathbf{p}, \omega; \mathbf{R}, T), \Sigma^<(\mathbf{p}, \omega; \mathbf{R}, T)]$$
$$= -\Sigma^>(\mathbf{p}, \omega; \mathbf{R}, T)g^<(\mathbf{p}, \omega; \mathbf{R}, T) + \Sigma^<(\mathbf{p}, \omega; \mathbf{R}, T)g^>(\mathbf{p}, \omega; \mathbf{R}, T)$$

$$(10.30)$$

provides an exact description of the response to slowly varying disturbances. All the quantities appearing in this equation may be expressed in terms of $g^>$ and $g^<$. In particular, $\Sigma^>$ and $\Sigma^<$ are defined by Green's function approximation, which gives the self-energy in terms of $g^>$ and $g^<$. The lowest-order approximation of this kind is given by Eq. (10.8). Both $g^>$ and $g^<$ are related to g by

$$\int \frac{d\omega}{2\pi}\frac{g^>(\mathbf{p}, \omega; \mathbf{R}, T) \mp g^<(\mathbf{p}, \omega; \mathbf{R}, T)}{z - \omega}$$
$$= g(\mathbf{p}, \omega; \mathbf{R}, T)$$
$$= \left[z - \left(\frac{p^2}{2m}\right) - U(\mathbf{R}, T) - \Sigma(\mathbf{p}, \omega; \mathbf{R}, T)\right]^{-1} \quad (10.31)$$

which is exactly the same relation as defines the equilibrium Green's functions.

To go from Eq. (10.30) back to the ordinary Boltzmann equation with Born approximation cross sections, we replace $\Sigma^>$ and $\Sigma^<$ on the right side of Eq. (10.30) by the approximations (10.9). On the left side of Eq. (10.30), however, we must employ the approximations $\Sigma^> = \Sigma^< = \Sigma = 0$. Since the left side of Eq. (10.30) determines the result Eq. (10.28b) for a, we must, therefore, replace Σ and Γ in Eq. (10.31) by zero. Then we get $a = 2\pi\delta\left(\omega - \left(\frac{p^2}{2m}\right) - U\left(\mathbf{R}, T\right)\right)$, so that we recover the ordinary Boltzmann equation (7.2).

The ordinary Boltzmann equation emerges then from an approximation in which the self-energies that appear on the left side of Eq. (10.30) are handled differently from those that appear on the right. One can see that these two appearances of the self-energy Σ play a very different physical role in the description of transport phenomena. The $\Sigma^>$ and $\Sigma^<$ on the right side of Eq. (10.30) describe the dynamical effect of collisions, i.e., how the collisions transfer particles from one energy–momentum configuration to another. On the other hand, the Σ's on the left side of Eq. (10.30) describe the kinetic effects of the potential, i.e., how the potential changes the energy–momentum relation from that of free particles, $\omega = \left(\frac{p^2}{2m}\right) + U$, to the more complex spectrum, Eq. (10.31). Because these two effects of Σ are physically so different, we should not be surprised to find that we can independently approximate the kinetic effects of Σ and the dynamical effects of Σ.

In the derivation of the ordinary Boltzmann equation, we completely neglect all the kinetic effects of Σ and retain the dynamic effects. In this way, we get to the familiar Boltzmann equation, which describes the particles as free particles in between collisions. The more general equation (10.30) includes the effects of the potential on the motion of particles even between collisions. These effects arise from several different sources. When the system is fairly dense, the particles never get away from the other particles in the system. Therefore, we cannot ever really think of the particles as being "in between collisions." Quantum mechanically, the wavefunctions of the particles are sufficiently smeared out so that there is always some overlap of wavefunctions; the particle is always colliding. Also the particle always retains some memory of collisions it has experienced through its correlations with other particles in the system. This memory is also contained in its energy–momentum relation.

Equations (10.30) and (10.31) can be used to describe all types of transport phenomena. In Chapter 11, we shall use these equations to describe the simplest transport process, ordinary sound propagation. In Chapter 12, these equations will be applied to a discussion of the behavior of low-temperature fermion systems.

Chapter 11

Quasi-Equilibrium Behavior: Sound Propagation

11.1 Complete Equilibrium Solutions

It is interesting to see how the nonequilibrium theory leads, as a special case, to the equilibrium theory of Chapters 2–5. There are two situations in which we expect an equilibrium solution to come out of the generalized Boltzmann equation. The first and most obvious case is when $U(\mathbf{R}, T)$ vanishes for all T previous to the time of observation. Then the system has never felt the disturbance, and it remains in its initial state of equilibrium. The second case is when $U(\mathbf{R}, T) = U_0$, a constant, for all times after some time, say T_0. Then if we observe the system at some time much later than T_0, we should expect that the system will have had sufficient time to relax to complete equilibrium.

In an equilibrium situation, the functions $g^>(\mathbf{p}, \omega; \mathbf{R}, T)$ and $g^<(\mathbf{p}, \omega; \mathbf{R}, T)$ are completely independent of $\mathbf{R}, T$. Since we are looking when $U(\mathbf{R}, T)$ is also independent of $\mathbf{R}$ and T, the left side of Eq. (10.30) vanishes. Therefore, $g^>$ and $g^<$ obey

$$0 = \Sigma^>(\mathbf{p}, \omega)\, g^<(\mathbf{p}, \omega) - \Sigma^<(\mathbf{p}, \omega)\, g^>(\mathbf{p}, \omega) \tag{11.1}$$

Annotations to Quantum Statistical Mechanics
In-Gee Kim
Copyright © 2018 Pan Stanford Publishing Pte. Ltd.
ISBN 978-981-4774-15-4 (Hardcover), 978-1-315-19659-6 (eBook)
www.panstanford.com

To see the consequences of Eq. (11.1), we consider, as an example, the Born collision approximation. Then Eq. (11.1) becomes

$$
\begin{aligned}
0 = \int & \frac{d\mathbf{p}'d\omega'}{(2\pi)^4} \frac{d\bar{\mathbf{p}}d\bar{\omega}}{(2\pi)^4} \frac{d\bar{\mathbf{p}}'d\bar{\omega}'}{(2\pi)^4} \left(\frac{1}{2}\right) \left[v\left(\mathbf{p}-\bar{\mathbf{p}}\right) \pm v\left(\mathbf{p}-\bar{\mathbf{p}}'\right)\right]^2 \\
& \times a\left(\mathbf{p}, \omega\right) a\left(\mathbf{p}', \omega\right) a\left(\bar{\mathbf{p}}, \bar{\omega}\right) a\left(\bar{\mathbf{p}}', \bar{\omega}'\right) \delta\left(\omega+\omega'-\bar{\omega}-\bar{\omega}'\right) \\
& \times \delta\left(\mathbf{p}+\mathbf{p}'-\bar{\mathbf{p}}-\bar{\mathbf{p}}'\right) (2\pi)^4 \left\{f\left(\mathbf{p}, \omega\right) f\left(\mathbf{p}', \omega'\right) \left[1 \pm f\left(\bar{\mathbf{p}}, \bar{\omega}\right)\right]\right. \\
& \times \left.\left[1 \pm f\left(\bar{\mathbf{p}}', \bar{\omega}'\right)\right] - \left[1 \pm f\left(\mathbf{p}, \omega\right)\right] \left[1 \pm f\left(\mathbf{p}', \omega'\right)\right] f\left(\bar{\mathbf{p}}, \bar{\omega}\right) f\left(\bar{\mathbf{p}}', \bar{\omega}'\right)\right\}
\end{aligned}
$$

$$(11.2)$$

where we have written

$$
\begin{aligned}
g^>\left(\mathbf{p}, \omega\right) &= \left[1 \pm f\left(\mathbf{p}, \omega\right)\right] a\left(\mathbf{p}, \omega\right) \\
g^<\left(\mathbf{p}, \omega\right) &= f\left(\mathbf{p}, \omega\right) a\left(\mathbf{p}, \omega\right)
\end{aligned}
$$

$$(11.3)$$

The expression in braces in Eq. (11.2) will vanish if $f\left(\mathbf{p}, \omega\right)$ is of the form

$$
f\left(\mathbf{p}, \omega\right) = \left\{\exp\left[\beta\left(\omega - \mathbf{p}\cdot\mathbf{v} + \frac{1}{2}mv^2 - \mu'\right)\right] \mp 1\right\}^{-1} \quad (11.4)
$$

where $\mathbf{v}$ is an arbitrary vector. In fact, it is possible to prove that Eq. (11.4) is the most general f for which Eq. (11.2) vanishes. The proof is quite analogous to the proof of the H theorem for the ordinary Boltzmann equation.

Therefore, to determine the possible equilibrium limits of $g^>\left(\mathbf{p}, \omega; \mathbf{R}, T\right)$ and $g^<\left(\mathbf{p}, \omega; \mathbf{R}, T\right)$, we must solve Eq. (10.31):

$$
g^{-1}\left(\mathbf{p}, z\right) = z - \left(\frac{p^2}{2m}\right) - U_0 - \Sigma\left(\mathbf{p}, z\right) \quad (10.31)
$$

using the relationships (11.3) and (11.4). These two may be written as

$$
g^>\left(\mathbf{p}, \omega\right) = e^{\beta\left(\omega - \mathbf{p}\cdot\mathbf{v} + \frac{1}{2}mv^2 - \mu'\right)} g^<\left(\mathbf{p}, \omega\right) \quad (11.5)
$$

Since $\Sigma\left(\mathbf{p}, z\right)$ is a function of $g^>$ and $g^<$, Eqs. (10.31) and (11.5) provide two relations between the two unknown functions $g^>\left(\mathbf{p}, \omega\right)$ and $g^<\left(\mathbf{p}, \omega\right)$.

When $U_0 = \mathbf{v} = 0$, Eqs. (10.31) and (11.5) are identical to the equations in Chapter 5 to determine the equilibrium Green's functions for chemical potential μ' and inverse temperature β. Writing these equilibrium functions as $G^>(p, \omega; \beta, \mu')$ and $G^>(p, \omega; \beta, \mu')$, we find

$$g^>(\mathbf{p}, \omega) = G^>(p, \omega; \beta, \mu')$$
$$g^<(\mathbf{p}, \omega) = G^<(p, \omega; \beta, \mu')$$
(11.6)

In this case, the nonequilibrium Green's functions reduce to their equilibrium counterparts, and the whole equilibrium theory of Chapters 2 through 5 emerges as a special case of the nonequilibrium theory developed in Chapters 9 and 10.

Consider next the case $U_0 \neq 0$, $\mathbf{v} = 0$. U_0 then represents a constant term added to the energy of every particle in the system. We expect that U_0 should have two effects on the equilibrium solution: First, the frequency should go into $\omega - U_0$, and second, the chemical potential should go into $\mu' - U_0$. If we define

$$\bar{g}^{\gtrless}(\mathbf{p}, \omega) = g^{\gtrless}(p, \omega + U_0)$$
(11.7)

we then expect that the solution to Eqs. (10.31) and (11.5) at $\mathbf{v} = 0$ is

$$\bar{g}^{\gtrless}(\mathbf{p}, \omega) = G^{\gtrless}(p, \omega; \beta, \mu)$$
(11.8)

where $\mu = \mu' - U_0$.

To verify this conjecture, we let $z \to z + U_0$ in Eq. (10.31). This then becomes

$$g^{-1}(\mathbf{p}, z + U_0) = \left[\int \frac{d\omega'}{2\pi} \frac{g^>(\mathbf{p}, \omega) \mp g^<(\mathbf{p}, \omega)}{z + U_0 - \omega'} \right]^{-1}$$
$$= \left[\int \frac{d\omega'}{2\pi} \frac{\bar{g}^>(\mathbf{p}, \omega) \mp \bar{g}^<(\mathbf{p}, \omega)}{z - \omega'} \right]^{-1}$$

and also

$$g^{-1}(\mathbf{p}, z + U_0) = z - \left(\frac{p^2}{2m} \right) - \Sigma(\mathbf{p}, z + U_0; g^>, g^<)$$

$$= z - \left(\frac{p^2}{2m} \right) - \Sigma_{\mathrm{HF}}(\mathbf{p}, g^<)$$

$$- \int \frac{d\omega}{2\pi} \frac{\Sigma^>(\mathbf{p}, \omega + U_0; g^>, g^<) - \Sigma^<(\mathbf{p}, \omega + U_0; g^>, g^<)}{z - \omega}$$

We may express the self-energies as functionals of $\bar{g}$. We first note that

$$\Sigma_{\mathrm{HF}}(\mathbf{p}, g^<) = \int \cdots \int \frac{d\omega'}{2\pi} g^<(\mathbf{p}, \omega') = \int \cdots \int \frac{d\omega'}{2\pi} \bar{g}^<(\mathbf{p}, \omega')$$
$$= \Sigma_{\mathrm{HF}}(\mathbf{p}, \bar{g}^<)$$

In the Born collision approximation,

$$\Sigma^> \left(\mathbf{p}, \omega + U_0\, g^>, g^<\right) \sim \int d\omega' d\bar{\omega} d\bar{\omega}'\; \delta \left(\omega + U_0 + \omega' - \bar{\omega} - \bar{\omega}'\right)$$
$$\times g^< \left(\mathbf{p}', \omega'\right) g^> \left(\bar{\mathbf{p}}, \bar{\omega}\right) g^> \left(\bar{\mathbf{p}}', \bar{\omega}'\right)$$
$$\sim \int d\omega' d\bar{\omega} d\bar{\omega}'\; \delta \left(\omega + U_0 + \omega' - \bar{\omega} - \bar{\omega}'\right)$$
$$\times \bar{g}^< \left(\mathbf{p}', \omega'\right) \bar{g}^> \left(\bar{\mathbf{p}}, \bar{\omega}\right) \bar{g}^> \left(\bar{\mathbf{p}}', \bar{\omega}'\right)$$
$$= \Sigma^> \left(\mathbf{p}, \omega; \bar{g}^>, \bar{g}^<\right)$$

Thus, we see that the $\bar{g}$'s obey

$$\left[\int \frac{d\omega'}{2\pi} \frac{\bar{g}^> \left(\mathbf{p}, \omega\right) \mp \bar{g}^< \left(\mathbf{p}, \omega\right)}{z - \omega'}\right]^{-1} = z - \left(\frac{p^2}{2m}\right) - \Sigma \left(\mathbf{p}, z; \bar{g}^>, \bar{g}^<\right)$$

which is exactly the same equation as is obeyed by the equilibrium $G^>$ and $G^<$. Furthermore, when $\mathbf{v} = 0$, the boundary condition (11.5) can be written in terms of the $\bar{g}$'s as

$$\bar{g}^> \left(\mathbf{p}, \omega\right) = e^{\beta \left(\omega - \mu' + U_0\right)} \bar{g}^< \left(\mathbf{p}, \omega\right)$$

The $\bar{g}$'s must then be the equilibrium G's, since they are both determined by the same equation.

The equilibrium state that results when $U = U_0$ and $\mathbf{v} = 0$ is thus the initial equilibrium state. The only difference is that the zero point of the particle energies has been shifted by an amount U_0.

We shall now see that the equilibrium state that occurs with $\mathbf{v} \neq 0$ is one in which the system as a whole is moving with a uniform velocity $\mathbf{v}$. If the entire system is moving, Green's function should be the same as the equilibrium Green's functions that would be "seen" by an observer moving with velocity $-\mathbf{v}$ past a fixed system. A particle moving with momentum $\mathbf{p}$ and energy ω in the fixed system would appear to the moving observer to have the extra momentum $-m\mathbf{v}$ and the extra kinetic energy $\left(\frac{1}{2m}\right) \left(\mathbf{p} - m\mathbf{v}\right)^2 - \left(\frac{p^2}{2m}\right)$. Therefore, if $\mathbf{v}$ does in fact represent the velocity of the system, $g^>$ and $g^<$ should be related to the equilibrium functions by

$$\bar{g}^{\gtrless} \left(\mathbf{p}, \omega\right) = G^{\gtrless} \left(p, \omega; \beta, \mu\right) \tag{11.9}$$

where

$$\bar{g}^{\gtrless} \left(\mathbf{p}, \omega\right) = g^{\gtrless} \left(\mathbf{p} + m\mathbf{v}, \omega + \mathbf{p} \cdot \mathbf{v} + \frac{1}{2}mv^2 + U_0\right) \tag{11.10}$$

To verify this, we must show that the $\bar{g}$'s satisfy the same equations as the G's. First, the boundary condition. From Eq. (11.5), we see that

$$\bar{g}^{>}(\mathbf{p}, \omega) = e^{\beta\left[\omega + \mathbf{p}\cdot\mathbf{v} + \frac{1}{2}mv^2 - (\mathbf{p}+m\mathbf{v})\cdot\mathbf{v} + \frac{1}{2}mv^2 - \mu\right]}\bar{g}^{<}(\mathbf{p}, \omega)$$

$$= e^{\beta(\omega - \mu)}\bar{g}^{<}(\mathbf{p}, \omega)$$

Thus, the $\bar{g}$'s satisfy the same boundary condition as the equilibrium G's. The other equation that determines $g^{>}$ and $g^{<}$ is Eq. (10.31). We can rewrite this equation in terms of the $\bar{g}$'s by letting $\mathbf{p} \to \mathbf{p} - m\mathbf{v}$ and $z \to z + U_0 + \mathbf{p} \cdot \mathbf{v} - \frac{1}{2}mv^2$. Then, it becomes

$$g^{-1}\left(\mathbf{p} + m\mathbf{v}, z + U_0 + \mathbf{p} \cdot \mathbf{v} + \frac{1}{2}mv^2\right)$$

$$= \left[\int \frac{d\omega}{2\pi} \frac{\bar{g}^{>}(\mathbf{p}, \omega) \mp \bar{g}^{<}(\mathbf{p}, \omega)}{z - \omega}\right]^{-1}$$

$$= \left(z + \mathbf{p} \cdot \mathbf{v} + \frac{1}{2}mv^2\right) - \frac{(\mathbf{p} + m\mathbf{v})^2}{2m}$$

$$- \Sigma\left(\mathbf{p} + m\mathbf{v}, z + \mathbf{p} \cdot \mathbf{v} + \frac{1}{2}mv^2 + U_0; g^{>}, g^{<}\right)$$

$$= z - \left(\frac{p^2}{2m}\right) - \Sigma\left(\mathbf{p} + m\mathbf{v}, z + \mathbf{p} \cdot \mathbf{v} + \frac{1}{2}mv^2 + U_0; g^{>}, g^{<}\right)$$

By essentially the same argument as we gave before, we can show that

$$\Sigma(\mathbf{p} + m\mathbf{v}, z + \mathbf{p} \cdot \mathbf{v} + U_0; g^{>}, g^{<}) = \Sigma(\mathbf{p}, z; \bar{g}^{>}, \bar{g}^{<})$$

Since the $\bar{g}$'s obey the same equations as the equilibrium Green's functions, Eq. (11.9) is, in fact, correct, and $\mathbf{v}$ is the average velocity of the system.

Such an equilibrium state would be reached if the potential $U(\mathbf{R}, T)$, when it acted, transferred a net momentum $mN\mathbf{v}$ to the system.

11.2 Local Equilibrium Solutions

A very simple extension of the results of Section 11.1 can be applied to a discussion of sound propagation. This is the primary reason for having described the equilibrium solutions to the generalized Boltzmann equation.

The arguments we shall use to find sound propagation will be very closely analogous to those used to find sound propagation from the ordinary Boltzmann equation. The left side of the generalized Boltzmann equation, (10.30), involves space and time derivatives; the right side does not. Therefore, when U ($\mathbf{R}$, T) varies very slowly in space and time, the left side of Eq. (10.30) is necessarily very small. Hence, in this limit, we can neglect the left side of Eq. (10.30) entirely. We then have to solve

$$\Sigma^> (\mathbf{p}, \omega; \mathbf{R}, T)\, g^< (\mathbf{p}, \omega; \mathbf{R}, T) - \Sigma^< (\mathbf{p}, \omega; \mathbf{R}, T)\, g^> (\mathbf{p}, \omega; \mathbf{R}, T) = 0 \tag{11.11}$$

In the Born collision approximation (11.11) becomes[a]

$$\int \frac{d\mathbf{p}'}{(2\pi)^3} \frac{d\omega'}{2\pi} \frac{d\bar{\mathbf{p}}}{(2\pi)^3} \frac{d\bar{\omega}}{2\pi} \frac{d\bar{\mathbf{p}}'}{(2\pi)^3} \frac{d\bar{\omega}'}{2\pi}$$
$$\times (2\pi)^4\, \delta \left(\mathbf{p} + \mathbf{p}' - \bar{\mathbf{p}} - \bar{\mathbf{p}}'\right) \delta \left(\omega + \omega' - \bar{\omega} - \bar{\omega}'\right)$$
$$\times \left(\frac{1}{2}\right) \left[v \left(\mathbf{p} - \bar{\mathbf{p}}\right) \pm v \left(\mathbf{p} - \bar{\mathbf{p}}'\right)\right]^2$$
$$\times \{g^< (\mathbf{p}, \omega; \mathbf{R}, T) g^< (\mathbf{p}', \omega'; \mathbf{R}, T) g^> (\bar{\mathbf{p}}, \bar{\omega}; \mathbf{R}, T) g^> (\bar{\mathbf{p}}', \bar{\omega}'; \mathbf{R}, T)$$
$$- g^> (\mathbf{p}, \omega; \mathbf{R}, T) g^> (\mathbf{p}', \omega'; \mathbf{R}, T) g^< (\bar{\mathbf{p}}, \bar{\omega}; \mathbf{R}, T) g^> (\bar{\mathbf{p}}', \bar{\omega}'; \mathbf{R}, T)\}$$
$$= 0 \tag{11.12}$$

From the discussion in Section 11.1, we know that the solution to Eq. (11.12) is

$$\frac{g^> (\mathbf{p}, \omega; \mathbf{R}, T)}{g^> (\mathbf{p}, \omega; \mathbf{R}, T)} = \exp \left\{ - \beta(\mathbf{R}, T) \left[\omega - \mathbf{p} \cdot \mathbf{v}(\mathbf{R}, T) + \frac{1}{2} m v^2 (\mathbf{R}, T) \right.\right.$$
$$\left.\left. - \mu(\mathbf{R}, T) + U (\mathbf{R}, T)\right] \right\} \tag{11.13}$$

where β^{-1} ($\mathbf{R}$, T), μ ($\mathbf{R}$, T), and $\mathbf{v}$ ($\mathbf{R}$, T) now represent the *local* temperature, chemical potential, and mean velocity of the particles in the system.

To determine $g^>$ and $g^<$, we make use of Eq. (10.31),

$$g^{-1} (\mathbf{p}, z; \mathbf{R}, T) = z - \left(\frac{p^2}{2m}\right) - U (\mathbf{R}, T) - \Sigma (\mathbf{p}, z; \mathbf{R}, T) \tag{10.31}$$

---

[a] The equality $= 0$ in the last line was omitted in the original text.

Since all the quantities in Eq. (10.31) depend on the values of $g^>$ and $g^<$ at only the space–time point **R**, T, we can directly carry over the discussion of Section 11.1 to establish the solution to Eqs. (10.31) and (11.13). In analogy to Eq. (11.9), we find

$$g^{\gtrless} \left(\mathbf{p} + m\mathbf{v}(\mathbf{R}, T), \, \omega + \mathbf{p} \cdot \mathbf{v}(\mathbf{R}, T) + \frac{1}{2} m v^2 (\mathbf{R}, T) + U(\mathbf{R}, T); \mathbf{R}, T \right)$$
$$= G^{\gtrless} (\mathbf{p}, \omega; \beta (\mathbf{R}, T), \mu (\mathbf{R}, T))$$

or

$$g^{\gtrless} (\mathbf{p}, \omega; \mathbf{R}, T) = G^{\gtrless} (\mathbf{p} - m\mathbf{v} (\mathbf{R}, T), \bar{\omega}; \beta (\mathbf{R}, T), \mu (\mathbf{R}, T))$$
$$\bar{\omega} = \omega - \mathbf{p} \cdot \mathbf{v} (\mathbf{R}, T) + \frac{1}{2} m v^2 (\mathbf{R}, T) - U (\mathbf{R}, T)$$

$$(11.14)$$

Here $G^{\gtrless} (\mathbf{p}, \omega; \beta.\mu)$ are the equilibrium Green's functions determined by the equilibrium Born collision approximation at the temperature β^{-1} and the chemical potential μ.

Therefore, when the disturbance varies very slowly in space and time, the nonequilibrium Green's functions $g^{\gtrless} (\mathbf{p}, \omega; \mathbf{R}, T)$ reduced to the equilibrium functions defined at the local temperature, chemical potential, and average velocity. Each portion of the system is very close to thermodynamic equilibrium, but the whole system is not in equilibrium because the temperature, chemical potential, and velocity vary from point to point.

We have derived this local-equilibrium result from the Born collision approximation. The result Eq. (11.14) is, in fact, much more generally valid. However, it is important to notice that Eq. (11.14) emerges from the application of Green's function approximations to a specific situation; it is not an extra assumption inserted into the theory. Equation (11.14) is not always correct; it is wrong in superfluid helium and in a superconductor, where the local-equilibrium state cannot be described by five parameters only. It is probably also wrong in a Coulomb system because of the long interaction range. The general theory is capable of predicting when Eq. (11.14) is correct, and when it is wrong.

To obtain a solution to Green's function equations of motion, we have to determine that local temperature, chemical potential, and velocity. Just as in the discussion of the ordinary Boltzmann equation, these parameters will be determined with the aid of the conservation laws for particle number, energy, and momentum.

11.3 Conservation Laws

The conservation laws can all be derived from the generalized Boltzmann equation (10.30). It is much more convenient, however, to derive from our starting point: Green's function equations of motion, Eqs. (7.28a) and (7.28b). We shall use only the difference of these two equations

$$\left[i \left(\frac{\partial}{\partial t_1} + \frac{\partial}{\partial t_{1'}} \right) + \frac{\nabla_1^2}{2m} - \frac{\nabla_{1'}^2}{2m} - U(1) + U(1') \right] g(1, 1'; U)$$
$$= \pm i \int d2 \left[V(1-2) - V(1'-2) \right] g_2(12^-, 1'2^+; U)$$

(11.15)

Eventually, we will employ the form of Eq. (11.15) in which $g_2(U)$ is determined by the Born collision approximation, but for now, we shall make use of only some rather general properties of $g_2(U)$.

If we set $1' = 1^+$ in Eq. (11.15), we derive the number conservation law

$$\frac{\partial}{\partial t_1} \left[\pm i g^<(1, 1; U) \right] + \nabla_{\mathbf{r}} \cdot \left[\frac{\nabla_1 - \nabla_{1'}}{2m} (\pm i) g^<(1, 1'; U) \right]_{1'=1} = 0$$

or

$$\frac{\partial}{\partial T} \langle \hat{n}(\mathbf{R}, T) \rangle_U + \nabla \cdot \left\langle \hat{\mathbf{j}}(\mathbf{R}, T) \right\rangle_U = 0 \qquad (11.16)$$

To find a differential momentum conservation law, we apply $\pm \frac{(\nabla_1 - \nabla_{1'})}{2i}$ to Eq. (11.15) and set $1' = 1^+$. In this way, we find[b]

$$\frac{\partial}{\partial t_1} \left[\frac{(\nabla_1 - \nabla_{1'})}{2i} (\pm i) g^<(1, 1'; U) \right]_{1'=1}$$
$$= m \frac{\partial}{\partial t_1} \left\langle \hat{\mathbf{j}}(1) \right\rangle_U$$
$$= - \left[\nabla_{\mathbf{r}_1} U(1) \right] \langle \hat{n}(1) \rangle_U$$
$$\quad - \nabla_{\mathbf{r}_1} \cdot \left\{ \frac{(\nabla_1 - \nabla_{1'})}{2i} \frac{(\nabla_1 - \nabla_{1'})}{2im} (\pm i) g^<(1, 1'; U) \right\}_{1'=1}$$
$$\quad + \int d\mathbf{r}_2 \left[\nabla_{\mathbf{r}_2} V(\mathbf{r}_1 - \mathbf{r}_2) \right] g_2(12; 1^+2^+; U) \big|_{t_2 = t_1^+}$$

(11.17)

---

[b]The 1^{++} symbol in the original text is corrected by 1^+.

So far this equation does not even have the structure of a conservation law because the term involving g_2 is not proportional to a divergence. However, in the limit as the disturbance varies slowly in space, this term may be approximately converted into a divergence. The point is that $g_2 \left(12; 1^+2^+\right)\big|_{t_2=t_1^+}$ can be written as $g_2 \left[\mathbf{r}_1 - \mathbf{r}_2; (\mathbf{r}_1 + \mathbf{r}_2)/2, t_1\right]$ and if the disturbance varies slowly in space, g_2 varies slowly as a function of $(\mathbf{r}_1 + \mathbf{r}_2)/2$. In fact, we may now write

$$\int d\mathbf{r}_2 \left[\nabla_{\mathbf{r}_2} v\left(\mathbf{r}_1 - \mathbf{r}_2\right)\right] g_2 \left(12; 1^+2^+; U\right)\big|_{t_2=t_1^+}$$

$$= \int d\mathbf{r} \left[\nabla v(r)\right] g_2 \left(\mathbf{r}; \mathbf{r}_1 - \frac{\mathbf{r}}{2}, t_1\right)$$

$$= \frac{1}{2} \int d\mathbf{r} \left[\nabla v(r)\right] \left[g_2 \left(\mathbf{r}; \mathbf{r}_1 - \frac{\mathbf{r}}{2}, t_1\right) - g_2 \left(-\mathbf{r}; \mathbf{r}_1 + \frac{\mathbf{r}}{2}, t_1\right)\right]$$

$$\tag{11.18}$$

Because of the symmetry

$$g_2 \left(12; 1^+2^+; U\right) = g_2 \left(21; 2^+1^+; U\right)$$

of both the exact $g_2(U)$ and any approximate $g_2(U)$ that obeys condition B, it follows that

$$g_2 \left(-\mathbf{r}; \mathbf{r}_1 - \frac{\mathbf{r}}{2}, t_1\right) = g_2 \left(\mathbf{r}; \mathbf{r}_1 + \frac{\mathbf{r}}{2}, t_1\right)$$

Thus, expression (11.18) becomes

$$\frac{1}{2} \int d\mathbf{r} \left[\nabla v(r)\right] \left[g_2 \left(\mathbf{r}; \mathbf{r}_1 - \frac{\mathbf{r}}{2}, t_1\right) - g_2 \left(\mathbf{r}; \mathbf{r}_1 + \frac{\mathbf{r}}{2}, t_1\right)\right]$$

If the disturbance varies very slowly over the force range, we can expand the g_2 to first order in $\mathbf{r}$, getting

$$\int d\mathbf{r}_2 \, \nabla_{\mathbf{r}_1} v\left(\mathbf{r}_1 - \mathbf{r}_2\right) g_2 \left(12; 1^+2^+; U\right)\big|_{t_2=t_1^+}$$

$$\approx -\int d\mathbf{r} \left[\nabla v(r)\right] \frac{\mathbf{r} \cdot \nabla_{\mathbf{r}_1}}{2} g_2 \left(\mathbf{r}; \mathbf{r}_1, t_1\right)$$

$$= -\sum_{j=1}^{3} \left(\nabla_{\mathbf{r}_1}\right)_j \left[\int d\mathbf{r}_2 \, \frac{\nabla v \left(|\mathbf{r}_1 - \mathbf{r}_2'|\right)}{2} \left(\mathbf{r}_1 - \mathbf{r}_2\right)_j g_2 \left(12; 1^+2^+; U\right)\big|_{t_2=t_1^+}\right]$$

Therefore, for slowly varying disturbance, the momentum conservation law has the structure

$$m\frac{\partial}{\partial T} \left\langle \hat{\mathbf{j}}\left(\mathbf{R}, T\right)\right\rangle_U = -\left[\nabla U\left(\mathbf{R}, T\right)\right] \left\langle \hat{n}\left(\mathbf{R}, T\right)\right\rangle_U - \nabla \cdot \mathfrak{T}\left(\mathbf{R}, T\right)$$

$$\tag{11.19}$$

where

$$\mathfrak{T}_{ij}\,(\mathbf{R},\,T) = \left[\left(\frac{\nabla_1 - \nabla_{1'}}{2i}\right)_i \left(\frac{\nabla_1 - \nabla_{1'}}{2im}\right)_j (\pm i)\,g^<\,(1,\,1';U)\right]_{1'=1=\mathbf{R},\,T}$$
$$+ \frac{1}{2}\int d\mathbf{r}_2 \frac{(\mathbf{r}_1 - \mathbf{r}_2)_i\,(\mathbf{r}_1 - \mathbf{r}_2)_j}{|\mathbf{r}_1 - \mathbf{r}_2|}\frac{\partial v\,(|\mathbf{r}_1 - \mathbf{r}_2|)}{\partial\,|\mathbf{r}_1 - \mathbf{r}_2|}$$
$$\times\,g_2\,(12;1^+2^+;U)\big|_{t_2=t_1^+,\,t_1=T,\,\mathbf{r}_1=\mathbf{R}} \qquad (11.20)$$

$\mathfrak{T}_{ij}$ is usually called the stress tensor. It is the momentum current; but since the momentum is a vector, its current is a tensor.

An exactly similar argument leads to a differential energy conservation law in which the time derivative of the energy density is

$$\frac{\partial}{\partial t_1}\,\langle\mathcal{E}(1)\rangle_U = \frac{\partial}{\partial t_1}\left\{\pm i\frac{\nabla_1\cdot\nabla_{1'}}{2m}g^<\,(1,\,1';U)\right.$$
$$\left. -\frac{1}{2}\int d\mathbf{r}_2\,v\,(\mathbf{r}_1 - \mathbf{r}_2)\,g_2\,(12;1^+2^+;U)\big|_{t_2=t_1^+}\right\}$$
$$= -\,[\nabla U\,(1)]\cdot\left\langle\hat{\mathbf{j}}(1)\right\rangle_U - \nabla\cdot\mathbf{j}_\varepsilon(1) \qquad (11.21)$$

where the energy current, for slowly varying disturbance, is[c]

$$\mathbf{j}_\varepsilon(1) = \pm i\left[\frac{\nabla_1 - \nabla_{1'}}{2im}\frac{\nabla_1\cdot\nabla_{1'}}{m}g^<\,(1,\,1';U)\right]_{1'=1}$$
$$-\int d\mathbf{r}_2\,v\,(\mathbf{r}_1 - \mathbf{r}_2)\frac{\nabla_1 - \nabla_{1'}}{2im}\,g_2\,(12,\,1'2^+;U)\big|_{t_2=t_1^+,\,1'=1^+}$$
$$+\int d\mathbf{r}_2\,\frac{v\,(\mathbf{r}_1 - \mathbf{r}_2)}{2}\,(\mathbf{r}_1 - \mathbf{r}_2)$$
$$\times\left\{\frac{\nabla_2\cdot(\nabla_1 - \nabla_{1'})}{2im}\,g_2\,(12,\,1'2^+;U)\big|_{t_2=t_1^+,\,1'=1^+}\right\}$$
$$\qquad (11.22)$$

11.4 Application of Conservation Laws to the Quasi-Equilibrium Situation

These conservation laws are true not only for the exact $g(U)$ and $g_2(U)$ but also for any conserving approximation for these functions.

---

[c]Here the redundant vertical bar on the right of the first term is omitted.

In particular, these laws hold in the Born collision approximation. Therefore, we may determine functions $\beta\,(\mathbf{R},\,T)$, $\mu\,(\mathbf{R},\,T)$, and $\mathbf{v}\,(\mathbf{R},\,T)$ in Eq. (11.14) by substituting the local equilibrium solutions into the conservation laws.

This is most simply done for the number conservation law (11.16), which can be expressed as

$$\int \frac{d\omega}{2\pi} \int \frac{d\mathbf{p}}{(2\pi)^3} \left[\frac{\partial}{\partial T} + \frac{\mathbf{p}\cdot\nabla_{\mathbf{R}}}{m} \right] g^<\,(\mathbf{p},\,\omega;\,\mathbf{R},\,T) = 0$$

For the local equilibrium solution (11.14), this is

$$\int \frac{d\omega}{2\pi} \int \frac{d\mathbf{p}}{(2\pi)^3} \left[\frac{\partial}{\partial T} + \frac{\nabla_{\mathbf{R}}\cdot\mathbf{p}}{m} \right] G^<(\mathbf{p}-m\mathbf{v}(\mathbf{R},\,T),\,\omega;\,\beta(\mathbf{R},\,T),\,\mu(\mathbf{R},\,T)) = 0$$

We now let $\mathbf{p} \to \mathbf{p}+m\mathbf{v}\,(\mathbf{R},\,T)$. Since the rotational invariance of the equilibrium Green's function implies

$$\int \frac{d\mathbf{p}}{(2\pi)^3}\,\mathbf{p}G^<\,(p,\,\omega;\,\beta,\,\mu) = 0$$

we find

$$\int \frac{d\omega}{2\pi} \int \frac{d\mathbf{p}}{(2\pi)^3} \left[\frac{\partial}{\partial T} + \nabla_{\mathbf{R}}\cdot\mathbf{v}(\mathbf{R},\,T) \right] G^<\,(p,\,\omega;\,\beta\,(\mathbf{R},\,T),\,\mu\,(\mathbf{R},\,T)) = 0$$

Hence, the number conservation law becomes

$$\frac{\partial}{\partial T}n\,(\beta\,(\mathbf{R},\,T),\,\mu\,(\mathbf{R},\,T))+\nabla_{\mathbf{R}}\cdot[\mathbf{v}\,(\mathbf{R},\,T)\,n\,(\beta\,(\mathbf{R},\,T),\,\mu\,(\mathbf{R},\,T))] = 0$$

$$(11.23)$$

where

$$n\,(\beta,\,\mu) = \int \frac{d\mathbf{p}}{(2\pi)^3}\frac{d\omega}{2\pi}\,G^<\,(p,\,\omega;\,\beta,\,\mu)$$

is the equilibrium density derived from the Born collision approximation, expressed as a function of the inverse temperature and the chemical potential. Similarly, in the local-equilibrium situation, the momentum conservation law (11.19) becomes

$$m\frac{d}{dT}\,[\mathbf{v}\,(\mathbf{R},\,T)\,n\,(\beta\,(\mathbf{R},\,T),\,\mu\,(\mathbf{R},\,T))]$$

$$= -n\,(\beta\,(\mathbf{R},\,T),\,\mu\,(\mathbf{R},\,T))\,\nabla_{\mathbf{R}}U\,(\mathbf{R},\,T) - \nabla_{\mathbf{R}}\cdot\mathfrak{T}\,(\mathbf{R},\,T)$$

$$(11.24)$$

while from Eq. (11.20), the stress tensor is seen to have the form

$$\mathfrak{T}_{ij}(\mathbf{R}, T) = \int \frac{d\omega}{e\pi} \int \frac{d\mathbf{p}}{(2\pi)^3} \frac{p_i p_j}{m} G^<(|\mathbf{p} - m\mathbf{v}(\mathbf{R}, T)|, \omega; \beta(\mathbf{R}, T), \mu(\mathbf{R}, T))$$

$$+ \int d\mathbf{r}_2 \frac{(\mathbf{r}_1 - \mathbf{r}_2)_i (\mathbf{r}_1 - \mathbf{r}_2)_j}{|\mathbf{r}_1 - \mathbf{r}_2|} \frac{\partial v(|\mathbf{r}_1 - \mathbf{r}_2|)}{\partial |\mathbf{r}_1 - \mathbf{r}_2|}$$

$$\times \, g_2 \left(12; 1^+ 2^+; U \right) \big|_{t_2 = t_1^+, t_1 - T, \mathbf{r} = \mathbf{R}} \tag{11.25}$$

By exactly the same argument that we used to evaluate the Born collision approximation $\Sigma(\mathbf{p}, z; \mathbf{R}, T)$ in terms of the local temperature, velocity, and chemical potential, it is easy to show that

$$g_2 \left(12, 1'2^+; U \right) = e^{im\mathbf{v}(\mathbf{R}, T) \cdot (\mathbf{r}_1 - \mathbf{r}_{1'})} G_2 \left(12, 1'2^+; \beta(\mathbf{R}, T), \mu(\mathbf{R}, T) \right) \tag{11.26}$$

for

$$\mathbf{R} = \frac{\mathbf{r}_1 + \mathbf{r}_{1'}}{2}$$

$$T = t_1, t_2 = t_1^+, t_{1'} = t_1^{++}$$

where $G_2(12; 1'2'; \beta, \mu)$ is the equilibrium two-particle Green's function, in the Born collision approximation. The rotational invariance of this function and $G(p, \omega)$ implies that Eq. (11.25) becomes

$$\mathfrak{T}_{ij}(\mathbf{R}, T) = m v_i(\mathbf{R}, T) v_j(\mathbf{R}, T) n(\beta(\mathbf{R}, T), \mu(\mathbf{R}, T))$$

$$+ \delta_{ij} P(\beta(\mathbf{R}, T), \mu(\mathbf{R}, T)) \tag{11.27}$$

where

$$P(\beta, \mu) = \int \frac{d\omega}{2\pi} \frac{d\mathbf{p}}{(2\pi)^3} \frac{p^2}{3m} G^<(p, \omega; \beta, \mu)$$

$$+ \frac{1}{6} \int d\mathbf{r}_2 \, |\mathbf{r}_1 - \mathbf{r}_2| \frac{\partial v(|\mathbf{r}_1 - \mathbf{r}_2|)}{\partial |\mathbf{r}_1 - \mathbf{r}_2|} G_2(12; 1^+ 2^+; \beta, \mu) \big|_{t_2 = t_1^+} \tag{11.28}$$

Thus, the momentum conservation law, Eq. (11.24), reduces to

$$m \frac{\partial}{\partial T} [\mathbf{v}(\mathbf{R}, T) n(\beta(\mathbf{R}, T), \mu(\mathbf{R}, T))]$$

$$= - n(\beta(\mathbf{R}, T), \mu(\mathbf{R}, T)) \nabla U(\mathbf{R}, T) \tag{11.29}$$

$$- \nabla \cdot [m\mathbf{v}(\mathbf{R}, T) \mathbf{v}(\mathbf{R}, T) n(\beta(\mathbf{R}, T), \mu(\mathbf{R}, T))]$$

$$- \nabla P(\beta(\mathbf{R}, T), \mu(\mathbf{R}, T))$$

We have used the symbol P for the part of the stress tensor proportional to the unit tensor δ_{ij}, in anticipation of the fact that this quantity is actually the pressure in the many-particle system. The most elementary reason for the appearance of the pressure in the momentum conservation law is that the pressure and the stress tensor have quite parallel meanings: The pressure is the average flux of momentum up to a surface of the system, whereas the stress tensor $\mathfrak{T}_{ij}$ gives the flux of the i-th direction momentum through a surface perpendicular to the j-th direction.

We can make this identification of the pressure mathematically as follows. Let us go back to the original Boltzmann equation (10.30):

$$\left[\omega - \frac{p^2}{2m} - U - \mathfrak{R}\Sigma, g^< \right] + [\mathfrak{R}g, \Sigma^<] = -\Sigma^> g^< + \Sigma^< g^> \quad (10.30)$$

For the case in which $U(\mathbf{R}, T)$ is independent of T, the time-independent local-equilibrium form

$$g^< (\mathbf{p}, \omega; \mathbf{R}, T) = f(\mathbf{p}, \omega; \mathbf{R}) \, a\, (\mathbf{p}, \omega; \mathbf{R})$$

$$\Sigma^< (\mathbf{p}, \omega; \mathbf{R}, T) = f(\mathbf{p}, \omega; \mathbf{R}) \, \Gamma\, (\mathbf{p}, \omega; \mathbf{R})$$

$$f(\mathbf{p}, \omega; \mathbf{R}) = \frac{1}{\exp[\beta(\mathbf{R})(\omega - \mathbf{p} \cdot \mathbf{v}(\mathbf{R}) + \frac{1}{2}mv^2(\mathbf{R}) - \mu(\mathbf{R}) - U(\mathbf{R}))] \mp 1}$$

is an exact solution to the Boltzmann equation, since then the right side of the Boltzman equation vanishes, and the left side becomes

$$\left[\omega - \left(\frac{p^2}{2m}\right) - U - \mathfrak{R}\Sigma, fa\right] + [\mathfrak{R}g, f\Gamma] = 0 \quad (11.30)$$

Like ordinary Poisson brackets, the generalized Poisson brackets satisfy

$$[A, BC] = C[A, B] + B[A, C]$$

Therefore, Eq. (11.30) may also be written as

$$f\left\{\left[\omega - \left(\frac{p^2}{2m}\right) - U - \mathfrak{R}\Sigma, a\right] + [\mathfrak{R}g, \Gamma]\right\}$$

$$+ a\left[\omega - \left(\frac{p^2}{2m}\right) - U - \mathfrak{R}\Sigma, f\right] + \Gamma[\mathfrak{R}g, f] = 0$$

However, the term in the braces must vanish, because a is evaluated by demanding that this term be zero. We are then left with

$$a\left[\omega - \left(\frac{p^2}{2m}\right) - U - \mathfrak{R}\Sigma, f\right] + \Gamma[\mathfrak{R}g, f] = 0 \quad (11.31)$$

which has the simple solution

$$\mathbf{v} = 0$$

$$\beta\,(\mathbf{R}) = \beta, \text{ independent of } \mathbf{R} \tag{11.32}$$

$$\mu\,(\mathbf{R}) + U\,(\mathbf{R}) = \mu', \text{ independent of } \mathbf{R}$$

For these values of β, μ, and $\mathbf{v}$, the function $f\,(\mathbf{p}, \omega; \mathbf{R})$ is independent of $\mathbf{p}$, and $\mathbf{R}$:

$$f\,(\mathbf{p}, \omega; \mathbf{R}) = \frac{1}{e^{\beta(\omega - \mu')} \mp 1}$$

and hence,

$$a\left[\omega - \left(\frac{p^2}{2m}\right) - U - \Re\Sigma, f\right] + \Gamma\,[\Re g, f]$$

$$= -\left[a\left(\frac{\partial}{\partial T}\right)\left(\omega - \left(\frac{p^2}{2m}\right) - U - \Re\Sigma\right) + \Gamma\left(\frac{\partial \Re g}{\partial T}\right)\right]\frac{\partial f}{\partial \omega}$$

Since neither

$$\omega - \left(\frac{p^2}{2m}\right) - U - \Re\Sigma$$

nor g depends on time, the choice Eq. (11.32) indeed gives a solution to Eq. (11.31).

Now we consider Eq. (11.29), the momentum conservation law, in the case $U\,(\mathbf{R}, T) = U\,(\mathbf{R})$. Using Eq. (11.32), we find

$$0 = -n\,(\beta, \mu\,(\mathbf{R}))\,\nabla U\,(\mathbf{R}) - \nabla P\,(\beta, \mu\,(\mathbf{R}))$$

or

$$0 = -\left[n\,(\beta, \mu\,(\mathbf{R})) - \frac{\partial P\,(\beta, \mu\,(\mathbf{R}))}{\partial \mu}\right]\nabla U\,(\mathbf{R})$$

But $\nabla U\,(\mathbf{R})$ is arbitrary, so that

$$\left(\frac{\partial}{\partial \mu}\right) P\,(\beta, \mu) = n\,(\beta, \mu) \tag{11.33}$$

This is identical with one of the thermodynamic definitions of the pressure that we used in Chapter 3. The identification of the P in Eq. (11.29) as the pressure is, therefore, correct. Incidentally, Eq. (11.28) is a useful expression for calculating the pressure.

Finally, we consider the energy conservation law (11.21). The substitution of the local-equilibrium solutions into this law yields

$$\left(\frac{\partial}{\partial T}\right)\left\{\mathcal{E}\left(\beta\left(\mathbf{R},\,T\right),\,\mu\left(\mathbf{R},\,T\right)\right)+\frac{1}{2}m\left[\mathbf{v}\left(\mathbf{R},\,T\right)\right]^2 n\left(\beta\left(\mathbf{R},\,T\right),\,\mu\left(\mathbf{R},\,T\right)\right)\right\}$$
$$= -\nabla\cdot\mathbf{j}_{\mathcal{E}}\left(\mathbf{R},\,T\right)-n\left(\beta\left(\mathbf{R},\,T\right),\,\mu\left(\mathbf{R},\,T\right)\right)\mathbf{v}\left(\mathbf{R},\,T\right)\cdot\nabla U\left(\mathbf{R},\,T\right)$$

$$(11.34)$$

where the equilibrium energy density is

$$\mathcal{E}\left(\beta,\,\mu\right)=\int\frac{d\mathbf{p}}{(2\pi)^3}\frac{d\omega}{2\pi}\frac{p^2}{2m}G^<\left(p,\,\omega;\,\beta,\,\mu\right)$$
$$-\frac{1}{2}\int d\mathbf{r}_2\,v\left(\mathbf{r}_1-\mathbf{r}_2\right)G_2\left(12;\,1^+2^+;\,\beta,\,\mu\right)\Big|_{t_2=t_1^+}$$

$$(11.35)$$

The energy current is given by Eqs. (11.22) and (11.26) as

$$j_{\mathcal{E}}(\mathbf{R},\,T)=\int\frac{d\mathbf{p}}{(2\pi)^3}\frac{d\omega}{2\pi}\mathbf{p}\frac{p^2}{2m}G^<\left(|\mathbf{p}-m\mathbf{v}(\mathbf{R},\,T)|,\,\omega;\,\beta(\mathbf{R},\,T),\,\mu(\mathbf{R},\,T)\right)$$
$$-\int d\mathbf{r}_2\,v\left(|\mathbf{r}_1-\mathbf{r}_2|\right)\left[\mathbf{v}\left(\mathbf{R},\,T\right)+\frac{\nabla_1-\nabla_{1'}}{2im}\right]$$
$$\times\,G\left(12;\,1'2^+;\,\beta\left(\mathbf{R},\,T\right),\,\mu\left(\mathbf{R},\,T\right)\right)\Big|_{1'=1^+,t_2=t_1^+}$$
$$+\int d\mathbf{r}_2\,v\left(|\mathbf{r}_1-\mathbf{r}_2|\right)\frac{\mathbf{r}_1-\mathbf{r}_2}{2}\left[\frac{\nabla_1-\nabla_{1'}}{2im}+v\left(\mathbf{R},\,T\right)\right]\cdot\nabla_2$$
$$\times\,G\left(12;\,1'2^+;\,\beta\left(\mathbf{R},\,T\right),\,\mu\left(\mathbf{R},\,T\right)\right)\Big|_{1'=1^+,t_2=t_1^+}$$

The rotational invariance of the equilibrium solution may be used to reduce the energy current to the form

$$j_{\mathcal{E}}\left(\mathbf{R},\,T\right)=v\left(\mathbf{R},\,T\right)\{\}_{\beta(\mathbf{R},T),\mu(\mathbf{R},T)}$$

where

$$\{\}_{\beta,\mu}=\frac{m}{2}v^2\left(\mathbf{R},\,T\right)n\left(\beta,\,\mu\right)+\int\frac{d\mathbf{p}}{(2\pi)^3}\frac{d\omega}{2\pi}\left(1+\frac{2}{3}\right)\frac{p^2}{2m}G^<\left(p,\,\omega;\,\beta,\,\mu\right)$$
$$-\int d\mathbf{r}_2\,v\left(|\mathbf{r}_1-\mathbf{r}_2|\right)G\left(12;\,1'2^+;\,\beta,\,\mu\right)\Big|_{1'=1^+,t_2=t_1^+}$$
$$+\frac{1}{6}\int d\mathbf{r}_2\,v\left(|\mathbf{r}_1-\mathbf{r}_2|\right)\left(\mathbf{r}_1-\mathbf{r}_2\right)\cdot\nabla_2\,G\left(12;\,1'2^+;\,\beta,\,\mu\right)\Big|_{1'=1^+,t_2=t_1^+}$$

When we integrate the last term in the braces by parts, it becomes

$$\int d\mathbf{r}_2\left[\nabla_{\mathbf{r}_1}v\left(\mathbf{r}_1-\mathbf{r}_2\right)\right]\cdot\left(\mathbf{r}_1-\mathbf{r}_2\right)G_2\,()+\frac{1}{2}\int d\mathbf{r}_2\,v\left(\mathbf{r}_1-\mathbf{r}_2\right)G_2\,()$$

We now see that the energy current may be expressed in terms of pressure and energy density, defined, respectively, by Eqs. (11.28) and (11.35):

$$j_{\mathcal{E}}\,(\mathbf{R},\,T) = v\,(\mathbf{R},\,T)\left[\frac{1}{2}mv^2\,(\mathbf{R},\,T)\,n\,(\beta\,(\mathbf{R},\,T),\,\mu\,(\mathbf{R},\,T))\right.$$

$$\left. + P\,(\beta\,(\mathbf{R},\,T),\,\mu\,(\mathbf{R},\,T)) + \mathcal{E}\,(\beta\,(\mathbf{R},\,T),\,\mu\,(\mathbf{R},\,T))\right]$$

$$(11.36)$$

The energy current is thus the local mean velocity times the sum of the mean increase in kinetic energy due to the local mean velocity and the enthalpy density, $\mathcal{E} + P$.

11.5 Sound Propagation

To derive the existence of sound propagation from these conservation laws, we consider the case in which $U\,(\mathbf{R},\,T)$ is small. At time $T = -\infty$, we consider the system to be in equilibrium with $\beta\,(\mathbf{R},\,T) = \beta;\, \mu\,(\mathbf{R},\,T) = \mu,\, v\,(\mathbf{R},\,T) = 0$. Then, for all later times $\beta\,(\mathbf{R},\,T) - \beta,\, \mu\,(\mathbf{R},\,T) - \mu$, and $v\,(\mathbf{R},\,T)$ will be small.

In the conservation laws, we consider only first-order terms. Then the number conservation law (11.23) is

$$\frac{\partial}{\partial T}n\,(\beta\,(\mathbf{R},\,T),\,\mu\,(\mathbf{R},\,T)) + n\nabla\cdot\mathbf{v}\,(\mathbf{R},\,T) = 0 \qquad (11.37)$$

The energy conservation law (11.34) is

$$\frac{\partial}{\partial T}\mathcal{E}\,(\beta\,(\mathbf{R},\,T),\,\mu\,(\mathbf{R},\,T)) + (\mathcal{E} + P)\nabla\cdot\mathbf{v}\,(\mathbf{R},\,T) = 0 \qquad (11.38)$$

and the momentum conservation law (11.29) is

$$n\frac{\partial}{\partial T}\mathbf{v}\,(\mathbf{R},\,T) + \nabla P\,(\beta\,(\mathbf{R},\,T),\,\mu\,(\mathbf{R},\,T)) = -n\nabla U\,(\mathbf{R},\,T) \quad (11.39)$$

We eliminate $\mathbf{v}$ from Eqs. (11.37) and (11.38) to find:

$$\frac{1}{n}\frac{\partial}{\partial T}n\,(\beta\,(\mathbf{R},\,T),\,\mu\,(\mathbf{R},\,T)) = \frac{1}{\mathcal{E} + P}\frac{\partial}{\partial T}\mathcal{E}\,(\beta\,(\mathbf{R},\,T),\,\mu\,(\mathbf{R},\,T))$$

$$(11.40)$$

We also eliminate $\mathbf{v}$ from Eqs. (11.37) and (11.39) by taking the divergence of the latter and then substituting the time derivative of the former. This gives

$$m \frac{\partial^2}{\partial T^2} n(\beta(\mathbf{R}, T), \mu(\mathbf{R}, T)) - \nabla^2 P(\beta(\mathbf{R}, T), \mu(\mathbf{R}, T))$$
$$= -n\nabla^2 U(\mathbf{R}, T) \tag{11.41}$$

These equations are almost identical to those that arose in the discussion of sound propagation based on the ordinary Boltzmann equation.

Equation (11.40) relates the permissible variations in $\mu(\mathbf{R}, T)$ and $\beta(\mathbf{R}, T)$. Since all the quantities that appear in this equation are thermodynamic functions, we can give a thermodynamic interpretation of Eq. (11.40). This equation demands that the change in β and μ be such that

$$\frac{dn}{n} - \frac{d\mathcal{E}}{\mathcal{E} + P} = 0 \tag{11.42}$$

where dn and $d\mathcal{E}$ are the local changes in n and $\mathcal{E}$. We recall the thermodynamic identities

$$TS = E + PV - \mu N$$

and

$$SdT = -Nd\mu + VdP$$

where S is the entropy, E is the total energy, and N is the total number of particles. These identities may be expressed solely in terms of intensive quantities by dividing both sides of each by N. Then

$$T\left(\frac{S}{N}\right) = \frac{\mathcal{E} + P}{n} - \mu \tag{11.43}$$

and

$$\frac{S}{N} dT = -d\mu + \frac{dP}{n} \tag{11.44}$$

If we take the differential of Eq. (11.43) and use the relation (11.44), we find the equation

$$Td\left(\frac{S}{N}\right) = \frac{d\mathcal{E}}{n} - \frac{\mathcal{E} + P}{n} \frac{dn}{n}$$

It follows that

$$\frac{dn}{n} - \frac{d\mathcal{E}}{\mathcal{E} + P} = -\frac{Tn}{\mathcal{E} + P} d\left(\frac{S}{N}\right)$$

Therefore, the restriction (11.42) may be written as

$$d\left(\frac{S}{N}\right) = 0 \tag{11.45}$$

This restriction means that β (**R**, T) and μ (**R**, T) change so that the entropy per particle, a local quantity, is constant.

Because of this restriction, the change in pressure must be related to the change in the density by

$$dP = \left(\frac{\partial P}{\partial n}\right)_{S/N} dn$$

Therefore, the momentum conservation law (11.41) becomes simply the sound-propagation equation

$$\left[\frac{\partial^2}{\partial T^2} - C^2 \nabla^2\right] n \left(\beta \, (\mathbf{R}, \, T), \, \mu \, (\mathbf{R}, \, T)\right) = -\nabla^2 U \, (\mathbf{R}, \, T) \, \frac{n}{m} \tag{11.46}$$

where C, the sound velocity, is determined by

$$mC^2 = \left(\frac{\partial P}{\partial n}\right)_{S/N} \tag{11.47}$$

For a perfect gas, the result (11.47) agrees with the sound velocity derived from the ordinary Boltzmann equation. When the potential is nonzero, these results differ. The sound velocity[d] (11.47) is amply verified by experiment.

This formula for the sound velocity can be obtained much more directly, assuming only local thermodynamic equilibrium and applying the conservation laws. The main justification for our rather elaborate Green's function arguments is that they provide a means of describing transport phenomena in a self-contained way, starting from a dynamical approximation, i.e., an approximation for $G_2(U)$ in terms of $G(U)$. These calculations require no extra assumptions. The existence of local thermodynamic equilibrium is derived from the

---

[d]The equation number was originally (11.31), which is the time-independent local-equilibrium Boltzmann equation, the wrong reference.

Green's function approximations. The various quantities that appear in the conservation laws are determined by the approximation. The theory provides at the same time a description of what transport processes occur, in this case sound propagation, and a determination of the numerical quantities that appear in the transport equation, in this case $(\partial P/\partial n)_{S/N}$.

Chapter 12

The Landau Theory of the Normal Fermi Liquid

12.1 The Boltzmann Equation

The nonequilibrium theory described in the previous chapters reduces to a particularly simple form for a system of fermions very close to zero temperature. To see this, let us define a "local occupation number" $f(\mathbf{p}, \omega; \mathbf{R}, T)$ by writing

$$g^<(\mathbf{p}, \omega; \mathbf{R}, T) = a(\mathbf{p}, \omega; \mathbf{R}, T) f(\mathbf{p}, \omega; \mathbf{R}, T) \qquad (12.1)$$

where

$$a(\mathbf{p}, \omega; \mathbf{R}, T) = g^>(\mathbf{p}, \omega; \mathbf{R}, T) \mp g^<(\mathbf{p}, \omega; \mathbf{R}, T)$$

In equilibrium, at zero temperature

$$f(\mathbf{p}, \omega; \mathbf{R}, T) \to f(\omega) = \begin{cases} 0 & \text{for } \omega > \mu \\ 1 & \text{for } \omega < \mu \end{cases}$$

Therefore, all "states" with $\omega < \mu$ are occupied, and all "states" with $\omega > \mu$ are empty. At very low temperatures, f differs from 0 or 1 only for ω very near to μ. We shall assume that f still has this form at low temperatures, even in the presence of a disturbance. That is, we assume that there exists a $\mu(\mathbf{R}, T)$ such that $f(\mathbf{p}, \omega; \mathbf{R}, T) = 0$

Annotations to Quantum Statistical Mechanics
In-Gee Kim
Copyright © 2018 Pan Stanford Publishing Pte. Ltd.
ISBN 978-981-4774-15-4 (Hardcover), 978-1-315-19659-6 (eBook)
www.panstanford.com

for ω appreciably greater than μ (**R**, T), and f (**p**, ω; **R**, T) $= 1$ for ω appreciably less than μ. The only frequencies for which the local occupation number, f (**p**, ω; **R**, T), is different from zero or one are those within an infinitesimal range of μ (**R**, T). This dependence of f on ω is essentially what we mean by "low temperature" for a non-equilibrium system. We shall show in a moment that this hypothesis about the behavior of f leads to a perfectly consistent solution to our basic nonequilibrium equations.

There is one simplification that makes this low-temperature system rather tractable: for ω near μ, the lifetime Γ becomes vanishingly small. We have mentioned in Chapter 5 that at zero temperature in equilibrium

$$\Sigma^> (p, \omega) = 0 \quad \text{for } \omega < \mu$$
$$\Sigma^< (p, \omega) = 0 \quad \text{for } \omega > \mu$$

The proof of this relations depends only on the fact that $f = 0$ for $\omega > \mu$ and $1 - f = 0$ for $\omega < \mu$. Since we are assuming that f has a similar behavior in the nonequilibrium case, it follows that

$$\Sigma^> (\textbf{p}, \omega; \textbf{R}, T) = 0 \quad \text{for } \omega < \mu (\textbf{R}, T)$$
$$\Sigma^< (\textbf{p}, \omega; \textbf{R}, T) = 0 \quad \text{for } \omega > \mu (\textbf{R}, T) \tag{12.2}$$

when the system is very little excited from its zero-temperature state. Therefore, for situations near zero temperature, we shall take both $\Sigma^>$ and $\Sigma^<$ to be very small at those frequencies, near μ, for which the occupation numbers f (**p**, ω; **R**, T) differ from 0 or 1. This approximation involves an assumption about the continuity of $\Sigma^>$ and $\Sigma^<$ at $\omega = \mu$. The continuity can be proved in all orders of perturbation theory, but it is not necessarily true for situations, such as the superconducting state, in which perturbation theory is not valid. Therefore, the discussion in the remainder of this chapter applies only to so-called "normal" fermion systems and not to the superconductor.

If $\Sigma^>$ and $\Sigma^<$ are both negligible for ω near μ, then this region the Boltzmann equation (10.30) becomes

$$\left[\omega - \left(\frac{p^2}{2m}\right) - U (\textbf{R}, T) - \Re\Sigma(\textbf{p}, \omega; \textbf{R}, T), a(\textbf{p}, \omega; \textbf{R}, T)f(\textbf{p}, \omega; \textbf{R}, T)\right] = 0$$

$$\text{for } \omega \approx \mu$$

$$\tag{12.3}$$

We may verify that out assumptions about f for ω appreciably greater or less than $\mu\,(\mathbf{R},\,T)$ lead to a consistent solution to the Boltzmann equation. First if ω is appreciably less than μ, then from the assumption $f = 1$, we have

$$\Sigma^> = 0,\; \Sigma^< = \Gamma,\; g^> = 0,\; \text{and}\; g^< = a \qquad (12.4)$$

When we substitute this solution into the Boltzmann equation (10.30), we find

$$\left[\omega - \left(\frac{p^2}{2m}\right) - U\,(\mathbf{R},\,T) - \Re\Sigma\,(\mathbf{p},\,\omega;\mathbf{R},\,T),\, a\,(\mathbf{p},\,\omega;\mathbf{R},\,T)\right]$$
$$- \left[\Gamma\,(\mathbf{p},\,\omega;\mathbf{R},\,T),\, \Re g\,(\mathbf{p},\,\omega;\mathbf{R},\,T)\right] = 0 \quad \text{for } \omega < \mu$$
$$(12.5)$$

Since this equation is, in fact, just Eq. (10.27) satisfied by a, Eq. (12.4) for ω appreciably less than μ. For ω appreciably greater than μ, the solution $g^< = \Sigma^< = 0$, which follows from the assumption $f = 0$, trivially satisfies Eq. (10.30).

We have shown in Chapter 10 that for all ω, $a\,(\mathbf{p},\,\omega;\mathbf{R},\,T)$ is given by[a] Eq. (10.28b) with the aid of Eq. (10.31)

$$a\,(\mathbf{p},\,\omega;\mathbf{R},\,T) = \frac{\Gamma\,(\mathbf{p},\,\omega;\mathbf{R},\,T)}{\left[\omega - \left(\frac{p^2}{2m}\right) - U\,(\mathbf{R},\,T) - \Re\Sigma\,(\mathbf{p},\,\omega;\mathbf{R},\,T)\right]^2 + \left[\frac{\Gamma(\mathbf{p},\omega;\mathbf{R},T)}{2}\right]^2}$$

Thus, when ω is close to μ, so that

$$\Gamma\,(\mathbf{p},\,\omega;\mathbf{R},\,T) = \Sigma^>\,(\mathbf{p},\,\omega;\mathbf{R},\,T) \mp \Sigma^<\,(\mathbf{p},\,\omega;\mathbf{R},\,T) \to 0$$

a becomes just the delta function

$$a\,(\mathbf{p},\,\omega;\mathbf{R},\,T) = 2\pi\left(\omega - \left(\frac{p^2}{2m}\right) - U\,(\mathbf{R},\,T) - \Re\Sigma\,(\mathbf{p},\,\omega;\mathbf{R},\,T)\right)$$
$$(12.6)$$

Note that at $\omega = \mu\,(\mathbf{R},\,T)$

$$\frac{\partial}{\partial\omega}\Re\Sigma\,(\mathbf{p},\,\omega;\mathbf{R},\,T) = \frac{\partial}{\partial\omega}\int\frac{d\omega'}{2\pi}\frac{\Gamma\,(\mathbf{p},\,\omega';\mathbf{R},\,T)}{\omega - \omega'}$$
$$= -\int\frac{d\omega'}{2\pi}\frac{\Gamma\,(\mathbf{p},\,\omega';\mathbf{R},\,T)}{(\mu - \omega')^2} < 0$$

since Γ is a positive function. By continuity, $\frac{\partial\Re\Sigma}{\partial\omega} < 0$ for all ω near μ. Therefore, the argument of the delta function in Eq. (12.6) is a

---

[a] In the original text, the reference equation number was only (10.31), but this is not enough to explain the following equation.

monotonically increasing function of ω for all ω near μ. It follows then that for every $\mathbf{p}$, $\mathbf{R}$, T, there exists just one root of

$$\omega = \left(\frac{p^2}{2m}\right) + U\ (\mathbf{R},\ T) + \Re\Sigma\ (\mathbf{p},\ \omega;\mathbf{R},\ T)$$

Let us write this solution as

$$\omega = E\ (\mathbf{p},\ \mathbf{R},\ T) + U\ (\mathbf{R},\ T) \tag{12.7}$$

where

$$E\ (\mathbf{p},\ \mathbf{R},\ T) = \left(\frac{p^2}{2m}\right) + \Re\Sigma\ (\mathbf{p},\ \omega;\mathbf{R},\ T)|_{\omega=E(\mathbf{p},\mathbf{R},T)+U(\mathbf{R},T)}$$

In equilibrium, $E\ (\mathbf{p},\ \mathbf{R},\ T)$ reduces to $E(p)$. Because the response to the disturbance is primarily a change in the occupation of single-particle levels with ω near μ, the response manifests itself mostly for momenta such that $E\ (p) \approx \mu$. We shall assume that there exists a unique momentum p_F, called the Fermi momentum, such that $E\ (p_F) = \mu$.

The two basic assumptions that go into this theory are the existence of a unique Fermi momentum and the smooth variation of $\Sigma^>$ and $\Sigma^<$ near $\omega \approx \mu$. Whenever these two assumptions are satisfied, the rest of our statements will hold for a fermion system at sufficiently low temperatures, in which the disturbance varies very slowly in space and time.

We can combine Eqs. (12.3) and (12.6) into the form

$$\left[\omega - \left(\frac{p^2}{2m}\right) - U - \Re\Sigma,\ 2\pi\delta\left(\omega - \left(\frac{p^2}{2m}\right) - U - \Re\Sigma\right)f(\mathbf{p},\ \omega;\mathbf{R},\ T)\right] = 0$$

$$\text{for } \omega \approx \mu$$

$$\tag{12.8}$$

Clearly, we need not consider the general $f\ (\mathbf{p},\ \omega;\mathbf{R},\ T)$ but only the simpler distribution function[b]

$$n\ (\mathbf{p},\ \mathbf{R},\ T) = f\ (\mathbf{p},\ \omega;\mathbf{R},\ T)|_{\omega=E(\mathbf{p},\mathbf{R},T)+U(\mathbf{R},T)} \tag{12.9}$$

We shall interpret $n\ (\mathbf{p},\ \mathbf{R},\ T)$ as the density of quasi-particles with momentum $\mathbf{p}$ at the space–time point $\mathbf{R}$, T. As we proceed, we

---

[b](*Original*) ‡The symbol $n\ (\mathbf{p},\ \mathbf{R},\ T)$ for the quasi-particle distribution function, rather than $f\ (\mathbf{p},\ \mathbf{R},\ T)$, is conventional in the literature of low-temperature fermion systems.

shall find that these quasi-particles behave very much like a system of weakly interacting particles.

For example, the quasi-particle distribution function obeys a simple Boltzmann equation. To derive this, we use the fact that

$$\left[\omega - \left(\frac{p^2}{2m} \right) - U - \Re \Sigma, 2\pi \delta \left(\omega - \left(\frac{p^2}{2m} \right) - U - \Re \Sigma \right) \right] = 0$$

to rewrite Eq. (12.8) in the form[c]

$$2\pi \delta \left(\omega - \left(\frac{p^2}{2m} \right) - U - \Sigma \right) \left[\omega - \left(\frac{p^2}{2m} \right) - U - \Sigma, n \right] = 0$$

$$(12.10)$$

It is possible to effect a considerable simplification in Eq. (12.10). First note that

$$\delta \left(\omega - \left(\frac{p^2}{2m} \right) - U - \Sigma \right) = \delta \left([\omega - U - E(\mathbf{p}, \mathbf{R}, T)] \left[1 - \frac{\partial \Sigma(\mathbf{p}, \omega; \mathbf{R}, T)}{\partial \omega} \right] \right)$$

$$= \frac{\delta(\omega - U(\mathbf{R}, T) - \Sigma(\mathbf{p}, \mathbf{R}, T))}{1 - \frac{\partial \Sigma(\mathbf{p}, \omega; \mathbf{R}, T)}{\partial \omega}} \qquad (12.11)$$

Thus, Eq. (12.10) can be written[d]

$$\frac{2\pi \delta(\omega - U(\mathbf{R}, T) - \Sigma(\mathbf{p}, \mathbf{R}, T))}{1 - \frac{\partial \Sigma(\mathbf{p}, \omega; \mathbf{R}, T)}{\partial \omega}} \left\{ \left[1 - \frac{\partial \Sigma(\mathbf{p}, \omega; \mathbf{R}, T)}{\partial \omega} \right] \frac{\partial n(\mathbf{p}, \mathbf{R}, T)}{\partial T} \right.$$

$$+ \nabla_{\mathbf{p}} \left[\left(\frac{p^2}{2m} \right) + U(\mathbf{R}, T) + \Sigma(\mathbf{p}, \omega; \mathbf{R}, T) \right] \cdot \nabla_{\mathbf{R}} n(\mathbf{p}, \mathbf{R}, T)$$

$$\left. - \nabla_{\mathbf{R}} \left[\left(\frac{p^2}{2m} \right) + U(\mathbf{R}, T) + \Sigma(\mathbf{p}, \omega; \mathbf{R}, T) \right] \cdot \nabla_{\mathbf{p}} n(\mathbf{p}, \mathbf{R}, T) \right\} = 0$$

$$(12.12)$$

Now

$$\left\{ \nabla_{\mathbf{p}} \left[\left(\frac{p^2}{2m} \right) + U(\mathbf{R}, T) + \Sigma(\mathbf{p}, \omega; \mathbf{R}, T) \right] \right\}_{\omega = U(\mathbf{R}, T) + E(\mathbf{p}, \mathbf{R}, T)}$$

$$= \nabla_{\mathbf{p}} E(\mathbf{p}, \mathbf{R}, T) - \left[\frac{\partial \Sigma(\mathbf{p}, \omega; \mathbf{R}, T)}{\partial \omega} \right]_{\omega = U(\mathbf{R}, T) + E(\mathbf{p}, \mathbf{R}, T)} \nabla_{\mathbf{p}} E(\mathbf{p}, \mathbf{R}, T)$$

---

[c](*Original*) §Since we are assuming that $\Sigma(\mathbf{p}, z; \mathbf{R}, T)$ is real near $z = \mu$, we shall drop the $\Re$ in $\Re \Sigma(\mathbf{p}, \omega; \mathbf{R}, T)$ henceforth in this chapter.

[d]In the original text, the equation number was (12.7).

and also

$$\left\{ \nabla_{\mathbf{R}} \left[\left(\frac{p^2}{2m} \right) + U \left(\mathbf{R}, T \right) + \Sigma \left(\mathbf{p}, \omega; \mathbf{R}, T \right) \right] \right\}_{\omega=U+E}$$
$$= \nabla_{\mathbf{R}} \left(E + U \right) \left(1 - \frac{\partial \Sigma}{\partial \omega} \right)_{\omega=U+E}$$

Therefore, Eq. (12.12) can be written in the much simpler form

$$2\pi\delta \left(\omega - E \left(\mathbf{p}, \mathbf{R}, T \right) \right) \left[\frac{\partial n \left(\mathbf{p}, \mathbf{R}, T \right)}{\partial T} + \nabla_{\mathbf{p}} E \left(\mathbf{p}, \mathbf{R}, T \right) \cdot \nabla_{\mathbf{R}} n \left(\mathbf{p}, \mathbf{R}, T \right) \right.$$
$$\left. - \nabla_{\mathbf{R}} E \left(\mathbf{p}, \mathbf{R}, T \right) \cdot \nabla_{\mathbf{p}} n \left(\mathbf{p}, \mathbf{R}, T \right) \right] = 0 \quad \text{for } \omega \approx \mu$$
$$(12.13)$$

Consequently, the quasi-particle distribution function satisfies the Boltzmann equation

$$\frac{\partial n}{\partial T} + \nabla_{\mathbf{p}} E \cdot \nabla_{\mathbf{R}} n - \nabla_{\mathbf{R}} E \cdot \nabla_{\mathbf{p}} n - \nabla_{\mathbf{R}} U \cdot \nabla_{\mathbf{p}} n = 0 \qquad (12.14)$$

12.2 Conservation Laws

The response of the system to a slowly varying external disturbance can be described in terms of the quasi-particles, whose distribution function is determined by the Boltzmann equation (12.14). From this Boltzmann equation, we can derive the forms of the conservation laws appropriate to a very low-temperature fermion system. These conservation laws will provide an identification of physical quantities like the number density, the momentum density, and the energy density in terms of the quasi-particle distribution function. Moreover, they will give a further confirmation of the quasi-particle picture.

We recall that the differential number conservation law is

$$\frac{\partial}{\partial T} \langle \hat{n} \left(\mathbf{R}, T \right) \rangle_U + \nabla \cdot \left\langle \hat{\mathbf{j}} \left(\mathbf{R}, T \right) \right\rangle_U = 0 \qquad (12.15)$$

To obtain a result that we can identify with this number conservation law, we integrate Eq. (12.14) over all momenta $\mathbf{p}$, and find

$$\int \frac{d\mathbf{p}}{(2\pi)^3} \frac{\partial n\,(\mathbf{p}, \mathbf{R}, T)}{\partial T} + \int \frac{d\mathbf{p}}{(2\pi)^3} \nabla_{\mathbf{p}} E\,(\mathbf{p}, \mathbf{R}, T) \cdot \nabla_{\mathbf{R}} n\,(\mathbf{p}, \mathbf{R}, T)$$

$$- \int \frac{d\mathbf{p}}{(2\pi)^3} \nabla_{\mathbf{R}} E\,(\mathbf{p}, \mathbf{R}, T) \cdot \nabla_{\mathbf{p}} n\,(\mathbf{p}, \mathbf{R}, T) = 0$$

$$(12.16)$$

The last term here can be converted into the form

$$\int \frac{d\mathbf{p}}{(2\pi)^3} \left[\nabla_{\mathbf{R}} \cdot \nabla_{\mathbf{p}} E\,(\mathbf{p}, \mathbf{R}, T)\right] n\,(\mathbf{p}, \mathbf{R}, T)$$

by an integration by parts, so that Eq. (12.16) becomes

$$\frac{\partial}{\partial T} \int \frac{d\mathbf{p}}{(2\pi)^3} n(\mathbf{p}, \mathbf{R}, T) + \nabla_{\mathbf{R}} \cdot \int \frac{d\mathbf{p}}{(2\pi)^3} \left[\nabla_{\mathbf{p}} E(\mathbf{p}, \mathbf{R}, T)\right] n(\mathbf{p}, \mathbf{R}, T) = 0$$

$$(12.17)$$

But the number density is the unique quantity constructible from $g^>$ and $g^<$ that satisfies a conservation law of the form of Eq. (12.15). Consequently, we can identify the first term in Eq. (12.17) with $\dfrac{\partial \langle \hat{n}\,(\mathbf{R}, T)\rangle_U}{\partial T}$ and the second term with $\nabla \cdot \left\langle \hat{\mathbf{j}}\,(\mathbf{R}, T)\right\rangle_U$. Thus

$$\langle \hat{n}\,(\mathbf{R}, T)\rangle_U = \int \frac{d\mathbf{p}}{(2\pi)^3}\, n\,(\mathbf{p}, \mathbf{R}, T) + n_0 \qquad (12.18)$$

$$\left\langle \hat{\mathbf{j}}\,(\mathbf{R}, T)\right\rangle_U = \int \frac{d\mathbf{p}}{(2\pi)^3} \left[\nabla_{\mathbf{p}} E\,(\mathbf{p}, \mathbf{R}, T)\right] n\,(\mathbf{p}, \mathbf{R}, T) + \mathbf{j}_0 \qquad (12.19)$$

The constants n_0 and $\mathbf{j}_0$ must be independent of time and space, respectively. Therefore, these constants must be independent of the distribution function $n\,(\mathbf{p}, \mathbf{R}, T)$. Since we shall only be interested in the variations in $\langle \hat{n}\rangle$ and $\langle \hat{\mathbf{j}}\rangle$ resulting from variations in the distribution function, we shall neglect these constants hereafter. Similarly, we can ignore the fact that $n\,(\mathbf{p}, \mathbf{R}, T)$ is ill-defined for p far from p_F. The only variations in $n\,(\mathbf{p}, \mathbf{R}, T)$ that we need consider are for p near p_F, and hence the integrals in Eqs. (12.18) and (12.19) will contribute only for p near p_F.

Equations (12.18) and (12.19) indicate the essential correctness of the quasi-particle picture. In Eq. (12.18), we see that the change in the density of particles is the integral over all momenta of the change in the density of quasi-particles with momentum $\mathbf{p}$. In Eq. (12.19), we see that the change in the total current is $\nabla_{\mathbf{p}} E (\mathbf{p}, \mathbf{R}, T)$, the velocity of a quasi-particle with momentum $\mathbf{p}$, times $n(\mathbf{p}, \mathbf{R}, T)$, the change in the density of quasi-particles with momentum $\mathbf{p}$, integrated over all momenta.

The momentum conservation law is

$$\frac{\partial}{\partial T} m \left\langle \hat{\mathbf{j}} (\mathbf{R}, T) \right\rangle_U + \nabla \cdot \mathfrak{T} (\mathbf{R}, T) = - \langle \hat{n} (\mathbf{R}, T) \rangle_U \nabla_{\mathbf{R}} U (\mathbf{R}, T)$$

(11.19)

To obtain the form of this law appropriate to the present situation, we multiply Eq. (12.14) by $\mathbf{p}$ and integrate over all momenta. Thus, we find

$$\frac{\partial}{\partial T} \int \frac{d\mathbf{p}}{(2\pi)^3} \mathbf{p} n(\mathbf{p}, \mathbf{R}, T) + \int \frac{d\mathbf{p}}{(2\pi)^3} \mathbf{p} \left\{ (\nabla_{\mathbf{p}} E) \cdot (\nabla_{\mathbf{R}} n) - (\nabla_{\mathbf{R}} E) \cdot (\nabla_{\mathbf{p}} n) \right\}$$
$$= - [\nabla_{\mathbf{R}} U (\mathbf{R}, T)] \langle \hat{n} (\mathbf{R}, T) \rangle_U$$

(12.20)

It is exceedingly plausible to identify the momentum density, $m \left\langle \hat{\mathbf{j}} (\mathbf{R}, T) \right\rangle_U$, with the integral of the momentum times the quasi-particle distribution function, i.e.,

$$\left\langle \hat{\mathbf{j}} (\mathbf{R}, T) \right\rangle_U = \int \frac{d\mathbf{p}}{(2\pi)^3} \frac{\mathbf{p}}{m} n (\mathbf{p}, \mathbf{R}, T)$$

(12.21)

This identification, as well as the identifications (12.18) and (12.19), can be put on a firm mathematical basis, but the arguments necessitate inquiring more deeply into the structure of the many-body perturbation theory than we care to at this point. We shall merely state that Eq. (12.21) can be shown to be a consequence of the momentum conservation law, while Eqs. (12.18) and (12.19) can be similarly derived from the number conservation law. Equation (12.21) is an alternative expression for the current, which should be compared with our earlier result, Eq. (12.19). Later we shall use the equality of these two expressions for the current in the calculation of the equilibrium value of $\nabla_{\mathbf{p}} E$.

Now let us consider the expression for the stress tensor that is derived by making use of identification (12.21) of the current. A comparison of the momentum conservation law (11.19) with Eq. (12.20) yields

$$\sum_{i=1}^{3} \frac{\partial}{\partial R_i} \mathfrak{T}_{ij} (\mathbf{R}, T) = \sum_{i=1}^{3} \int \frac{d\mathbf{p}}{(2\pi)^3} p_j \left[\frac{\partial E}{\partial p_i} \frac{\partial n}{\partial R_i} - \frac{\partial E}{\partial R_i} \frac{\partial n}{\partial p_i} \right]$$

(12.22)

By integrating the last term in Eq. (12.22) by parts, we can write

$$\sum_{i=1}^{3} \frac{\partial}{\partial R_i} \mathfrak{T}_{ij} (\mathbf{R}, T) = \sum_{i=1}^{3} \int \frac{d\mathbf{p}}{(2\pi)^3} \left[p_j \frac{\partial E}{\partial p_i} \frac{\partial n}{\partial R_i} + n \frac{\partial}{\partial p_i} \left(\frac{\partial E}{\partial R_i} p_j \right) \right]$$

$$= \sum_{i=1}^{3} \frac{\partial}{\partial R_i} \left[\int \frac{d\mathbf{p}}{(2\pi)^3} \left(p_j \frac{\partial E}{\partial p_i} + \delta_{ij} E \right) n \right]$$

$$- \int \frac{d\mathbf{p}}{(2\pi)^3} E \frac{\partial n}{\partial R_j} \tag{12.23}$$

If the right side of this equation is really to be the divergence of a tensor,

$$\int \frac{d\mathbf{p}}{(2\pi)^3} E (\mathbf{p}, \mathbf{R}, T) \nabla_R n (\mathbf{p}, \mathbf{R}, T)$$

must be the gradient of some scalar. Let us denote this scalar by the $\mathcal{E} (\mathbf{R}, T)$. Then $\mathcal{E} (\mathbf{R}, T)$ is defined by

$$\nabla_R \mathcal{E} (\mathbf{R}, T) = \int \frac{d\mathbf{p}}{(2\pi)^3} E (\mathbf{p}, \mathbf{R}, T) \nabla_R n (\mathbf{p}, \mathbf{R}, T) \tag{12.24}$$

$\mathcal{E} (\mathbf{R}, T)$ is a functional of $n (\mathbf{p}, \mathbf{R}, T)$ for all values of $\mathbf{p}$. And because

$$\Sigma (\mathbf{p}, \omega; \mathbf{R}, T) |_{\omega = U(\mathbf{R}, T) + E(\mathbf{p}, \mathbf{R}, T)}$$

can be expressed (as we saw in Chapter 11) as a functional of $n (\mathbf{p}'; \mathbf{R}, T)$ with no explicit dependence on $U (\mathbf{R}, T)$, $\mathcal{E} (\mathbf{R}, T)$ does not have any explicit dependence on U. Therefore, we can compute $\nabla_R \mathcal{E} (\mathbf{R}, T)$ as

$$\nabla_R \mathcal{E} (\mathbf{R}, T) = \int d\mathbf{p} \frac{\delta \mathcal{E} (\mathbf{R}, T)}{\delta n (\mathbf{p}, \mathbf{R}, T)} \nabla_R n (\mathbf{p}, \mathbf{R}, T) \tag{12.25}$$

By comparing Eqs. (12.24) and (12.25), we see that

$$E (\mathbf{p}, \mathbf{R}, T) = (2\pi)^3 \frac{\delta \mathcal{E} (\mathbf{R}, T)}{\delta n (\mathbf{p}, \mathbf{R}, T)} \tag{12.26}$$

Because the last term in Eq. (12.23) is the gradient of $\mathcal{E}$, we can solve this equation for $\mathfrak{T}$ to find

$$\mathfrak{T}_{ij} (\mathbf{R}, T) = \int \frac{d\mathbf{p}}{(2\pi)^3} \left[p_j \frac{\partial E (\mathbf{p}, \mathbf{R}, T)}{\partial p_i} + \delta_{ij} E (\mathbf{p}, \mathbf{R}, T) \right] n (\mathbf{p}, \mathbf{R}, T)$$

$$- \mathcal{E} (\mathbf{R}, T) \delta_{ij} \tag{12.27}$$

We can, by calculating $\frac{\partial \mathcal{E}}{\partial T}$, discover the physical interpretation of $\mathcal{E}$. From Eq. (12.26)

$$\frac{\partial \mathcal{E}\,(\mathbf{R},\,T)}{\partial T} = \int d\mathbf{p}\, \frac{\delta \mathcal{E}\,(\mathbf{R},\,T)}{\delta n\,(\mathbf{p},\,\mathbf{R},\,T)}\, \frac{\delta n\,(\mathbf{p},\,\mathbf{R},\,T)}{\partial T}$$

$$= \int \frac{d\mathbf{p}}{(2\pi)^3}\, E\,(\mathbf{p},\,\mathbf{R},\,T)\, \frac{\partial n\,(\mathbf{p},\,\mathbf{R},\,T)}{\partial T}$$

From the Boltzmann equation (12.14), we see that

$$\int \frac{d\mathbf{p}}{(2\pi)^3}\, E\,(\mathbf{p},\,\mathbf{R},\,T)\, \frac{\partial n\,(\mathbf{p},\,\mathbf{R},\,T)}{\partial T}$$

$$= - \int \frac{d\mathbf{p}}{(2\pi)^3}\, \left[E\nabla_\mathbf{p} E \cdot \nabla_\mathbf{R} n - E\nabla_\mathbf{R} \cdot \nabla_\mathbf{p} n \right] + \int \frac{d\mathbf{p}}{(2\pi)^3}\, E\nabla_\mathbf{R} U \cdot \nabla_\mathbf{p} n$$

$$= - \nabla_\mathbf{R} \cdot \int \frac{d\mathbf{p}}{(2\pi)^3}\, \left(E\nabla_\mathbf{p} E \right) n - \nabla_\mathbf{R} U \cdot \int \frac{d\mathbf{p}}{(2\pi)^3}\, \left(\nabla_\mathbf{p} E \right) n$$

so that[e]

$$\frac{\partial \mathcal{E}\,(\mathbf{R},\,T)}{\partial T} + \nabla_\mathbf{R} \cdot \int \frac{d\mathbf{p}}{(2\pi)^3}\, E\,(\mathbf{p},\,\mathbf{R},\,T)\, \left[\nabla_\mathbf{p} E\,(\mathbf{p},\,\mathbf{R},\,T) \right] n\,(\mathbf{p},\,\mathbf{R},\,T)$$

$$= - \nabla_\mathbf{R} U\,(\mathbf{R},\,T) \cdot \int \frac{d\mathbf{p}}{(2\pi)^3}\, \left[\nabla_\mathbf{p} E\,(\mathbf{p},\,\mathbf{R},\,T) \right] n\,(\mathbf{p},\,\mathbf{R},\,T)$$

$$(12.28)$$

This is exactly the form of an energy conservation law with an energy current

$$\mathbf{j}_\epsilon\,(\mathbf{R},\,T) = \nabla_\mathbf{R} \cdot \int \frac{d\mathbf{p}}{(2\pi)^3}\, E\,(\mathbf{p},\,\mathbf{R},\,T)\, \left[\nabla_\mathbf{p} E\,(\mathbf{p},\,\mathbf{R},\,T) \right] n\,(\mathbf{p},\,\mathbf{R},\,T)$$

$$(12.29)$$

equal to the sum over all momenta of the density of quasi-particles, times the energy of the quasi-particle E, times the quasi-particle velocity $\nabla_\mathbf{p} E$. The source term in the conservation law is

$$- \nabla_\mathbf{R} U\,(\mathbf{R},\,T) \cdot \int \frac{d\mathbf{p}}{(2\pi)^3}\, \left[\nabla_\mathbf{p} E\,(\mathbf{p},\,\mathbf{R},\,T) \right] n\,(\mathbf{p},\,\mathbf{R},\,T)$$

which is the power fed into the system. Hence, Eq. (12.28) becomes exactly the usual energy conservation law

$$\frac{\partial \mathcal{E}\,(\mathbf{R},\,T)}{\partial T} + \nabla \cdot \mathbf{j}_\epsilon\,(\mathbf{R},\,T) = - \nabla U\,(\mathbf{R},\,T) \cdot \left\langle \hat{\mathbf{j}}\,(\mathbf{R},\,T) \right\rangle_U \qquad (12.30)$$

---

[e]We insert the parenthesis $[\cdots]$ in the right-hand side for avoiding the differentiation confusion.

12.3 Thermodynamic Properties

It seems quite clear by this point that $\mathcal{E}(\mathbf{R}, T)$ is just the energy density. A final check on this point, we compute, in the case of equilibrium $[U(\mathbf{R}, T) = 0]$, the change in $\mathcal{E}(\mathbf{R}, T)$ resulting from a change in the chemical potential μ. In this situation

$$\delta\mathcal{E} = \int \frac{d\mathbf{p}}{(2\pi)^3} E(p)\delta n(p) \tag{12.31}$$

From the definition of $n(\mathbf{p}, \mathbf{R}, T)$ in equilibrium, at zero temperature

$$n(p) = f(E(p)) = \begin{cases} 0 & \text{for } E(p) > \mu \\ 1 & \text{for } E(p) < \mu \end{cases} \tag{12.32}$$

Therefore, all contributions to Eq. (12.31) come at $p = p_F$, where $E(p) = \mu$. Thus

$$\delta\mathcal{E} = \int \frac{d\mathbf{p}}{(2\pi)^3} \delta n(p)\mu = \mu\delta n$$

so that

$$\frac{d\mathcal{E}}{dn} = \mu \tag{12.33}$$

We, therefore, recover the thermodynamic relationship that at zero temperature, the derivative of the energy density with respect to the particle density is the chemical potential. This is but another indication that $\mathcal{E}$ is the energy density.

We would like to see how the other important thermodynamic quantities appear in this theory. To do this, let us note that the basic element of the theory, the quantity that can be calculated directly from Green's function, is $E(\mathbf{p}; \mathbf{R}, T)$, the quasi-particle energy expressed as a functional of the distribution function. From $E(\mathbf{p}, \mathbf{R}, T)$, we can calculate

$$f(\mathbf{p}, \mathbf{p}'; \mathbf{R}, T) = (2\pi)^3 \frac{\delta E(\mathbf{p}, \mathbf{R}, T)}{\delta n(\mathbf{p}', \mathbf{R}, T)} \tag{12.34}$$

Since $f(\mathbf{p}, \mathbf{p}'; \mathbf{R}, T)$ is a second variational derivative, and two such derivatives comment, it is symmetrical in $\mathbf{p}$ and $\mathbf{p}'$, i.e.,

$$f(\mathbf{p}, \mathbf{p}'; \mathbf{R}, T) = f(\mathbf{p}', \mathbf{p}; \mathbf{R}, T)$$

This second variational derivative of the energy is a kind of effective interaction. For example, in the Hartree approximation,

$$E\left(\mathbf{p}, \mathbf{R}, T\right) = \left(\frac{p^2}{2m}\right) + \int \frac{d\mathbf{p}'}{(2\pi)^3}\left[v - v\left(\left|\mathbf{p} - \mathbf{p}'\right|\right)\right] n\left(\mathbf{p}', \mathbf{R}, T\right)$$

Therefore,

$$\mathcal{E}\left(\mathbf{R}, T\right) = \int \frac{d\mathbf{p}}{(2\pi)^3}\frac{p^2}{2m} n\left(\mathbf{p}', \mathbf{R}, T\right)$$

$$+ \frac{1}{2}\int \frac{d\mathbf{p}}{(2\pi)^3}\frac{d\mathbf{p}'}{(2\pi)^3} n\left(\mathbf{p}, \mathbf{R}, T\right)\left[v - v\left(\left|\mathbf{p} - \mathbf{p}'\right|\right)\right]$$

and

$$f\left(\mathbf{p}, \mathbf{p}'; \mathbf{R}, T\right) = \left[v - v\left(\left|\mathbf{p} - \mathbf{p}'\right|\right)\right]$$

Unfortunately, this is the last case in which we can obtain any moderately simple forms for E, $\mathcal{E}$, and f. For example, in the Born collision approximation $\Sigma_c\left(\mathbf{p}, z; \mathbf{R}, T\right)$ is expressed as complicated integrals of products of $g^>\left(\mathbf{p}', \omega'; \mathbf{R}, T\right)$ and $g^>\left(\mathbf{p}'', \omega''; \mathbf{R}, T\right)$. Through the contribution of these integrals for frequencies near $\mu\left(\mathbf{R}, T\right)$, $\Sigma_c\left(\mathbf{p}, z; \mathbf{R}, T\right)$ gains a dependence on $n\left(\mathbf{p}', \mathbf{R}, T\right)$. Also Σ_c depends on a for all frequencies, and a in turn is expressed in terms of Σ. Thus, a and E turn out to depend on n in a very complex implicit fashion. But even though we cannot obtain simple expressions for $\mathcal{E}$, E, and f, we can use the theory to derive some interesting general relations between these quantities.

In equilibrium, $E\left(\mathbf{p}, \mathbf{R}, T\right) = E(p)$. All the interesting properties of the system are determined by the distribution function for momenta near p_F. To find these properties, we need to know the behavior of $E(p)$ near $p = p_F$. In particular, we should know the effective mass m^*, defined by

$$E(p) = \mu + \frac{p^2 - p_F^2}{2m^*} \quad \text{near } p = p_F \tag{12.35}$$

We can express this effective mass in terms of $f\left(\mathbf{p}, \mathbf{p}'\right)$ by making use of the fact that Eqs. (12.19) and (12.21) are both valid expressions for the current $\left\langle \hat{\mathbf{j}} (\mathbf{R}, T) \right\rangle$. We have[f]

$$\left\langle \hat{\mathbf{j}} (\mathbf{R}, T) \right\rangle = \int \frac{d\mathbf{p}}{(2\pi)^3} \frac{\mathbf{p}}{m} n (\mathbf{p}, \mathbf{R}, T) \tag{12.21}$$

$$= \int \frac{d\mathbf{p}}{(2\pi)^3} \nabla_{\mathbf{p}} E (\mathbf{p}, \mathbf{R}, T) n (\mathbf{p}, \mathbf{R}, T) \tag{12.19}$$

By taking the variational derivative of this equation with respect to $n (\mathbf{p}, \mathbf{R}, T)$, we find

$$\frac{\mathbf{p}}{m} = \nabla_{\mathbf{p}} E (\mathbf{p}, \mathbf{R}, T) + \int \frac{d\mathbf{p}'}{(2\pi)^3} \left[\nabla_{\mathbf{p}'} f (\mathbf{p}, \mathbf{p}'; \mathbf{R}, T) \right] n (\mathbf{p}', \mathbf{R}, T)$$

In equilibrium, this becomes[g]

$$\frac{\mathbf{p}}{m} = \frac{\mathbf{p}}{m^*} - \int \frac{d\mathbf{p}'}{(2\pi)^3} f (\mathbf{p}, \mathbf{p}') \nabla_{\mathbf{p}'} n (p') \tag{12.36}$$

But

$$n(p) = \begin{cases} 0 & p > p_F \\ 1 & p < p_F \end{cases}$$

so that

$$\frac{\mathbf{p}}{m} = \frac{\mathbf{p}}{m^*} - \int \frac{d\mathbf{p}'}{(2\pi)^3} f (\mathbf{p}, \mathbf{p}') \frac{\mathbf{p}'}{p'} \delta (p' - p_F) \tag{12.37}$$

At $p = p_F$ and $p' = p_F$, $f (\mathbf{p}, \mathbf{p}')$ depends only on $\cos \theta = \frac{\mathbf{p} \cdot \mathbf{p}'}{p_F^2}$. Thus, at $p = p_F$, we can write Eq. (12.37) as

$$\frac{1}{m} = \frac{1}{m^*} - \frac{1}{p_F} \int \frac{d\mathbf{p}'}{(2\pi)^3} f (\cos \theta) \cos \theta \delta (p' - p_F)$$

or

$$\frac{1}{m} = \frac{1}{m^*} - \frac{p_F}{2\pi^2} \int_{-1}^{1} \frac{d (\cos \theta)}{2} f (\cos \theta) \cos \theta \tag{12.38}$$

This expression relates the effective mass to a moment of the effective two-particle interaction. For example, in the Hartree–Fock approximation, this gives the effective mass as

$$\frac{1}{m} = \frac{1}{m^*} - \frac{p_F}{2\pi^2} \int_{-1}^{1} d (\cos \theta) \left\{ v(0) - v \left(p_F \sqrt{2 - 2 \cos \theta} \right) \right\} \cos \theta$$

---

[f]The equation numbers were not written in the original text. Here those numbers are written intentionally for providing better understanding.

[g]This can be done by the integration by parts on the integral term of the right-hand side.

Another thermodynamic quantity of some importance is the thermodynamic derivative $\frac{dn}{d\mu}$. From Eq. (12.18), the change in n can be written as

$$dn = \int \frac{d\mathbf{p}'}{(2\pi)^3} \, dn \, (\mathbf{p})$$

But from Eq. (12.32)

$$dn \, (\mathbf{p}) = -d \, [E(p) - \mu] \, \delta \, (E(p) - \mu)$$

Since $dn \, (\mathbf{p})$ depends only on p when we change μ, we have

$$dE(p) = \int \frac{d\mathbf{p}'}{(2\pi)^3} \, f \, (\mathbf{p}, \mathbf{p}') \, dn \, (p')$$

$$= \int_{-1}^{1} \frac{d \, (\cos\theta)}{2} \, f \, (\cos\theta) \int \frac{d\mathbf{p}'}{(2\pi)^3} \, dn \, (p')$$

$$= \int_{-1}^{1} \frac{d \, (\cos\theta)}{2} \, f \, (\cos\theta) \, dn$$

Thus

$$dn(p) = -\delta \, (E(p) - \mu) \left[dn \int_{-1}^{1} \frac{d \, (\cos\theta)}{2} f \, (\cos\theta) - d\mu \right]$$

and

$$dn = \int \frac{d\mathbf{p}}{(2\pi)^3} \, \delta \, (E(p) - \mu) \left[d\mu - \int_{-1}^{1} \frac{d \, (\cos\theta)}{2} f \, (\cos\theta) \, dn \right]$$

Because $\delta \, (E(p) - \mu) = \delta \left(\left(\frac{p^2}{2m^*} \right) - \left(\frac{p_F^2}{2m^*} \right) \right)$,

$$dn = \frac{m^* p_F}{2\pi^2} \left[d\mu - dn \int_{-1}^{1} \frac{d \, (\cos\theta)}{2} f \, (\cos\theta) \right]$$

and

$$\frac{dn}{d\mu} = \left(\frac{p_F}{2m} \right) \left[\frac{1}{m^*} + \frac{p_F}{2\pi^2} \int_{-1}^{1} \frac{d \, (\cos\theta)}{2} f \, (\cos\theta) \right]^{-1}$$

If we make use of expression (12.36) for m^*, we find

$$\frac{dn}{d\mu} = \left[\frac{2\pi^2}{m^* p_F} + \int_{-1}^{1} \frac{d \, (\cos\theta)}{2} f \, (\cos\theta) \, (1 - \cos\theta) \right]^{-1} \qquad (12.39)$$

as our expression for the thermodynamic derivative in terms of the effective two-particle interaction.

Expressions (12.38) and (12.39) were originally derived by Landau. He goes on to use the basic equations we have written here to derive all the properties of a low-temperature normal fermion system, including the existence of zero sound. Since we feel that we cannot hope to surpass the clarity and beauty of Landau's original presentation, we strongly suggest that the reader refer to his papers cited in References and Supplementary Reading.

Chapter 13

Shielded Potential

13.1 Green's Function Approximation for Coulomb Gas

In our discussion of the random phase approximation, we saw that the particles in a Coulomb system move so as to produce a decided shielding effect. They reduce the effect of slowly varying external forces applied to the system. In particular, the applied field $U\left(\mathbf{R}, T\right)$ produces the reduced total potential field[a]

$$U_{\text{eff}}\left(\mathbf{R}, T\right) = U\left(\mathbf{R}, T\right) + \int d\mathbf{R}' \ \frac{e^2}{|\mathbf{R} - \mathbf{R}'|} \left(\langle \hat{n}\left(\mathbf{R}', T\right)\rangle - n\right)$$

$$= U\left(\mathbf{R}, T\right) + \int d\mathbf{R}' \ \frac{e^2}{|\mathbf{R} - \mathbf{R}'|}$$
$$\times \left[\pm i \left(2S + 1\right) G^<\left(\mathbf{R}', T; \mathbf{R}', T; U\right) - n\right] \qquad (13.1)$$

The constant n, the average density, represents the subtraction of the uniform background. The $(2S + 1)$ comes from summing over the spin degree of freedom [cf. Eq. (8.3a)]. The main application of this chapter will be to an electron gas for which $(2S + 1) = 2$.

---

[a]The coordinate variable $\mathbf{R}'$ in $G^<$ was $\mathbf{R}$, which was a typographic error, in the original text.

Annotations to Quantum Statistical Mechanics
In-Gee Kim
Copyright © 2018 Pan Stanford Publishing Pte. Ltd.
ISBN 978-981-4774-15-4 (Hardcover), 978-1-315-19659-6 (eBook)
www.panstanford.com

The reduction in the applied field is measured by the dielectric response function

$$K(1, 2) = \left[\frac{\delta U_{\text{eff}}(1)}{\delta U(2)} \right] \tag{13.2}$$

In fact, the Fourier transform of K goes to zero in the low-wave-number, low-frequency limit [cf. Eq. (8.32) for example], implying that the applied field is completely shielded out in this limit.

Now all of the approximations we have discussed so far have been derived by expanding G_2 or Σ in a power series in V and G. In Chapter 6, these expansions were derived by considering quantities such as $\delta\Sigma(1, 1'; U)/\delta U(2)$ to be small in comparison with $\delta(2 - 1')\delta(1 - 2)$. This kind of approximation is certainly wrong in a Coulomb system. To see this, we should note that to lowest order

$$\frac{\delta\Sigma(1, 1'; U)}{\delta U(2)} = \frac{\delta\Sigma_{\text{Hartree}}(1, 1'; U)}{\delta U(2)}$$

$$= \frac{\delta}{\delta U(2)} [U_{\text{eff}}(1) - U(1)]\delta(1 - 1')$$

Then

$$\frac{\delta\Sigma(1, 1'; U)}{\delta U(2)} = \delta(1 - 1')[K(1, 2) - \delta(1 - 2)]$$

But we have already said that K can usually be considered to be a small quantity, in the sense that its Fourier transform is usually much less than one. Therefore, in the lowest approximation in a Coulomb system,

$$\frac{\delta\Sigma(1, 1'; U)}{\delta U(2)} \approx -\delta(1 - 1')\delta(1 - 2)$$

Clearly, then we cannot use approximations derived from the statement

$$\frac{\delta\Sigma(1, 1'; U)}{\delta U(2)} \ll \delta(1 - 1')\delta(1 - 2)$$

We shall instead derive approximations for the Coulomb system by considering how functions change when U_{eff} is changed. There is much physical sense in saying that the relevant quantity for a Coulomb system is the total field through which the particles move, and not the applied field. We can expect that physical quantities should vary rather slowly in their dependence on the total field U_{eff}.

To derive approximations, we begin from the exact equation (6.24a)

$$
\begin{aligned}
G^{-1}\left(1, 1'; U\right) = {}& \left[i\frac{\partial}{\partial t_1} + \frac{\nabla_1^2}{2m} - U_{\text{eff}}(1)\right]\delta\left(1 - 1'\right) \\
& - i\int_0^{-i\beta} d\bar{1}d\bar{2}\, V\left(1 - 2\right)\left[\frac{\delta}{\delta U(2)}G\left(1, \bar{1}; U\right)\right]G^{-1}\left(\bar{1}, 1'; U\right)
\end{aligned}
$$

$$(13.3)$$

which holds for the time arguments in the imaginary interval $[0, -i\beta]$. Since the only occurrence of U in this equation is in U_{eff}, we see that G depends on U only in so far as it depends on U_{eff}. We shall, therefore, regard G as a functional of U_{eff}. We may handle variational derivatives very much as ordinary derivatives. Thus, we may use the chain rule for differentiating $G\left(U_{\text{eff}}\right)$ with respect to U, i.e.,

$$
\frac{\delta G\left(1, 1'; U_{\text{eff}}\right)}{\delta U(2)} = \int_0^{-i\beta} d3\, \frac{\delta U_{\text{eff}}(3)}{\delta U(2)}\frac{\delta G\left(1, 1'; U_{\text{eff}}\right)}{\delta U_{\text{eff}}(3)}
$$

$$(13.4)$$

The $(\mathbf{r}_3, t_3)$ integral is over all space and all times in the interval $[0, -i\beta]$, since G depends on U_{eff} in that entire region. Then we can rewrite Eq. (13.3) as

$$
\begin{aligned}
G^{-1}\left(1, 1'; U\right) = {}& \left[i\frac{\partial}{\partial t_1} + \frac{\nabla_1^2}{2m} - U_{\text{eff}}(1)\right]\delta\left(1 - 1'\right) \\
& - i\int_0^{-i\beta} d\bar{1}d3\, V_S\left(1, 3\right)\left[\frac{\delta G\left(1, \bar{1}; U_{\text{eff}}\right)}{\delta U_{\text{eff}}(3)}\right]G^{-1}\left(\bar{1}, 1'; U_{\text{eff}}\right)
\end{aligned}
$$

$$(13.5)$$

The quantity

$$
\begin{aligned}
V_S\left(1, 3\right) &= \int_0^{-i\beta} d2\, V\left(1 - 2\right)\frac{\delta U_{\text{eff}}(3)}{\delta U(2)} \\
&= \int_0^{-i\beta} d2\, V\left(1 - 2\right)K\left(3, 2\right)
\end{aligned}
$$

$$(13.6)$$

occurring in the above equation is interpreted simply as an effective time-dependent interaction between particles at the points 1 and 3. A particle at 1 can affect a particle at 3 in two ways. First, the particle at 3 can feel the effects of the potential $V\left(1 - 3\right)$ directly. Also the potential V can effect particles at 2, which in turn will change the potential they exert at point 1. This intermediate polarization of the medium leads to the time dependence of the effective interaction. The first effect is represented in the delta-function part of K and the second effect in the remainder in K. Because of the dynamic shielding, V_S is of much smaller than V; we shall call it the shielded potential.

To the lowest order, we can approximate Eq. (13.5) by neglecting $\delta G/\delta U_{\text{eff}}$. This yields the Hartree approximation

$$G^{-1} = G_0^{-1} - U_{\text{eff}}$$

To obtain the next-order result, we define

$$G^{-1}\left(1, 1'; U_{\text{eff}}\right) = \left[i\frac{\partial}{\partial t_1} + \frac{\nabla_1^2}{2m} - U_{\text{eff}}(1)\right]\delta\left(1 - 1'\right) - \Sigma'(1, 1'; U_{\text{eff}})$$

$$(13.7)$$

where Σ' differs from Σ in that it does not contain the Hartree self-energy. From Eq. (13.5), we find

$$\Sigma'\left(1, 1'; U_{\text{eff}}\right) = -i\int V_S\left(1, 3\right) G\left(1, \bar{1}\right) \frac{\delta G^{-1}\left(\bar{1}, 1'\right)}{\delta U_{\text{eff}}(3)}$$

$$= i V_S\left(1, 1'\right) G\left(1, 1'\right) + i\int V_S\left(1, 3\right) G\left(1, \bar{1}\right) \frac{\delta \Sigma'\left(\bar{1}, 1'\right)}{\delta U_{\text{eff}}(3)}$$

$$(13.8)$$

Our approximation will be to neglect $\delta\Sigma'/\delta U_{\text{eff}}$. Thus

$$\Sigma'\left(1, 1'; U_{\text{eff}}\right) = i V_S\left(1, 1'; U_{\text{eff}}\right) G\left(1, 1'; U_{\text{eff}}\right) \qquad (13.9)$$

We then need an expression for V_S. From its definition, Eq. (13.6) and the definition of U_{eff}, we write the exact equation

$$V_S\left(1, 3\right) = \int d2\, V\left(1 - 2\right) \frac{\delta U_{\text{eff}}(3)}{\delta U\left(2\right)}$$

$$= V\left(1 - 3\right) \pm i\left(2S + 1\right)\int V\left(1 - 2\right) \frac{\delta G\left(4, 4^+; U_{\text{eff}}\right)}{\delta U\left(2\right)} V\left(4 - 3\right)$$

$$= V\left(1 - 3\right) \pm i\left(2S + 1\right)\int V_S\left(1, 2\right) \frac{\delta G\left(4, 4^+\right)}{\delta U_{\text{eff}}(2)} V\left(4 - 3\right)$$

$$= V\left(1 - 3\right) \pm i\left(2S + 1\right)\int V_S\left(1, 2\right) G\left(4, 2\right) G\left(2, 4^+\right) V\left(4 - 3\right)$$

$$\pm i\left(2S + 1\right)\int V_S\left(1, 2\right) G\left(4, 5\right) \frac{\delta \Sigma'\left(5, 5'\right)}{\delta U_{\text{eff}}(2)} G\left(5', 4\right) V\left(4 - 3\right)$$

Again we neglect $\delta\Sigma'/\delta U_{\text{eff}}$. Thus

$$V_S(1, 3) = V\left(1 - 3\right) \pm i(2S + 1)\int V_S(1, 2)G(4, 2)G(3, 4)V\left(4 - 3\right)$$

$$(13.10)$$

We shall use the approximate Eqs. (13.9) and (13.10) to describe the one-particle Green's function in an electron gas.

Incidentally, if we started from the random phase approximation for K, we would arrive at essentially the same equation as Eq. (13.10) for V_S, but the G's would be replaced by Hartree Green's function. To see this, we recall that to derive the random phase approximation, we began with the Hartree approximation for G in the presence of U. Then to find K, we differentiated

$$U_{\text{eff}}(1) = U(1) + \int V(1-2) \left[\pm i(2S+1)G_H(2, 2^+) - n\right].$$

with respect to U. Here G_H is the Hartree Green's function. Thus

$$\frac{\delta U_{\text{eff}}(1)}{\delta U(3)} = \delta(1-3) + \int V(1-2)\left[\pm i(2S+1)\right]\frac{\delta G_H(2, 2^+)}{\delta U_{\text{eff}}(4)}\frac{\delta U_{\text{eff}}(4)}{\delta U(3)}$$

or

$$K(1, 3) = \delta(1-3) \pm i(2S+1)\int V(1-2)G_H(2, 4)G_H(4, 2)K(4, 3)$$

$$(13.11)$$

Then using the definition (13.6) of V_S, we find for V_S in this approximation

$$V_S(1, 3) = V(1-3) \pm i(2S+1)\int V_S(1, 2)G_H(4, 2)G_H(2, 4)V(4-3)$$

$$(13.12)$$

In some ways, it is better to use the Hartree Green's functions than the real G's to determine V_S. The derivation of the plasma pole in V_S (or equivalently in K) from Eq. (13.12) depends rather critically on the use of the properties of the Hartree Green's functions. A calculation shows that the plasmon pole appears in V_S in the approximation (13.10) but only at relatively high wavenumber. Therefore, the low wavenumber form of V_S is not given too well by Eq. (13.10). One would need a fancier equation than Eq. (13.10) to get the correct low wavenumber behavior of V_S, using real G's. Nonetheless, we shall use Eq. (13.10) in the evaluation of G.

Let us proceed to the analysis of the equilibrium Green's function. Since we are finished taking functional derivatives with respect to U and U_{eff}, we may set $U = 0$ in Eqs. (13.9) and (13.10). Then $U_{\text{eff}} = 0$, because we have included a uniform positive background to guarantee over-all electric neutrality of the system. This background has the effect of canceling the Hartree field of the electrons. Had we not included the background, U_{eff} would be given by

$$U_{\text{eff}}(\mathbf{r}, t) = \int d\mathbf{r}' \, \frac{ne^2}{|\mathbf{r} - \mathbf{r}'|}$$

where the integral extends over the entire volume of the system. Thus, U_{eff} would become infinite as the system became infinite.

As in Chapter 5, we wish to determine[b]

$$A(\mathbf{p}, \omega) = \frac{\Gamma(\mathbf{p}, \omega)}{[\omega - E(p) - \Re\Sigma_c(\mathbf{p}, \omega)]^2 + \left[\frac{\Gamma(\mathbf{p}, \omega)}{2}\right]^2}$$

where

$$\Gamma(\mathbf{p}, \omega) = \Sigma^>(\mathbf{p}, \omega) - \Sigma^<(\mathbf{p}, \omega)$$

$$\Sigma_c(\mathbf{p}, z) = \int \frac{d\omega'}{2\pi} \, \frac{\Gamma(\mathbf{p}, \omega')}{\omega' - z}$$

and

$$E(p) = \frac{p^2}{2m} \pm (2S + 1) \int \frac{d\mathbf{p}'}{(2\pi)^3} \, \frac{4\pi e^2}{|\mathbf{p} - \mathbf{p}'|^2} \langle \hat{n}(\mathbf{p}') \rangle$$

To write down an expression for Σ_c, we must note a few simple facts about V_S. The shielded potential obeys the periodic boundary condition

$$V_S(1 - 1')\big|_{t_1=0} = V_S(1 - 1')\big|_{t_1=-i\beta} \tag{13.13}$$

The difference $V_S - V$, like G, is composed of two analytic functions

$$V_S(1 - 1') - V(1 - 1') = \begin{cases} V_S^>(1 - 1') & \text{for } it_1 > it_{1'} \\ V_S^<(1 - 1') & \text{for } it_1 < it_{1'} \end{cases}$$

Therefore, V_S may be written in terms of a Fourier series, where the Fourier coefficient is

$$V_S(\mathbf{k}, \Omega_v) = V(\mathbf{k}) + \int \frac{d\omega}{2\pi} \, \frac{V_S^>(\mathbf{k}, \omega) - V_S^<(\mathbf{k}, \omega)}{\Omega_v - \omega} \tag{13.14}$$

$$\Omega_v = \frac{\pi v}{-i\beta}$$

We may then take Fourier coefficients of Eq. (13.10) and obtain

$$V_S(\mathbf{k}, \Omega_v) = V(\mathbf{k})[1 + L_1(\mathbf{k}, \Omega_v) V_S(\mathbf{k}, \Omega_v)] \tag{13.15}$$

---

[b]The necessary vector symbols are explicitly given in this chapter for clarity. In the original text, those are written in scalar format.

where $L_1 (\mathbf{k}, \Omega_v)$, the Fourier coefficient of $\pm (2S + 1) G(4, 2)$ $G(2, 4)$ is given by

$$L_1 (\mathbf{k}, \Omega_v) = \int \frac{d\omega}{2\pi} \frac{L_1^> (\mathbf{k}, \Omega_v) - L_1^< (\mathbf{k}, \Omega_v)}{\Omega - \omega} \tag{13.16}$$

and

$$L_1^\gtrless (\mathbf{k}, \omega) = (2S + 1) \int \frac{d\mathbf{p}'}{(2\pi)^3} \frac{d\omega'}{2\pi} G^\gtrless \left(\mathbf{p} + \frac{\mathbf{k}}{2}, \omega' + \frac{\omega}{2}\right) G^\lessgtr \left(\mathbf{p} - \frac{\mathbf{k}}{2}, \omega' - \frac{\omega}{2}\right) \tag{13.17}$$

It is now simple algebra to convince oneself that

$$V_S^> (\mathbf{k}, \omega) - V_S^< (\mathbf{k}, \omega) = 2\Im V_S (\mathbf{k}, \omega - i\epsilon)$$

$$= 2\Im \left[\frac{V(\mathbf{k})}{1 - V(\mathbf{k}) L_1 (\mathbf{k}, \omega - i\epsilon)} \right]$$

$$= |V_S (\mathbf{k}, \omega - i\epsilon)|^2 \times \left[L_1^> (\mathbf{k}, \omega) - L_1^< (\mathbf{k}, \omega) \right]$$

Since

$$V_S^> (\mathbf{k}, \omega) = e^{\beta \omega} V_S^< (\mathbf{k}, \omega) \quad \text{and} \quad L_1^> (\mathbf{k}, \omega) = e^{\beta \omega} L_1^< (\mathbf{k}, \omega)$$

It follows that

$$V_S^\gtrless (\mathbf{k}, \omega) = |V_S (\mathbf{k}, \omega - i\epsilon)|^2 L_1^\gtrless (\mathbf{k}, \omega) \tag{13.18}$$

We shall first find $\Sigma_c^> (\mathbf{p}, \omega)$, the collision rate of a particle with momentum $\mathbf{p}$ and energy ω. The collisional part of the self-energy differs from Σ' by the single-particle exchange energy. Thus, from Eq. (13.9),

$$\Sigma_c (1 - 1') = i \left[V_S (1 - 1') - V (1 - 1') \right] G (1 - 1')$$

so that

$$\Sigma^\gtrless (1 - 1') = i V_S^\gtrless (1 - 1') G^\gtrless (1 - 1')$$

and

$$\Sigma^\gtrless (\mathbf{p}, \omega) = \int \frac{d\mathbf{p}'}{(2\pi)^3} \frac{d\omega'}{2\pi} V_S^\gtrless (\mathbf{p} - \mathbf{p}', \omega - \omega') G^\gtrless (\mathbf{p}', \omega')$$

Now from the result (13.18), we find that

$$\Sigma^> (\mathbf{p}, \omega) = (2S + 1) \int \frac{d\mathbf{p}'}{(2\pi)^3} \frac{d\omega'}{2\pi} \frac{d\bar{\mathbf{p}}}{(2\pi)^3} \frac{d\bar{\omega}}{2\pi} \frac{d\bar{\mathbf{p}}'}{(2\pi)^3} \frac{d\bar{\omega}'}{2\pi}$$

$$\times (2\pi)^3 \delta (\mathbf{p} + \mathbf{p}' - \bar{\mathbf{p}} - \bar{\mathbf{p}}') 2\pi \delta (\omega + \omega' - \bar{\omega} - \bar{\omega}')$$

$$\times |V_S(\mathbf{p} - \bar{\mathbf{p}}, \omega - \bar{\omega} + i\epsilon)|^2 G^< (\mathbf{p}', \omega') G^> (\bar{\mathbf{p}}, \bar{\omega}) G^> (\bar{\mathbf{p}}', \bar{\omega}') \tag{13.19a}$$

Similarly, $\Sigma^< (\mathbf{p}, \omega)$, the collision rate of an excitation produced by removing a particle with momentum $\mathbf{p}$ and energy ω is

$$\Sigma^< (\mathbf{p}, \omega) = (2S + 1) \int \frac{d\mathbf{p}'}{(2\pi)^3} \frac{d\omega'}{2\pi} \frac{d\bar{\mathbf{p}}}{(2\pi)^3} \frac{d\bar{\omega}}{2\pi} \frac{d\bar{\mathbf{p}}'}{(2\pi)^3} \frac{d\bar{\omega}'}{2\pi}$$
$$\times (2\pi)^3 \delta (\mathbf{p} + \mathbf{p}' - \bar{\mathbf{p}} - \bar{\mathbf{p}}') 2\pi \delta (\omega + \omega' - \bar{\omega} - \bar{\omega}')$$
$$\times |V_S (\mathbf{p} - \bar{\mathbf{p}}, \omega - \bar{\omega} + i\epsilon)|^2$$
$$\times G^> (\mathbf{p}', \omega') G^< (\bar{\mathbf{p}}, \bar{\omega}) G^< (\bar{\mathbf{p}}', \bar{\omega}') \qquad (13.19b)$$

Notice that these results are exactly the same as those that emerged from the Born collision approximation (without exchange) except that in the collision cross section,

$$|V_S (\mathbf{p} - \bar{\mathbf{p}}, \omega - \bar{\omega} + i\epsilon)|^2$$

replaces

$$[v (\mathbf{p} - \bar{\mathbf{p}})]^2$$

This replacement is absolutely necessary when dealing with the Coulomb interaction. In this case, the first Born approximation differential cross section is proportional to the non-integrable function

$$[v(k)]^2 = \left[\frac{4\pi e^2}{k^2} \right]^2$$

There is a very small-angle scattering from the long-ranged Coulomb force, the total cross section diverges, and the lifetime Γ is infinite. However, using the shielded potential in the form of Eq. (13.15):

$$V_S (\mathbf{k}, \Omega) =$$

$$\frac{4\pi e^2}{k^2 - 4\pi e^2 (2S + 1) \int \frac{d\mathbf{p}}{(2\pi)^3} \frac{d\omega}{2\pi} \frac{d\omega'}{2\pi} \frac{G^> \left(\mathbf{p}+\frac{\mathbf{k}}{2},\omega\right) G^< \left(\mathbf{p}-\frac{\mathbf{k}}{2},\omega'\right) - G^< \left(\mathbf{p}+\frac{\mathbf{k}}{2},\omega\right) G^> \left(\mathbf{p}-\frac{\mathbf{k}}{2},\omega'\right)}{\Omega - \omega + \omega'}}$$
$$(13.20)$$

The low-momentum transfer divergence disappears, and the total cross section is quite finite. Thus, it is essential to use the shielded potential in discussing the Coulomb gas.

Not only is it essential to describe the scattering of particles in the medium by shielded potential, but it is quite reasonable to do so. $V_S (\mathbf{k}, \Omega)$ represents the total potential field produced by an externally added charge distribution proportional to

$$e^{i\mathbf{k}\cdot\mathbf{R} - i\Omega T}$$

But the system should not be able to distinguish very well between external perturbations and the fields produced by the particles within the medium. Therefore, if one adds a particle to the medium, its scattering should be described by the average total field it produces, i.e., V_S.

Another way of stating the same result is to notice that a particle moving through the medium produces a rather complicated disturbance. It tends to repel other particles from its immediate neighborhood so that at large distances, the net disturbance produces a small to repel particles in its neighborhood. In some sense, the total disturbance—added particle plus lowered density in the neighborhood—moves as a single entity. This entity is called a quasi-particle. The elementary scattering processes are not the collisions of particles but the collisions of quasi-particles. The effective potential between quasi-particles is not V $(1 - 2)$ but the shielded potential V_S $(1 - 2)$.

To determine A, we must solve Eqs. (13.12) and (13.19) self-consistently. It is extremely difficult to get very far in carrying out this solution. Hence, we shall leave this aspect of the problem here and turn a discussion of the equation of state of the Coulomb gas in the shielded potential approximation.

13.2 Calculation of the Equation of State of a Coulomb Gas

In Chapter 3, we described a method for computing the pressure of a system by means of an integral, Eq. (3.15), of the interaction energy over an interaction strength[c] parameter. This integral is

$$P - P_0 = -\frac{1}{\Omega} \int_0^1 \frac{d\lambda}{\lambda} \, \langle \lambda V \rangle_\lambda \quad \Omega = \text{volume of system} \quad (13.21)$$

where P_0 is the pressure of a non-interacting gas with the same values of the chemical potential and temperature. The interaction

---

[c]There is a typographic error on the word "strength" by "strengthn" in the original text.

energy may be expressed in terms of G_2 as

$$\langle \lambda V \rangle_\lambda = \left\langle \left(\frac{1}{2}\right) \int d\mathbf{r}_1 d\mathbf{r}_2 \hat{\psi}^\dagger(\mathbf{r}_1)\hat{\psi}^\dagger(\mathbf{r}_2)\lambda v\,(\mathbf{r}_1 - \mathbf{r}_2)\,\hat{\psi}(\mathbf{r}_2)\hat{\psi}(\mathbf{r}_1) \right\rangle_\lambda$$

$$= \frac{1}{2} \int d\mathbf{r}_1 d\mathbf{r}_2 \; \lambda v\,(\mathbf{r}_1 - \mathbf{r}_2)\, G_2 \left(12, 1^+2^+; \lambda\right)_{t_2 = t_1^+} \qquad (13.22)$$

Thus

$$P = P_0 + \int_0^1 \frac{d\lambda}{\lambda} \int d\mathbf{r}_2 \; \lambda v\,(\mathbf{r}_1 - \mathbf{r}_2)\, G_2 \left(12, 1^+2^+; \lambda\right)_{t_2 = t_1^+} \qquad (13.23)$$

This equation can be used to obtain an implicit form for the equation of state. Since the density n is given as

$$n = \left(\frac{\partial P}{\partial \mu}\right)_\beta$$

it implies

$$n = n_0 + \int_0^1 \frac{d\lambda}{\lambda} \int d\mathbf{r}_2 \; \lambda v\,(\mathbf{r}_1 - \mathbf{r}_2) \left[\frac{\partial}{\partial \mu} G_2 \left(12, 1^+2^+; \lambda\right)\right]_{\lambda\beta}$$

$$(13.24)$$

Equations (13.23) and (13.24) lead to expressions for the pressure and the density in terms of the variables β and μ. We shall now indicate briefly the structure of this result for a Coulomb gas.

For the approximation in the last section, the total interaction energy is

$$-\sum_{\text{spin}} \int d\mathbf{r}_2 \; v\,(\mathbf{r}_1 - \mathbf{r}_2)\, G_2 \left(12, 1^+2^+; \lambda\right)_{t_2 = t_1^+}$$

$$= \pm i\,(2S + 1) \int_0^{-i\beta} d\bar{1}\; \Sigma'\,(1 - \bar{1})\, G\,(\bar{1} - 1^+)$$

$$= \pm (2S + 1) \int \frac{d\mathbf{p}}{(2\pi)^3} \frac{d\mathbf{p}'}{(2\pi)^3} \frac{4\pi e^2}{|\mathbf{p} - \mathbf{p}'|^2} \int \frac{d\omega\, d\omega'}{2\pi\; 2\pi} G^<(\mathbf{p}, \omega)\, G^<(\mathbf{p}', \omega')$$

$$\pm i\,(2S + 1) \int \frac{d\mathbf{p}}{(2\pi)^3} \left[\int_0^{t_1} d\bar{t}_1 \; \Sigma^>(\mathbf{p}, t_1 - \bar{t}_1)\, G^<(\mathbf{p}, \bar{t}_1 - t_1)\right.$$

$$\left. - \int_{t_1}^{-i\beta} d\bar{t}_1 \; \Sigma^<(\mathbf{p}, t_1 - \bar{t}_1)\, G^>(\mathbf{p}, \bar{t}_1 - t_1)\right] \qquad (13.25)$$

Since the left side is independent of t_1, we may, for convenience, choose $t_1 = 0$. Then, using the Fourier transforms of $\Sigma^<$ and $G^>$, we find that the last term is

$$-(2S+1)\int\frac{d\mathbf{p}}{(2\pi)^3}\frac{d\omega}{2\pi}\frac{d\omega'}{2\pi}\frac{e^{\beta(\omega-\omega')}-1}{\omega-\omega'}\Sigma^<(\mathbf{p},\omega)\,G^>(\mathbf{p},\omega')$$

$$=-(2S+1)\int\frac{d\mathbf{p}}{(2\pi)^3}\frac{d\omega}{2\pi}\frac{d\omega'}{2\pi}\frac{\Sigma^>(\mathbf{p},\omega)G^<(\mathbf{p},\omega')-\Sigma^<(\mathbf{p},\omega)G^>(\mathbf{p},\omega')}{\omega-\omega'}$$

Thus[d]

$$P=P_0\mp(2S+1)\int_0^1\frac{d\lambda}{\lambda}\int\frac{d\mathbf{p}}{(2\pi)^3}\frac{d\mathbf{p}'}{(2\pi)^3}\frac{4\pi\lambda e^2}{|\mathbf{p}-\mathbf{p}'|^2}\langle\hat{n}(\mathbf{p})\rangle_\lambda\langle\hat{n}(\mathbf{p}')\rangle_\lambda$$

$$+(2S+1)\int_0^1\frac{d\lambda}{\lambda}\int\frac{d\omega}{2\pi}\int\frac{d\omega'}{2\pi}\int\frac{d\mathbf{p}}{(2\pi)^3}$$

$$\times\frac{\Sigma^>(\mathbf{p},\omega)\,G^<(\mathbf{p},\omega')-\Sigma^<(\mathbf{p},\omega)\,G^>(\mathbf{p},\omega')}{\omega-\omega'}\qquad(13.26)$$

where $\langle\hat{n}(\mathbf{p})\rangle_\lambda$ is the density of particles with a particular spin direction.

When we substitute the result (13.19) for $\Sigma^\gtrless$ into Eq. (13.26), we find

$$P=P_0\mp(2S+1)\int_0^1\frac{d\lambda}{\lambda}\int\frac{d\mathbf{p}}{(2\pi)^3}\frac{d\mathbf{p}'}{(2\pi)^3}\frac{4\pi\lambda e^2}{|\mathbf{p}-\mathbf{p}'|^2}\langle\hat{n}(\mathbf{p})\rangle_\lambda\langle\hat{n}(\mathbf{p}')\rangle_\lambda$$

$$+(2S+1)^2\int_0^1\frac{d\lambda}{\lambda}\int\frac{d\mathbf{p}d\omega}{(2\pi)^4}\frac{d\mathbf{p}'d\omega'}{(2\pi)^4}\frac{d\bar{\mathbf{p}}d\bar{\omega}}{(2\pi)^4}\frac{d\bar{\mathbf{p}}'d\bar{\omega}'}{(2\pi)^4}$$

$$\times(2\pi)^4\frac{\delta(\mathbf{p}+\mathbf{p}'-\bar{\mathbf{p}}-\bar{\mathbf{p}}')}{\omega+\omega'-\bar{\omega}-\bar{\omega}'}\left|V_S\left(\lambda,\mathbf{p}'-\bar{\mathbf{p}}',\omega'-\bar{\omega}'+i\epsilon\right)\right|^2$$

$$\times\left[G^>(\mathbf{p},\omega;\lambda)\,G^>(\mathbf{p}',\omega';\lambda)\,G^<(\bar{\mathbf{p}},\bar{\omega};\lambda)\,G^<(\bar{\mathbf{p}}',\bar{\omega}';\lambda)\right.$$

$$\left.-G^<(\mathbf{p},\omega;\lambda)\,G^<(\mathbf{p}',\omega';\lambda)\,G^>(\bar{\mathbf{p}},\bar{\omega};\lambda)\,G^>(\bar{\mathbf{p}}',\bar{\omega}';\lambda)\right]$$

$$=P_0+P_1+P_2\qquad(13.27)$$

Note incidentally the detailed similarity between the last term in Eq. (13.27) and a typical quantum mechanical second-order perturbation theory calculation of an energy shift. The factor

$$|V_S|^2\,\delta\left(\mathbf{p}+\mathbf{p}'-\bar{\mathbf{p}}-\bar{\mathbf{p}}'\right)$$

is the matrix element for a process

$$\mathbf{p}\omega+\mathbf{p}'\omega'\to\bar{\mathbf{p}}\bar{\omega}+\bar{\mathbf{p}}'\bar{\omega}'$$

---

[d] The equation number (13.26) was missing in the original text. In addition, the prime symbol $'$ at the one of the integral symbol $\int\frac{d\omega}{2\pi}$ was also missing in the original text.

The $G^<$'s are densities of initial states, the $G^>$'s are densities of available final states, and the factor

$$\left[\omega + \omega' - \bar{\omega} - \bar{\omega}'\right]$$

is the typical energy denominator that enters such a calculation.

The reason for this similarity is that for the particular case of a zero-temperature system, the pressure is simply[e]

$$P = -\left(\frac{1}{\Omega}\right)\left[\langle \hat{H} \rangle - \mu \langle \hat{N} \rangle\right] \tag{13.28}$$

This can be seen from the thermodynamic relation

$$TS = \langle \hat{H} \rangle - \mu \langle \hat{N} \rangle + P\Omega$$

Therefore, Eq. (13.26) also determines the ground-state energy. When the G's in Eq. (13.26) are replaced by G_0's, Eq. (13.26) leads to a calculation of the ground-state energy of an electron gas similar to that done by Gell-Mann and Brueckner.[f]

In general, there is no guarantee that the pressure determined by Eq. (13.27) will be the same as that determined by Eq. (3.12), an integral of the density over the chemical potential. It is true that these alternative methods will lead to identical results for all the approximations for G we have discussed up to now.[g] However, these methods require solving for G self-consistently, i.e., as the solution to a nonlinear integral equation. The closer we come to self-consistency in the approximate solution to these nonlinear equations, the closer we will come to making the results of the μ' integrations for P outlined in Chapter 3 correspond to the result (13.27).

To carry the evaluation of the pressure further, we replace the G's that appear in Eq. (13.27) by G_0's. There are then two cases in which we can get results simply. The first is a zero-temperature electron gas, and the second is a classical system.

---

[e]There is no equation number (13.28) in the original text. The equation number (13.28) is assigned to this equation from the context.

[f]*(Original)* ‡M. Gell-Mann and K. Brueckner, *Phys. Rev.*, **106**, 364 (1957).

[g]*(Original)* §The proof of this result will be published shortly by one of us (GB) in the *Physical Review*. *(Author)* This article is indeed published in Gordon Baym, *Phys. Rev.* **127**, 1391 (1962).

For zero-temperature electrons, the Hartree–Fock term in the pressure becomes simply the negative of the exchange energy. Here $2S + 1 = 2$. Thus, setting $p_F^0 = \sqrt{2m\mu}$,

$$
P_1 = 2 \int_0^1 \frac{d\lambda}{\lambda} \int_{p < p_F^0} \frac{d\mathbf{p}}{(2\pi)^3} \int_{p' < p_F^0} \frac{d\mathbf{p}'}{(2\pi)^3} \frac{4\pi e^2 \lambda}{|\mathbf{p} - \mathbf{p}'|^2}
$$

$$
= \frac{e^2}{2\pi^3} \int_0^{p_F^0} p^2 dp \int_0^{p_F^0} p'^2 dp' \int_{-1}^1 d\alpha \, \frac{1}{p^2 + p'^2 - 2\alpha p p'}
$$

$$
= \frac{e^2 \left(p_F^0\right)^4}{4\pi^3} \tag{13.29}
$$

To the degree of accuracy to which we shall work, it makes no difference if we replace the p_F^0 in P_1 and P_2 by the Fermi momentum p_F, which is conventionally defined by[h] $p_F = (3\pi^2 n)^{1/3}$. Therefore, we can write the result (13.29) as

$$
P_1 = \frac{e^2 p_F^4}{4\pi^3} \tag{13.29a}
$$

The density n of an interacting gas with a certain value of μ is not equal to the density n_0 of a free gas with the same value of μ. Therefore, $p_F = (3\pi^2 n)^{1/3}$, different from[i] $p_F^0 = (3\pi^2 n_0)^{1/3}$. For example, in the Hartree approximation, $p_F^0 = \sqrt{2m\mu}$, whereas $p_F = \sqrt{2m\,(\mu - nv)}$. In replacing the G's by G_0's in the collision term, we write

$$
G^> (p, \omega) \rightarrow 2\pi\delta \left(\omega - \frac{p^2}{2m} \right) \left[1 \pm f \left(\frac{p^2}{2m} \right) \right]
$$

and

$$
G^< (p, \omega) \rightarrow 2\pi\delta \left(\omega - \frac{p^2}{2m} \right) f \left(\frac{p^2}{2m} \right)
$$

We make the change of variables

$$
\mathbf{p} \rightarrow \mathbf{p} - \frac{\mathbf{k}}{2} \equiv \mathbf{p}_- \quad \bar{\mathbf{p}} \rightarrow \mathbf{p} + \frac{\mathbf{k}}{2} \equiv \mathbf{p}_+
$$

$$
\mathbf{p}' \rightarrow \mathbf{p}' + \frac{\mathbf{k}}{2} \equiv \mathbf{p}'_+ \quad \bar{\mathbf{p}}' \rightarrow \mathbf{p} - \frac{\mathbf{k}}{2} \equiv \mathbf{p}'_-
$$

---

[h]The power factor $1/3$ was typographically wrong by $1/2$ in the original text.
[i]The subscript 0 at p_F^0, which was omitted in the original text, is necessary.

in the integral. Then the collision term in the pressure becomes

$$P_2 = 4 \int \frac{d\lambda}{2\lambda} \int \frac{d\mathbf{p}}{(2\pi)^3} \frac{d\mathbf{p}'}{(2\pi)^3} \frac{d\mathbf{k}}{(2\pi)^3}$$

$$\times f\left(\frac{p_+^2}{2m}\right) f\left(\frac{(p_-')^2}{2m}\right) \left[1 \pm f\left(\frac{p_-^2}{2m}\right)\right] \left[1 \pm f\left(\frac{(p_+')^2}{2m}\right)\right]$$

$$\times \frac{2\left|V_S\left(\mathbf{k}, \left(\frac{\mathbf{p}'\cdot\mathbf{k}}{m}\right) + i\epsilon\,\lambda\right)\right|^2}{\frac{(\mathbf{p}-\mathbf{p}')\cdot\mathbf{k}}{m}} \tag{13.30}$$

The extra factor two arises from the use of the symmetry of

$$\left|V_S\left(\mathbf{k}, \left(\frac{\mathbf{p}'\cdot\mathbf{k}}{m}\right) + i\epsilon\,\lambda\right)\right|^2$$

under $\mathbf{k} \to -\mathbf{k}$.

We recall that in the discussion of the random phase approximation, we found

$$K\left(k, \Omega = 0\right) = \frac{k^2}{k^2 + \left(\frac{1}{r_D}\right)^2}$$

for k small. Thus, in this approximation

$$V_S\left(k, \Omega = 0; \lambda\right) = \frac{4\pi e^2 \lambda}{k^2 + \left(\frac{1}{r_D(\lambda)}\right)^2}$$

We may expect that for k^{-1} much less than the screening radius r_D, $V_S\left(\lambda\right)$ is nearly equal to

$$V\left(\lambda\right) = \frac{4\pi e^2 \lambda}{k^2}$$

To see the qualitative effects of the shielding, we shall replace the shielding in Eq. (13.30) by a cutoff at low momentum transfer k. We take as a cutoff $k_{min} = \frac{1}{r_D}$. For $k > \frac{1}{r_D}$, we take $V_S = \frac{4\pi\lambda e^2}{k^2}$. Then Eq. (13.30) becomes

$$P_2 = 2\left(4\pi e^2\right) \int_0^1 d\lambda\lambda \int \frac{d\mathbf{p}}{(2\pi)^3} \frac{d\mathbf{p}'}{(2\pi)^3}$$

$$\times \int_{k > r_D^{-1}(\lambda)} \frac{d\mathbf{k}}{(2\pi)^3} \frac{1}{k^4} \frac{1}{(\mathbf{p}' - \mathbf{p})\cdot\left(\frac{\mathbf{k}}{m}\right)}$$

$$\times f\left(\frac{p_+^2}{2m}\right) f\left(\frac{(p_-')^2}{2m}\right) \left[1 \pm f\left(\frac{p_-^2}{2m}\right)\right] \left[1 \pm f\left(\frac{(p_+')^2}{2m}\right)\right] \tag{13.31}$$

If the $\mathbf{k}$ integral were not cut off below, it would be divergent.

Let us evaluate this for fermions at zero temperature. For large r_D, $\frac{1}{r_D} \ll p_F$, the main contribution to this integral comes from $k \ll p_F$. Therefore, we can cut off the above integral at $k = p_F$ and make approximations appropriate to small k within the integrand. In particular, we note that the factor

$$ f\left(\frac{p_+^2}{2m}\right)\left[1 \pm f\left(\frac{(p'_+)^2}{2m}\right)\right] $$

is nonzero only when $\mathbf{p} + \frac{\mathbf{k}}{2}$ is within the Fermi sphere, $\left|\mathbf{p} + \frac{\mathbf{k}}{2}\right| < p_F$ and when $\mathbf{p} - \frac{\mathbf{k}}{2}$ is outside the Fermi sphere, $\left|\mathbf{p} - \frac{\mathbf{k}}{2}\right| > p_F$. This can only happen if p is close to p_F. Therefore, we can approximately write

$$ \mathbf{p} \cdot \mathbf{k} = p_F k \alpha $$

$$ \frac{\left(\mathbf{p} \pm \frac{\mathbf{k}}{2}\right)^2}{2m} = \frac{p^2}{2m} \pm \frac{p_F k \alpha}{2m} $$

where α is direction cosine between $\mathbf{k}$ and $\mathbf{p}$. We can also approximately write

$$ \int \frac{d\mathbf{p}}{(2\pi)^3} = \frac{1}{2\pi^2}\int_{-1}^{1}\frac{d\alpha}{2}\int_{0}^{\infty} dp\, p^2 \approx \frac{m p_F}{2\pi^2}\int\frac{d\alpha}{2}\int_{0}^{\infty} d\left(\frac{p^2}{2m}\right) $$

Thus, Eq. (13.31) becomes

$$ P_2 = \frac{4\pi e^2}{\pi^2}\left(\frac{m p_F}{2\pi^2}\right)^2 \int_{-\infty}^{\infty} dE_p \int_{-\infty}^{\infty} dE_{p'} \int_{-1}^{1}\frac{d\alpha}{2}\int_{-1}^{1}\frac{d\alpha'}{2}\int_{1/r_D}^{p_F}\frac{dk}{k^2} $$

$$ \times f\left(E_p + \frac{k p_F \alpha}{2m}\right)\left[1 - f\left(E_p - \frac{k p_F \alpha}{2m}\right)\right] $$

$$ \times f\left(E_{p'} - \frac{k p_F \alpha'}{2m}\right)\left[1 - f\left(E_{p'} + \frac{k p_F \alpha'}{2m}\right)\right]\frac{1}{\frac{p_F k}{m}(\alpha - \alpha')} $$

where $E_p = \frac{p^2}{2m}$. Now the integrals over E are easily evaluated, since f is either 1 or 0. In particular,

$$ \int dE_p\, f\left(E_p + \frac{x}{2}\right)\left[1 - f\left(E_p - \frac{x}{2}\right)\right] = \begin{cases} 0 & \text{for } x > 0 \\ -x & \text{for } x < 0 \end{cases} $$

so that

$$P_2 = 4 \left(\frac{m p_F e^2}{\pi^2} \right) \int_{-1}^0 \frac{d\alpha}{2} \int_0^1 \frac{d\alpha'}{2} \int_{1/r_D}^{p_F} \frac{dk}{k^2} \frac{\left(\frac{p_F k}{m} \right) |\alpha| \left(\frac{p_F k}{m} \right) |\alpha'|}{\frac{p_F k}{m} (\alpha' - \alpha)}$$

$$= \left(\frac{m p_F e^2}{\pi^2} \right) \frac{p_F}{m} \int_{1/r_D}^{p_F} \frac{dk}{k} \int_0^1 d\alpha \int_0^1 d\alpha' \frac{\alpha \alpha'}{\alpha' + \alpha}$$

$$= \left(\frac{m p_F e^2}{\pi^2} \right) \frac{p_F}{m} \ln (p_F r_D) \frac{2}{3} (1 - \ln 2)$$

From Eq. (8.35)

$$r_D{}^2 = \frac{\pi \hbar}{4 p_F} a_0 \sim \frac{1}{e^2}$$

Thus,

$$P_2 = -\frac{p_F^2}{3\pi^4} m e^4 (1 - \log 2) \left[\ln \frac{m e^2}{p_F} + \mathcal{O}(1) \right] \qquad (13.32)$$

Note the appearance of the $e^4 \ln e^2$ in this term.

Since $P_0 = (p_F^0)^5 / 15 m \pi^2$ for zero-temperature fermions (with spin), we find for Eq. (13.27):

$$P = \frac{(p_F^0)^5}{15 m \pi^2} + \frac{e^2 p_F^4}{4\pi^3} - \frac{p_F^2}{3\pi^4} m e^4 (1 - \log 2) \log \frac{m e^2}{p_F} + \cdots \quad (13.33)$$

To find an equation of state, we must now express P in terms of n by eliminating p_F^0 in Eq. (13.33). Using the thermodynamic identity $n = \left(\frac{\partial P}{\partial \mu} \right)_\beta$, we have

$$n = \frac{\partial P}{\partial p_F^0} \frac{\partial p_F^0}{\partial \mu}$$

$$= \frac{m}{p_F^0} \left\{ \frac{(p_F^0)^4}{3 m \pi^2} + \frac{\partial p_F}{\partial p_F^0} \left[\frac{e^2 p_F^3}{\pi^3} - \frac{p_F^2}{\pi^4} m e^4 (1 - \ln 2) \ln \frac{m e^2}{p_F} \right] \right\}$$

$$\approx \frac{(p_F^0)^3}{3\pi^2} + \frac{e^2 m p_F^2}{\pi^3} - \frac{m^2 p_F e^4}{\pi^4} (1 - \ln 2) \ln \left(\frac{m e^2}{p_F} \right)$$

The last two terms in this equation represent the change in the density from that of a non-interacting gas with the same value of the chemical potential. We must solve this equation for p_F^0 in terms of n. Since $p_F = (3\pi n)^{1/3}$,

$$p_F^0 = \left[1 - \frac{3 e^2 m}{\pi p_F} + \frac{3 m^2 e^4}{\pi^2 p_F^2} (1 - \ln 2) \ln \frac{e^2 m}{p_F} \right]^{1/3} p_F \qquad (13.34)$$

Substituting Eq. (13.34) into Eq. (13.33) and writing $(1 - X)^{5/3}$ as $1 - \frac{5}{3}X$, we discover the equation of state for the Coulomb gas:

$$P = \frac{np_F^2}{5m} - \frac{ne^2}{4\pi} p_F \qquad (13.35)$$

When the pressure is expressed as a function of n, instead of p_F^0, the $e^4 \ln e^2$ term fortuitously cancels out.

We can now use this equation of state to find the ground-state energy of the Coulomb gas. From Eq. (13.28), $E/\Omega = \mu n - P$. We evaluate μ in terms of n from Eq. (13.34) as

$$\mu = \frac{\left(p_F^0\right)^2}{2m} \approx \frac{p_F^2}{2m} - e^2 p_F + \frac{me^4}{p_F}(1 - \ln 2) \ln \frac{me^2}{p_F} \qquad (13.36)$$

so that

$$\frac{E}{\Omega} = \left(\frac{3}{10}\right) np_F^2 - \frac{3}{4}\frac{e^2}{4\pi} np_F + \frac{nme^4}{\pi^2}(1 - \ln 2) \ln \frac{me^2}{p_F} \qquad (13.37)$$

It is customary in the literature to express results like this in terms of the Rydberg unit of energy,

$$\frac{e^2}{2a_0} = \frac{me^4}{2\hbar^2}$$

(a_0 = Bohr radius), and the dimensionless parameter r_s, which is essentially the ratio of the inter-particle spacing to the Bohr radius,

$$r_s = \left(\frac{3}{4\pi n}\right)^{1/3} \frac{me^2}{\hbar^2} = \left(\frac{9\pi}{4}\right)^{1/3} \frac{1}{p_F a_0}$$

Thus,

$$\frac{E}{\Omega} = \left[\frac{3}{5}\left(\frac{9\pi}{4}\right)^{2/3}\frac{1}{r_s^2} - \frac{3}{2\pi}\left(\frac{9\pi}{4}\right)^{1/3}\frac{1}{r_s} + \frac{2}{\pi^2}(1 - \ln 2)\ln r_s + \mathcal{O}(1)\right]\frac{me^4}{2\hbar^2}n$$

$$\approx \left[\frac{2.21}{r_s^2} - \frac{0.916}{r_s} + 0.0622 \ln r_s\right]\frac{me^4}{2\hbar^2}n \qquad (13.38a)$$

and

$$P = n\frac{me^4}{2\hbar^2}\left[\frac{2}{5}\left(\frac{9\pi}{4}\right)^{2/3}\frac{1}{r_s^2} - \frac{1}{2\pi}\left(\frac{9\pi}{4}\right)^{1/3}\frac{1}{r_s}\right] \qquad (13.38b)$$

These expressions are the first few terms in expansions of the energy and pressure in terms of r_s—expansions that are increasingly accurate in the high-density ($r_s \to 0$) limit. It is important to notice the appearance of the $e^4 \ln e^2$ term in the energy. It means that these expansions can only be asymptotic; they are not power series expansion. Such logarithms will appear in the expansion of any physical quantity in the Coulomb gas. Therefore, no physical quantity can be expanded in a power series in e^2.

There is Dyson's old argument why this should be so. If physical quantities could be expanded in a power series in e^2, the expansion would be just valid for negative e^2, an alternative Coulomb interaction, as for $e^2 > 0$. However, a purely attractive Coulomb interaction is indeed a very strange interaction; the system would be able to undergo extremely coherent processes.

One indication of this is the plasma pole, which we found near

$$\Omega^2 = \omega_p^2 = \frac{4\pi e^2}{m}$$

When e^2 becomes negative, this becomes a complex pole at $z = \pm i\sqrt{n|e^2|/m}$. Such a complex pole, as we have discussed in Chapter 8, leads to unstable behavior of the system, and this means that the Green's function analysis that we have given cannot be correct for $e^2 < 0$.

The next term in the expansion of the pressure is of the form (const) $\times n\left(me^4/2\hbar^2\right)$. Our expression, Eq. (13.27), gives only part of this term. The remainder comes from the term $\delta\Sigma'/\delta U_{\text{eff}}$, which we neglected in Eq. (13.8). To find the contribution to order e^4 from this term, we take $\Sigma'(1, 1') = iV_S(1, 1')G(1, 1')$ in the right side of Eq. (13.8) and keep only the $\delta G/\delta U_{\text{eff}}$ term. Then to order e^4, the correction term (13.9) is

$$\delta\Sigma'(1, 1'; U_{\text{eff}}) = -\int_0^{-i\beta} d2\,d3\,V(1-3)V(2-1')$$
$$\times G_0(1, 2)G_0(2, 3)G_0(3, 1') \qquad (13.39)$$

The contribution of this term to the pressure must be evaluated numerically.

This highly quantum mechanical formalism leads to reasonable results in the $\hbar \to 0$ limit. We could calculate P_2 directly from Eq. (13.30), but it is somewhat simpler to go back to our original equation (13.23). We wrote $P = P_0 + P_2$, where

$$P = \sum_{\text{integral variables 1 and 2}} \int \frac{d\lambda}{\lambda} d\mathbf{r}_2 \, \lambda v \, (\mathbf{r}_1 - \mathbf{r}_2) \qquad (13.40)$$
$$\times \left[G_2 \left(1, 2; 1^+, 2^+ \right) - G \left(1, 1^+ \right) G \left(2, 2^+ \right) \right]_{t_2 = t_1^+}$$

and again make use of the shielded potential approximation for G_2. We find

$$P_2 = \pm (2S + 1) \int_0^{-i\beta} d\bar{1} d\bar{2} K (1 - \bar{1}) V (\bar{1} - \bar{2}) G (\bar{2} - 1^+) G (1 - \bar{2}^+) \qquad (13.41)$$

where the dielectric function K is defined by

$$K (1 - 2) = \delta (1 - 2) \pm i (2S + 1) \int_0^{-i\beta} d\bar{1} d\bar{2} K (1 - \bar{1}) V (\bar{1} - \bar{2})$$
$$\times G (\bar{2} - 2) G (2 - \bar{2}) \qquad (13.42)$$

By comparing Eq. (13.41) with Eq. (13.42), we see that vG_2 may be simply expressed in terms of $K - 1$.

There is one complication. In Eq. (13.41), 1^+ and $\bar{2}^+$ signify that the $\delta (1 - \bar{1})$ term in $K (1 - \bar{1})$ should reproduce the exchange term

$$\cdots \int d\mathbf{r}_2 \, v \, (\mathbf{r}_1 - \mathbf{r}_2) \, G \left(\mathbf{r}_2 - \mathbf{r}_1, t_1 - t_1^+ \right) G \left(\mathbf{r}_1 - \mathbf{r}_2, t_1 - t_1^+ \right)$$

But in the integral in Eq. (13.42), the $\delta (1 - \bar{1})$ term in K yields

$$\lim_{t_2 \to t_1} \int d\mathbf{r}_2 \, v \, (\mathbf{r}_1 - \mathbf{r}_2) \, G \, (\mathbf{r}_2 - \mathbf{r}_1, t_1 - t_2) \, G \, (\mathbf{r}_1 - \mathbf{r}_2, t_2 - t_1)$$

which, because of the different equal-time limit of the G's, is not the same as the exchange term. Thus, to express vG_2 in terms of $K - 1$, we write

$$G(\bar{2} - 1^+)G(1 - \bar{2}^+) = \left[G(\bar{2} - 1^+)G(1 - \bar{2}^+) - G(\bar{2} - 1)G(1 - \bar{2}) \right]$$
$$+ G(\bar{2} - 1)G(1 - \bar{2})$$

Substituting this in the right side of Eq. (13.41) and using Eq. (13.42) give

$$\sum \int v(G_2 - GG) = \pm (2S + 1) \int_0^{-i\beta} d\bar{1} d\bar{2} K (1 - \bar{1}) v(\bar{1} - \bar{2})$$
$$\times \left[G(\bar{2} - 1^+)G(1 - \bar{2}^+) - G(\bar{2} - 1)G(1 - \bar{2}) \right]$$
$$- i \lim_{\substack{\mathbf{r}_2 \to \mathbf{r}_1 \\ t_2 \to t_1}} \left[K(1 - 2) - \delta(1 - 2) \right]$$

$$(13.43)$$

The difference $G\left(\bar{2}-1^{+}\right)G\left(1-\bar{2}^{+}\right)-G\left(\bar{2}-1\right)G\left(1-\bar{2}\right)$ contributes only when $\bar{t}_2 = t_1$. Hence, only the $\delta\left(1-\bar{1}\right)$ term in $K\left(1-\bar{1}\right)$ contributes to the first term in Eq. (13.43). We may, therefore, replace $K\left(1-\bar{1}\right)$ by $\delta\left(1-\bar{1}\right)$ in this term. Thus

$$\sum \int v\left(G_2 - GG\right) = \pm\left(2S+1\right)\lim_{t_2 \to t_1}\int d\mathbf{r}_2\, v\left(\mathbf{r}_1 - \mathbf{r}_2\right)$$
$$\times\left[G\left(\mathbf{r}_2 - \mathbf{r}_1, 0^-\right)G\left(\mathbf{r}_1 - \mathbf{r}_2, 0^-\right)\right.$$
$$\left.- G\left(\mathbf{r}_2 - \mathbf{r}_1, t_2 - t_1\right)G\left(\mathbf{r}_1 - \mathbf{r}_2, t_1 - t_2\right)\right]$$
$$- i \lim_{\substack{\mathbf{r}_2 \to \mathbf{r}_1 \\ t_2 \to t_1}}\left[K\left(1-2\right)-\delta\left(1-2\right)\right]$$

We get the same result whether we let $t_2 \to t_1^+$ or $t_2 \to t_{\bar{1}}$; we consider the latter case. Then

$$\sum \int v(G_2 - GG) = \pm(2S+1)\int d\mathbf{r}v(\mathbf{r})G^<(-\mathbf{r}, 0)(G^<(\mathbf{r}, 0) - G^>(\mathbf{r}, 0))$$
$$- iK^>(\mathbf{r} = 0, t = 0) \qquad (13.44)$$

From the equal-time commutation relations of $\hat{\psi}$ and $\hat{\psi}^\dagger$, we have

$$G^<\left(\mathbf{r}, 0\right) - G^>\left(\mathbf{r}, 0\right) = -i\delta\left(\mathbf{r}\right)$$

so that the right side of Eq. (13.44) is $nv\left(\mathbf{r} = 0\right) - iK^>\left(\mathbf{r} = 0, t = 0\right)$. These two terms are individually divergent in the Coulomb case, but their difference is finite. Writing them in terms of their Fourier transforms, we find

$$\sum \int v\left(G_2 - GG\right) = \int \frac{d\mathbf{k}}{(2\pi)^3}\left[nv\left(\mathbf{k}\right) - i\int_{-\infty}^{\infty}\frac{d\omega}{2\pi}K^>\left(\mathbf{k}, \omega\right)\right]$$

Now we know from the boundary condition on K that

$$K^>\left(\mathbf{k}, \omega\right) = \frac{1}{i}\frac{Q\left(\mathbf{k}, \omega\right)}{1 - e^{-\beta\omega}}$$

where $Q\left(\mathbf{k}, \omega\right)$ is the discontinuity of the function $K\left(\mathbf{k}, z\right)$ across the real axis:

$$Q\left(\mathbf{k}, \omega\right) = -i\left[K\left(\mathbf{k}, \omega - i\epsilon\right) - K\left(\mathbf{k}, \omega + i\epsilon\right)\right] \qquad (13.45)$$

Thus, P_2 becomes

$$P_2 = \int_0^1 \frac{d\lambda}{\lambda}\int \frac{d\mathbf{k}}{(2\pi)^3}\left(\lambda n\left(\lambda\right)v\left(\mathbf{k}\right) - \int \frac{d\omega}{2\pi}\frac{Q\left(\mathbf{k}, \omega\right)}{1 - e^{-\beta\omega}}\right) \qquad (13.46)$$

The weight function $Q(\mathbf{k}, \omega)$ contributes appreciably to the ω integral only in the neighborhoods of density excitations of the system, e.g., for $\omega \sim \omega_p$. In the classical limit, these contributions are for $\hbar\beta\omega \ll 1$, so that we may replace $(1 - e^{-\hbar\beta\omega})^{-1}$ in the integral by $\hbar\beta\omega$:

$$\int \frac{d\omega}{2\pi} \frac{Q(\mathbf{k}, \omega)}{1 - e^{-\beta\omega}} \rightarrow \int \frac{d\omega}{2\pi} \frac{Q(\mathbf{k}, \omega)}{\beta\omega}$$

Now in the high-frequency ($|\Omega| \rightarrow \infty$) limit, $K(\mathbf{k}, \Omega) \rightarrow 1$, so that from Cauchy's integral theorem, $K(\mathbf{k}, \Omega)$ may be written

$$K(\mathbf{k}, \Omega) - 1 = \int \frac{d\omega}{2\pi} \frac{1}{\Omega - \omega} i\,[K(\mathbf{k}, \omega - i\epsilon) - K(\mathbf{k}, \omega + i\epsilon)]$$

$$= \int \frac{d\omega}{2\pi} \frac{1}{\Omega - \omega} Q(\mathbf{k}, \omega)$$

Therefore, we see that

$$\int \frac{d\omega}{2\pi} \frac{Q(\mathbf{k}, \omega)}{\beta\omega} = -\frac{1}{\beta}[K(\mathbf{k}, \Omega = 0) - 1]$$

(using the fact that $K(\mathbf{k}, \Omega = 0)$ is real so that the $\Omega \rightarrow 0$ limit may be taken uniquely). Thus, P_2 assumes the rather simpler form

$$P_2 = \int_0^1 \frac{d\lambda}{\lambda} \int \frac{d\mathbf{k}}{(2\pi)^3} \left(\lambda n(\lambda) v(\mathbf{k}) + \frac{1}{\beta}[K(\mathbf{k}, \Omega = 0) - 1] \right)$$

$$(13.47)$$

To evaluate this, we recall that in the classical limit, for $\Omega = 0$, we found

$$K^{-1}(\mathbf{k}, 0; \lambda) = 1 + \beta\lambda n(\lambda) v(\mathbf{k})$$

$$= 1 + [kr_D(\lambda)]^{-2}$$

where the λ-dependent screening radius is defined by

$$r_D(\lambda) = \frac{1}{\sqrt{4\pi \lambda e^2 \beta n(\lambda)}}$$

Substituting this evaluation of K into Eq. (13.47), we find

$$P_2 = \frac{1}{\beta} \int_0^1 \frac{d\lambda}{\lambda} \int \frac{d\mathbf{k}}{(2\pi)^3} \frac{1}{k^2 [r_D(\lambda)]^2} \frac{1}{1 + k^2 [r_D(\lambda)]^2}$$

Doing the $\mathbf{k}$ integral gives

$$P_2 = \frac{1}{\beta} \int_0^1 \frac{d\lambda}{8\pi\lambda} \frac{1}{[r_D(\lambda)]^3}$$

The lowest-order contribution to this term may be evaluated by replacing $n(\lambda)$ in $r_D(\lambda)$ by n_0, the density of a non-interacting gas with the same value of the chemical potential. Thus, finally,

$$P_2 = \frac{1}{9} n_0 k_B T \frac{1}{n_0 \left(\frac{4\pi}{3}\right) r_D^3} \tag{13.48}$$

It is clear from this form that the dimensions are correct.

What is the physical interpretation of the calculation that we have just done for P_2? Let us go back to our starting point, Eq. (13.21), which relates the pressure to the interaction energy. The interaction energy is a perfectly reasonable classical concept. We can express it classically as

$$\frac{1}{2} \int d\mathbf{r}_1 d\mathbf{r}_2 \, v(\mathbf{r}_1 - \mathbf{r}_2) \, \rho(\mathbf{r}_1, \mathbf{r}_2)$$

where the density correlation function $\rho(\mathbf{r}_1, \mathbf{r}_2)$ is the probability for finding a particle at $\mathbf{r}_1$ and a (different) particle at $\mathbf{r}_2$, in an equal time measurement. To the lowest order, the density correlation function is just the product of the densities $n_0 n_0$. However, since this interaction energy diverges for the Coulomb system, we have added a background charge that cancels it out. Therefore, we must estimate $\rho(\mathbf{r}_1, \mathbf{r}_2)$ more accurately to find the lowest-order order change in the pressure in a Coulomb gas.

We notice that when there is a particle present at $\mathbf{r}_2$, the density of particles in the immediate neighborhood will be lowered, since the particle repels its neighbors. According to the Maxwell–Boltzmann distribution, the density of particle at $\mathbf{r}_2$ in the potential field $v(\mathbf{r}_1 - \mathbf{r}_2)$, will be proportional to $e^{-\beta v(\mathbf{r}_1 - \mathbf{r}_2)}$. Therefore, we might guess that

$$\rho(\mathbf{r}_1, \mathbf{r}_2) \approx n_0^2 e^{-\beta v(\mathbf{r}_1 - \mathbf{r}_2)}$$

and the interaction energy will be

$$\frac{1}{2} \int d\mathbf{r}_1 d\mathbf{r}_2 \, v(\mathbf{r}_1 - \mathbf{r}_2) \left[e^{-\beta v(\mathbf{r}_1 - \mathbf{r}_2)} - 1 \right] n_0^2$$

If βv is usually much less than one, we may expand the exponential to find an interaction energy

$$-\beta \frac{\Omega}{2} n_0^2 \int d\mathbf{r} \, \left[v^2(r) \right]^2 = -\beta \frac{\Omega}{2} n_0^2 \int \frac{d\mathbf{k}}{(2\pi)^3} \left[v^2(k) \right]^2$$

This second-order interaction energy leads to exactly the same second-order pressure as we would have obtained had we replaced V_S by V in the last term of Eq. (13.27) and take the classical limit. However, this result diverges for a Coulomb gas since $[v(r)]^2 \sim \frac{1}{r^2}$. But in a Coulomb system, the shielding effect will decrease the amount that a particle repels the other particles in the system, so that more realistically, $\rho\,(\mathbf{r}_1, \mathbf{r}_2)$ should be estimated by

$$\rho\,(\mathbf{r}_1, \mathbf{r}_2) \approx n_0^2 e^{-\beta V_S(\mathbf{r}_1 - \mathbf{r}_2)} \qquad (13.49)$$

Therefore, the interaction energy will be

$$\frac{1}{2} \int d\mathbf{r}_1 d\mathbf{r}_2\, v\,(\mathbf{r}_1 - \mathbf{r}_2) \left[e^{-\beta V_S(\mathbf{r}_1 - \mathbf{r}_2)} - 1 \right] n_0^2$$

which, when βV_S is usually much less than one, is

$$-\beta \frac{\Omega}{2} n_0^2 \int d\mathbf{r}\, v(r) V_S(r) = -\beta \frac{\Omega}{2} n_0^2 \int \frac{d\mathbf{k}}{(2\pi)^3}\, v(k) V_S(k)$$

Taking

$$V_S(k) = \frac{4\pi e^2}{k^2 + r_D^{-2}}$$

yields a P_2 identical to Eq. (13.48).

We can use Eq. (13.48) to get an equation of state for the Coulomb gas. We calculated that the pressure is

$$P = P_0 \left[1 + \frac{1}{9}\frac{1}{n_0 \left(\frac{4\pi}{3}\right) r_D^3} \right] = n_0 k_B T + \frac{k_B T}{12\pi} \left(4\pi e^2 n_0 \beta \right)^{3/2}$$

$$(13.50)$$

where[j] $P_0 = n_0 k_B T$ is the pressure of an ideal gas with temperature T and chemical potential μ:

$$n_0 = \int \frac{d\mathbf{p}}{(2\pi\hbar)^3}\, e^{-\beta\left(\frac{p^2}{2m} - \mu\right)}$$

We remember that the real density is not n_0 but $\left.\frac{\partial P}{\partial \mu}\right|_T$. If we use Eq. (13.50) and $\frac{\partial n_0}{\partial \mu} = \beta n_0$, we see that

$$n = n_0 + \frac{3}{2}\frac{1}{12\pi} \left(4\pi e^2 \beta n_0 \right)^{3/2}$$

---

[j]The Boltzmann constant k_B was written as k in the original text.

so that

$$P = nk_B T \left(1 - \frac{1}{18} \frac{1}{\left(\frac{4\pi}{3}\right) r_D^3 n} \right) \qquad (13.51)$$

Equation (13.51) indicates that the first-order effect of the correlations is to reduce the pressure. To understand this, we need only note that the direct effect of the average Coulomb force would be to produce an (infinite) increase in the pressure. As each particle got near the wall, all its fellows would push against it and help it along. We have explicitly eliminated this infinite helping effect by including the background of charges. The shielding tends to further reduce this helping effect by reducing the forces felt by the particles. Therefore, the shielding acts to reduce the pressure.

Equation (13.51) represents the first few terms in the expansion of the pressure in terms of the shielded potential. The parameter that we consider small is

$$\frac{1}{n \left(\frac{4\pi}{3}\right) r_D^3}$$

the inverse of the number of particles within a sphere with radius r_D. This number of particles has to be large in order that the description of shielding that we are using be sensible. If the number is less than one, there are no particles available to shield. Notice that this expansion is certainly not an expansion in the potential strength e^2. The first term we have here is of order e^3. Therefore, in this high-temperature limit, as in the low-temperature limit, a Coulomb force seems highly unamenable to expansion in a power series of e^2. Nevertheless, there exists a well-defined asymptotic expansion for the limit of small e^2.

One final point. Equations (13.9) and (13.10) can be used as the basis of a description of nonequilibrium phenomena in plasmas. It is easy to verify that they are a conserving approximation. Eventually, they lead to a Boltzmann equation for a plasma in which the left side is the same as in the collision-less Boltzmann equation, and the collision term involves scattering cross sections proportional to $|V_S|^2$.

Chapter 14

T Approximation

14.1 Structure of the *T* Matrix

All our Green's function approximations so far have been based on the idea that the potential is small. Even the shielded potential approximation depends on there being a dimensionless parameter, proportional to the strength of the interaction, which is small. For zero-temperature fermions, this parameter is $r_s = \frac{1}{a_0} \left(\frac{3}{4\pi n} \right)^{1/3}$, and in the classical limit, it is $\frac{1}{r_D} \left(\frac{3}{4\pi n} \right)^{1/3}$. However, in many situations of practical interest, the potential is not small, but nonetheless the effects of the potential are small because the potential is very short-ranged. For example, a gas composed of hard spheres with radius r_0 has the potential

$$v(r) = \begin{cases} 0 & \text{for } r > r_0 \\ 1 & \text{for } r < r_0 \end{cases} \qquad (14.1)$$

but when $r_0 \to 0$, the properties of this gas are essentially identical with the properties of a free gas.

We can make a first estimate of the properties of such a gas by adding up an infinite sequence of terms in the expansion of

Annotations to Quantum Statistical Mechanics
In-Gee Kim
Copyright © 2018 Pan Stanford Publishing Pte. Ltd.
ISBN 978-981-4774-15-4 (Hardcover), 978-1-315-19659-6 (eBook)
www.panstanford.com

$G_2 \left(12; 1'2'\right)$. In the Born approximation,

$$\pm \text{ (exchange terms)}$$

Only processes in which two particles propagate independently or come together and interact only once are considered. If the potential is strong, we have to take into account that the particles feel the effect of the potential many, many times as they approach one another, i.e., that

$$\pm \text{ (exchange terms)} \tag{14.2}$$

Equation (14.2) represents the power-series expansion of the integral equation

$$G_2 \left(12; 1'2'\right) = G \left(1, 1'\right) G \left(2, 2'\right) \pm G \left(1, 2'\right) G \left(2, 1'\right)$$
$$+ i \int_0^{-i\beta} d\bar{1} d\bar{2} G \left(1, \bar{1}\right) G \left(2, \bar{2}\right) V \left(\bar{1} - \bar{2}\right) G_2 \left(\bar{1}\bar{2}; 1'2'\right)$$

$$\tag{14.3}$$

This should be compared with Eq. (5.6).

To see the consequences of Eq. (14.3), we introduce the auxiliary quantity T, which satisfies[a]

$$\langle 12| \, \hat{T} \, |1'2' \rangle = V \, (1 - 2) \, \delta \, (1 - 1') \, \delta \, (2 - 2')$$
$$+ i \int d\bar{1} d\bar{2} \, \langle 12| \, \hat{T} \, |\bar{1}\bar{2} \rangle \, G \left(\bar{1}1' \right) G \left(\bar{2}, 2' \right) V \left(1' - 2' \right)$$

$$(14.4)$$

We shall see that in the low-density limit, T reduces to the T matrix of conventional scattering theory. The T matrix defined in Eq. (14.4) is related to the G_2 defined in Eq. (14.3) by

$$V \, (1 - 2) \, G_2 \left(12; 1'2' \right) = \int d\bar{1} d\bar{2} \, \langle 12| \, \hat{T} \, |\bar{1}\bar{2} \rangle$$
$$\times \left[G \left(\bar{1}, 1' \right) G \left(\bar{2}, 2' \right) \pm G \left(\bar{1}, 2' \right) G \left(\bar{2}, 1' \right) \right]$$

$$(14.5)$$

This is easiest to see if we write Eqs. (14.3) and (14.4) in matrix notation:

$$[1 - iGGV] \, G_2 = GG \pm GG$$
$$T \, [1 - iGGV] = V \qquad (14.3a)$$

$$V G_2 = V \, \frac{1}{1 - iGGV} \, [GG \pm GG] \qquad (14.4a)$$

Thus,

$$V G_2 = T \, [GG \pm GG]$$

which is just the right side of Eq. (14.5). The combination $V \, (1 - 2) \, G_2 \, (12, 1'2')$ appears in the equation of motion for G.

Even when the potential is infinite, e.g., v is of the form (14.1), T can be finite. The reason is that the correlation between particles ensures that there can be no particles closer together than r_0. This is reflected in the vanishing of the $G_2 \left(\mathbf{r}t, \mathbf{r}'t; \mathbf{r}t^+, \mathbf{r}'t^+ \right)$ defined by Eq. (14.3) when $|\mathbf{r} - \mathbf{r}'|$ is less than r_0.

---

[a]Here one should not be confused with the notations for the quantity T and the operator $\hat{T}$. In the context, the quantity T means $T(12; 34) = \langle 12| \, \hat{T} \, |34 \rangle$, while the operator $\hat{T}$ should not be confused with the time ordering operator $\hat{T}$ in the previous chapters.

Let us see how T may be determined. From Eq. (14.4), it follows that T has the structure

$$\langle 1, 2 | \hat{T} | 1', 2' \rangle = \delta (t_1 - t_2) \delta (t_{1'} - t_{2'}) \langle \mathbf{r}_1, \mathbf{r}_2 | T (t_1 - t_{1'}) | \mathbf{r}_{1'}, \mathbf{r}_{2'} \rangle$$

$$\langle \mathbf{r}_1, \mathbf{r}_2 | T (t_1 - t_{1'}) | \mathbf{r}_{1'}, \mathbf{r}_{2'} \rangle = \begin{cases} \langle \mathbf{r}_1, \mathbf{r}_2 | T^> (t_1 - t_{1'}) | \mathbf{r}_{1'}, \mathbf{r}_{2'} \rangle & \text{for } i t_1 > i t_{1'} \\ \langle \mathbf{r}_1, \mathbf{r}_2 | T^< (t_1 - t_{1'}) | \mathbf{r}_{1'}, \mathbf{r}_{2'} \rangle & \text{for } i t_1 < i t_{1'} \\ \langle \mathbf{r}_1, \mathbf{r}_2 | T_0 (t_1 - t_{1'}) | \mathbf{r}_{1'}, \mathbf{r}_{2'} \rangle & \text{for } i t_1 = i t_{1'} \end{cases}$$

$$(14.6)$$

where $T^>$ and $T^<$ are analytic functions of the time arguments. T satisfies the same boundary conditions as $G (t_1 - t_{1'}) G (t_1 - t_{1'})$, i.e.,[b]

$$\langle | T (t_1 - t_{1'}) | \rangle |_{t_1 = 0} = \langle |^< T (t_1 - t_{1'}) | \rangle |_{t_1 = 0}$$
$$= e^{2\beta\mu} \langle | T^> (t_1 - t_{1'}) | \rangle |_{t_1 = -i\beta}$$
$$= e^{2\beta\mu} \langle | T (t_1 - t_{1'}) | \rangle |_{t_1 = -i\beta}$$

so that $T^>$ and $T^<$ are related by

$$\langle \mathbf{r}_1, \mathbf{r}_2 | T^> (\omega) | \mathbf{r}_{1'}, \mathbf{r}_{2'} \rangle = e^{\beta(\omega - 2\mu)} \langle \mathbf{r}_1, \mathbf{r}_2 | T^< (\omega) | \mathbf{r}_{1'}, \mathbf{r}_{2'} \rangle \quad (14.7)$$

where

$$\langle | T^> (\omega) | \rangle = \int_{-\infty}^{\infty} dt \, e^{-i\omega t} i \langle | T^> (t) | \rangle$$

$$\langle | T^< (\omega) | \rangle = \int_{-\infty}^{\infty} dt \, e^{-i\omega t} i \langle | T^< (t) | \rangle$$

We can represent this boundary condition by writing T as the Fourier series

$$\langle | T (t_1 - t_{1'}) | \rangle = \frac{1}{-i\beta} \sum_{\nu} e^{-i z_\nu (t_1 - t_{1'})} \langle | T (z_\nu) | \rangle \quad (14.8)$$

where

$$z_\nu = \frac{\pi \nu}{-i\beta} + 2\mu \quad \nu = \text{even integer}$$

Essentially, the same calculation as we went through in Chapter 9 [c.f. Eq. (9.5a)] indicates that the Fourier coefficient of T is

$$\langle | T (z) | \rangle = \langle | T_0 (z) | \rangle + \int \frac{d\omega}{2\pi} \frac{\langle | T^> (\omega) | \rangle - \langle | T^< (\omega) | \rangle}{z - \omega} \quad (14.9)$$

---

[b]*(Author)* Here and after, we understand the symbol $\langle |A| \rangle$ to be an abbreviation of $\langle \mathbf{r}_1, \mathbf{r}_2 | A | \mathbf{r}_{1'}, \mathbf{r}_{2'} \rangle$.

where

$$\langle| \, T_0 \, (z) \, |\rangle = \int_{-i\epsilon}^{i\epsilon} dt \, e^{-izt} \, \langle| \, T_0 \, (z) \, |\rangle$$

The other function of time that appears in Eq. (14.4) for T is

$$\langle \mathbf{r}_1, \mathbf{r}_2 | \mathcal{G}(t_1 - t_{1'}) | \mathbf{r}_{1'}, \mathbf{r}_{2'} \rangle = i G(\mathbf{r}_1 - \mathbf{r}_{1'}, t_1 - t_{1'}) G(\mathbf{r}_2 - \mathbf{r}_{2'}, t_1 - t_{1'})$$

$$(14.10)$$

We can similarly expand $\langle| \, \mathcal{G} \, |\rangle$ in a Fourier series and find that its Fourier coefficient is

$$\langle \mathbf{r}_1, \mathbf{r}_2 | \, G \, (z) \, | \mathbf{r}_{1'}, \mathbf{r}_{2'} \rangle = \int \frac{d\omega}{2\pi} \, \frac{\langle| \mathcal{G}^> (\omega) |\rangle - \langle| \mathcal{G}^< (\omega) |\rangle}{z - \omega}$$

$$= \int \frac{d\omega \, d\omega'}{2\pi \, 2\pi}$$

$$\times \frac{G^> (\mathbf{r}_1 - \mathbf{r}_{1'}, \omega) G^> (\mathbf{r}_2 - \mathbf{r}_{2'}, \omega') - G^< (\mathbf{r}_1 - \mathbf{r}_{1'}, \omega) G^< (\mathbf{r}_2 - \mathbf{r}_{2'}, \omega')}{z - \omega - \omega'}$$

$$(14.11)$$

Now we can write Eq. (14.4) as

$$\langle \mathbf{r}_1, \mathbf{r}_2 | T \, (t_1 - t_{1'}) | \mathbf{r}_{1'}, \mathbf{r}_{2'} \rangle = \delta(\mathbf{r}_1 - \mathbf{r}_{1'}) \delta(\mathbf{r}_2 - \mathbf{r}_{2'}) \delta(t_1 - t_{1'}) v(\mathbf{r}_{1'} - \mathbf{r}_{2'})$$

$$+ \int_0^{-i\beta} d\bar{t} \int d\bar{\mathbf{r}}_1 d\bar{\mathbf{r}}_2 \langle \mathbf{r}_1, \mathbf{r}_2 | T \, (t_1 - t_{1'}) | \bar{\mathbf{r}}_1, \bar{\mathbf{r}}_2 \rangle$$

$$\times \langle \bar{\mathbf{r}}_1, \bar{\mathbf{r}}_2 | \mathcal{G}(\bar{t} - t_{1'}) | \mathbf{r}_{1'}, \mathbf{r}_{2'} \rangle v(\mathbf{r}_{1'} - \mathbf{r}_{2'})$$

We take Fourier coefficients of this equation by multiplying by $e^{iz_v(t_1 - t_{1'})}$ and integrating over all t_1 in $[0, -i\beta]$. Then we find

$$\langle \mathbf{r}_1, \mathbf{r}_2 | \, T \, (z) \, | \mathbf{r}_{1'}, \mathbf{r}_{2'} \rangle = \delta \, (\mathbf{r}_1 - \mathbf{r}_{1'}) \, \delta \, (\mathbf{r}_2 - \mathbf{r}_{2'}) \, v \, (\mathbf{r}_{1'} - \mathbf{r}_{2'})$$

$$+ \int d\bar{\mathbf{r}}_1 d\bar{\mathbf{r}}_2 \langle \mathbf{r}_1, \mathbf{r}_2 | T \, (z) | \bar{\mathbf{r}}_1, \bar{\mathbf{r}}_2 \rangle \langle \bar{\mathbf{r}}_1, \bar{\mathbf{r}}_2 | \mathcal{G}(z) | \mathbf{r}_{1'}, \mathbf{r}_{2'} \rangle v(\mathbf{r}_{1'} - \mathbf{r}_{2'})$$

$$(14.12)$$

Equation (14.12) is originally only derived for

$$z = z_v = \frac{\pi v}{-i\beta} + \mu \quad v = \text{ even integer}$$

but both sides may be continued to all complex values of z. This complex variable corresponds to the total energy of the particles that take part in the scattering process. We can also Fourier transform with respect to the center of mass variables in Eq. (14.12). We write

$$\langle \mathbf{r}_1, \mathbf{r}_2 | T(z) | \mathbf{r}_{1'}, \mathbf{r}_{2'} \rangle = \int \frac{d\mathbf{P}}{(2\pi)^3} \, \exp\left[-\frac{i}{2} \mathbf{P} \cdot (\mathbf{r}_1 + \mathbf{r}_2 - \mathbf{r}_{1'} - \mathbf{r}_{2'}) \right]$$
$$\times \langle \mathbf{r}_1 - \mathbf{r}_2 | T(\mathbf{P}, z) | \mathbf{r}_{1'} - \mathbf{r}_{2'} \rangle$$

$$\langle \mathbf{r}_1, \mathbf{r}_2 | \mathcal{G}(z) | \mathbf{r}_{1'}, \mathbf{r}_{2'} \rangle = \int \frac{d\mathbf{P}}{(2\pi)^3} \, \exp\left[-\frac{i}{2} \mathbf{P} \cdot (\mathbf{r}_1 + \mathbf{r}_2 - \mathbf{r}_{1'} - \mathbf{r}_{2'}) \right]$$
$$\times \langle \mathbf{r}_1 - \mathbf{r}_2 | \mathcal{G}(\mathbf{P}, z) | \mathbf{r}_{1'} - \mathbf{r}_{2'} \rangle \qquad (14.13)$$

so that Eq. (14.12) becomes

$$\langle \mathbf{r} | T(\mathbf{P}, z) | \mathbf{r}' \rangle = \delta(\mathbf{r} - \mathbf{r}') v(r') + \int d\bar{r} \, \langle \mathbf{r} | T(\mathbf{P}, z) | \bar{\mathbf{r}} \rangle \langle \bar{\mathbf{r}} | \mathcal{G}(\mathbf{P}, z) | \mathbf{r}' \rangle v(r')$$
$$(14.14)$$

Equation (14.14) remains an integral equation in the radial variables. This integral equation cannot be solved exactly except in a very few special cases. To see the nature of this equation, let us assume that v is finite, so that it may be Fourier-transformed. We multiply this equation by $e^{-i\mathbf{p}\cdot\mathbf{r}+i\mathbf{r}'\cdot\mathbf{r}'}$ and integrate over all $\mathbf{r}$ and $\mathbf{r}'$. We then find

$$\langle \mathbf{p} | T(\mathbf{P}, z) | \mathbf{p}' \rangle$$
$$= v(\mathbf{p} - \mathbf{p}') + \int \frac{d\bar{\mathbf{p}}}{(2\pi)^3} \frac{d\bar{\mathbf{p}}'}{(2\pi)^3} \langle \mathbf{p} | T(\mathbf{P}, z) | \bar{\mathbf{p}} \rangle \langle \bar{\mathbf{p}} | \mathcal{G}(\mathbf{P}, z) | \bar{\mathbf{p}}' \rangle v(\mathbf{p}' - \bar{\mathbf{p}}')$$
$$(14.15)$$

Here, $\mathbf{p}$ represents the momentum of one of the initial particles in the center of mass system, $\mathbf{p}'$ is the momentum of this particle after the scattering, $\mathbf{P}$ is the center of mass momentum, and $\langle \mathbf{p} | T(\mathbf{P}, z) | \mathbf{p}' \rangle$ is the scattering amplitude for such a process.

To see the relation of T to the conventional scattering amplitude, let us consider the low-density limit in which

$$\beta\mu \to -\infty$$

and

$$A(p, \omega) \to A^0(p, \omega) = 2\pi\delta\left(\omega - \frac{p^2}{2m}\right)$$

Then

$$\langle \mathbf{r}_1, \mathbf{r}_2 | \mathcal{G}(z) | \mathbf{r}_{1'}, \mathbf{r}_{2'} \rangle = \int \frac{d\mathbf{p}_1}{(2\pi)^3} \frac{d\mathbf{p}_2}{(2\pi)^3} \frac{e^{i\mathbf{p}_1 \cdot (\mathbf{r}_1 - \mathbf{r}_{1'}) + i\mathbf{p}_2 \cdot (\mathbf{r}_2 - \mathbf{r}_{2'})}}{z - \left(\frac{p_1^2}{2m}\right) - \left(\frac{p_2^2}{2m}\right)}$$

$$\langle \mathbf{r} | \mathcal{G}(\mathbf{P}, z) | \mathbf{r}' \rangle = \int \frac{d\mathbf{p}}{(2\pi)^3} \frac{e^{i\mathbf{p} \cdot (\mathbf{r} - \mathbf{r}')}}{z - \frac{\left(\mathbf{p} + \frac{\mathbf{P}}{2}\right)^2}{2m} - \frac{\left(\mathbf{p} - \frac{\mathbf{P}}{2}\right)^2}{2m}}$$

and

$$\langle \mathbf{p}| \mathcal{G} \left(\mathbf{P}, z\right) |\mathbf{p}'\rangle = \frac{(2\pi)^3 \, \delta \left(\mathbf{p} - \mathbf{p}'\right)}{z - \left(\frac{P^2}{4m}\right) - \left(\frac{p^2}{2m}\right)} \tag{14.16}$$

With this value of $\mathcal{G}$, Eq. (14.15) becomes

$$\langle \mathbf{p}|T\left(\mathbf{P}, z\right)|\mathbf{p}'\rangle = v(\mathbf{p} - \mathbf{p}') + \int \frac{d\bar{\mathbf{p}}}{(2\pi)^3} \langle \mathbf{p}|T\left(\mathbf{P}, z\right)|\bar{\mathbf{p}}\rangle v(\bar{\mathbf{p}} - \mathbf{p}')$$

$$\times \frac{1}{z - \left(\frac{P^2}{4m}\right) - \left(\frac{p^2}{2m}\right)} \tag{14.17}$$

When the complex variable z is replaced by the total energy of the incident particles $\left(\frac{P^2}{4m}\right) - \left(\frac{p^2}{2m}\right) + i\epsilon$, Eq. (14.17) determines the scattering amplitude of conventional scattering theory. This scattering matrix is defined by

$$\langle \mathbf{p}| \hat{T} = \langle \varphi_{\mathbf{p}}| v \tag{14.18}$$

where $\langle \mathbf{p}|$ is a free two-particle state and $\langle \varphi_{\mathbf{p}}|$ is a two-particle scattering state with energy $\frac{p^2}{2m}$. The state $\langle \varphi_{\mathbf{p}}|$ satisfies

$$\langle \varphi_{\mathbf{p}}| \left(\hat{H}_0 + v - \left(\frac{p^2}{2m}\right)\right) = 0 \tag{14.19}$$

where $\hat{H}_0$ is the free-particle Hamiltonian. We may write the solution to this equation as

$$\langle \varphi_{\mathbf{p}}| = \langle \mathbf{p}| + \langle \varphi_{\mathbf{p}}| v \frac{1}{\left(\frac{p^2}{2m}\right) - \hat{H}_0 + i\epsilon}$$

where the $i\epsilon$ is chosen so that the solution to $\langle \varphi_{\mathbf{p}}|$ corresponds to an outgoing wave. Multiplying by v and using Eq. (14.18) then gives

$$\langle \mathbf{p}| T |\mathbf{p}'\rangle = \langle \mathbf{p}| v |\mathbf{p}'\rangle + \langle \mathbf{p}| T \frac{1}{\left(\frac{p^2}{2m}\right) - \hat{H}_0 + i\epsilon} v |\mathbf{p}'\rangle$$

which is Eq. (14.17), with $z = \left(\frac{p^2}{2m}\right) + i\epsilon + \left(\frac{P^2}{4m}\right)$.

In this conventional two-body scattering matrix, the particles may be thought of as propagating as free particles, between Born approximation scatterings, while in the many-body case, the particles feel the full effects of the medium between the scatterings with each other. Even if the interactions of the particles with the medium are neglected, $A\,(p,\omega) \rightarrow 2\pi\delta\left(\omega - \frac{p^2}{2m}\right)$, the weightings of the intermediate states between scatterings are changed by the presence of the medium. This is reflected in the factors of f and $1 \pm f$ that appear in Eq. (14.11). Also the many-body T matrix depends on the center of mass momentum of the two particles, whereas the conventional scattering matrix is independent of this momentum.

The many-particle T satisfies an optical theorem quite analogous to the one obeyed by the conventional scattering matrix. To derive this theorem, let us consider T to be a matrix in the variables $\mathbf{p}$ and $\mathbf{p}'$. Then Eq. (14.15) may be written with the momentum indices suppressed as

$$T\,(z) = v + T\,(z)\mathcal{G}(z)v$$

or as

$$T^{-1}(z) = v^{-1} - \mathcal{G}(z) \tag{14.20}$$

T and $\mathcal{G}$ are real functions of the complex variable z. We let $z = \omega - i\epsilon$. Then the imaginary part of T is given by

$$\begin{aligned}
\Im T\,(\omega - i\epsilon) &= - \left[T\,(\omega - i\epsilon)\right]^* \left[\Im T^{-1}\,(\omega - i\epsilon)\right] T\,(\omega - i\epsilon) \\
&= - T\,(\omega + i\epsilon) \left[\Im T^{-1}\,(\omega - i\epsilon)\right] T\,(\omega - i\epsilon)
\end{aligned}$$

Now from Eq. (14.20),

$$\begin{aligned}
\Im T^{-1}\,(\omega - i\epsilon) &= - \Im \mathcal{G}^{-1}\,(\omega - i\epsilon) \\
&= - \frac{1}{2}\left[\mathcal{G}^>\,(\omega) - \mathcal{G}^<\,(\omega)\right]
\end{aligned}$$

and

$$\Im T\,(\omega - i\epsilon) = \frac{1}{2}\left[T^>\,(\omega) - T^<\,(\omega)\right]$$

Thus,

$$T^>\,(\omega) - T^<\,(\omega) = T\,(\omega + i\epsilon)\left[\mathcal{G}^>\,(\omega) - \mathcal{G}^<\,(\omega)\right] T\,(\omega - i\epsilon)$$

or, with the matrices indices reinserted,

$$\begin{aligned}
\left\langle \mathbf{p}\left| T^>\,(\mathbf{P}, \omega) - T^<\,(\mathbf{P}, \omega)\right| \mathbf{p}'\right\rangle &= \int \frac{d\bar{\mathbf{p}}}{(2\pi)^3}\frac{d\bar{\mathbf{p}}'}{(2\pi)^3} \left\langle \mathbf{p}\left| T\,(\mathbf{P}, \omega + i\epsilon)\right| \bar{\mathbf{p}}\right\rangle \\
&\quad \times \left\langle \bar{\mathbf{p}}\left| \mathcal{G}^>\,(\mathbf{P}, \omega) - \mathcal{G}^<\,(\mathbf{P}, \omega)\right| \bar{\mathbf{p}}'\right\rangle \\
&\quad \times \left\langle \bar{\mathbf{p}}'\left| T\,(\mathbf{P}, \omega - i\epsilon)\right| \mathbf{p}'\right\rangle
\end{aligned}$$

$$\tag{14.21}$$

Since

$$T^> (\omega) = e^{\beta(\omega - 2\mu)} T^< (\omega)$$

and

$$\mathcal{G}^> (\omega) = e^{\beta(\omega - 2\mu)} \mathcal{G}^< (\omega)$$

we can derive from Eq. (14.21) that

$$\left\langle \mathbf{p} \left| T^{\gtrless} (\mathbf{P}, \omega) \right| \mathbf{p}' \right\rangle$$

$$= \int \langle \mathbf{p} | T (\mathbf{P}, \omega + i\epsilon) | \bar{\mathbf{p}} \rangle \frac{d\bar{\mathbf{p}}}{(2\pi)^3} \left\langle \bar{\mathbf{p}} \left| \mathcal{G}^{\gtrless} (\mathbf{P}, \omega) \right| \bar{\mathbf{p}}' \right\rangle$$

$$\times \frac{d\bar{\mathbf{p}}'}{(2\pi)^3} \langle \bar{\mathbf{p}}' | T (\mathbf{P}, \omega - i\epsilon) | \mathbf{p}' \rangle \qquad (14.22)$$

Equations (14.21) and (14.22) are generalizations of the optical theorem of ordinary scattering theory.

Let us now substitute the approximation (14.5) for G_2 into the equation of motion for G. Then

$$\left(i \frac{\partial}{\partial t_1} + \frac{\nabla_1^2}{2m} \right) G (1, 1') = \delta (1 - 1') \pm \int V (1 - 2) G_2 (12; 1'2^+)$$

$$= \delta (1 - 1') \pm \int \langle 12 | T | \bar{1}\bar{2} \rangle$$

$$\times \left[G (\bar{1}, 1') G (\bar{2}, 2^+) \pm G (\bar{1}, 2^+) G (\bar{2}, 1') \right]$$

$$\equiv \delta (1 - 1') + \int \Sigma (1, \bar{1}) G (\bar{1}, 1')$$

so that the self-energy is, in this approximation,

$$\Sigma (1, 1') = \pm \int d2 d\bar{2} \left[\langle 12 | T | 1'\bar{2} \rangle \pm \langle 12 | T | \bar{2}1' \rangle \right] G (\bar{2}, 2^+)$$

$$= \pm i \int d\mathbf{r}_2 d\bar{\mathbf{r}}_2 \left[\langle \mathbf{r}_1 \mathbf{r}_2 | T (t_1 - t_{1'}) | \mathbf{r}_{1'} \bar{\mathbf{r}}_2 \rangle \right.$$

$$\left. \pm \langle \mathbf{r}_1 \mathbf{r}_2 | T (t_1 - t_{1'}) | \bar{\mathbf{r}}_2 \mathbf{r}_{1'} \rangle \right] G (\bar{\mathbf{r}}_2 - \mathbf{r}_2, t_{1'} - t_1)$$

$$(14.23)$$

To understand the T approximation for G, let us compute[c] $\Sigma^> (\mathbf{p}, \omega)$, the average collision rate for a particle traveling through

[c]The momentum $\mathbf{p}$ of the arguments of $\Sigma^>$ and $G^{\gtrless}$, which was printed in scalar form in the original text, is vectorized for clarity.

the medium with momentum $\mathbf{p}$ and energy ω. From Eq. (14.23), we see that

$$\Sigma^> (\mathbf{p}, \omega) = \int \frac{d\mathbf{p}'}{(2\pi)^3} \frac{d\omega'}{2\pi} \left[\left\langle \frac{\mathbf{p} - \mathbf{p}'}{2} \left| T^> (\mathbf{p} + \mathbf{p}', \omega + \omega') \right| \frac{\mathbf{p} - \mathbf{p}'}{2} \right\rangle \right.$$

$$\left. \pm \left\langle \frac{\mathbf{p} - \mathbf{p}'}{2} \left| T^> (\mathbf{p} + \mathbf{p}', \omega + \omega') \right| \frac{\mathbf{p}' - \mathbf{p}}{2} \right\rangle \right] G^< (\mathbf{p}', \omega')$$

Using the optical theorem (14.22), we find

$$\Sigma^> (\mathbf{p}, \omega) = \int \frac{d\mathbf{p}' d\omega'}{(2\pi)^4} \frac{d\bar{\mathbf{p}}}{(2\pi)^3} \frac{d\bar{\mathbf{p}}'}{(2\pi)^3}$$

$$\times \left\langle \frac{\mathbf{p} - \mathbf{p}'}{2} \left| T (\mathbf{p} + \mathbf{p}', \omega + \omega' + i\epsilon) \right| \bar{\mathbf{p}} \right\rangle$$

$$\times \left\langle \bar{\mathbf{p}} \left| \mathcal{G}^> (\mathbf{p} + \mathbf{p}', \omega + \omega') \right| \bar{\mathbf{p}}' \right\rangle$$

$$\times \left[\left\langle \bar{\mathbf{p}}' \left| T (\mathbf{p} + \mathbf{p}', \omega + \omega' - i\epsilon) \right| \frac{\mathbf{p} - \mathbf{p}'}{2} \right\rangle \right.$$

$$\left. \mp \left\langle \bar{\mathbf{p}}' \left| T (\mathbf{p} + \mathbf{p}', \omega + \omega' - i\epsilon) \right| \frac{\mathbf{p}' - \mathbf{p}}{2} \right\rangle \right]$$

However,

$$\left\langle \bar{\mathbf{p}} \left| \mathcal{G}^> (\mathbf{P}, \omega) \right| \bar{\mathbf{p}}' \right\rangle = (2\pi)^3 \delta (\bar{\mathbf{p}} - \bar{\mathbf{p}}') \int \frac{d\omega'}{2\pi}$$

$$\times G^> \left(\bar{\mathbf{p}} + \frac{\mathbf{P}}{2}, \omega' + \frac{\omega}{2} \right) G^> \left(-\bar{\mathbf{p}} + \frac{\mathbf{P}}{2}, -\omega' + \frac{\omega}{2} \right)$$

so that $\Sigma^> (\mathbf{p}, \omega)$ has the form

$$\Sigma^> (\mathbf{p}, \omega) = \int \frac{d\mathbf{p}' d\omega'}{(2\pi)^4} \int \frac{d\bar{\mathbf{p}} d\bar{\omega}}{(2\pi)^4} \int \frac{d\bar{\mathbf{p}}' d\bar{\omega}'}{(2\pi)^4}$$

$$\times (2\pi)^4 \delta (\mathbf{p} + \mathbf{p}' - \bar{\mathbf{p}} - \bar{\mathbf{p}}') \delta (\omega + \omega' - \bar{\omega} - \bar{\omega}')$$

$$\times \left(\frac{1}{2} \right) \left| \left\langle \frac{\mathbf{p} - \mathbf{p}'}{2} \left| T (\mathbf{p} + \mathbf{p}', \omega + \omega' + i\epsilon) \right| \frac{\bar{\mathbf{p}} - \bar{\mathbf{p}}'}{2} \right\rangle \right.$$

$$\left. \pm \left\langle \frac{\mathbf{p} - \mathbf{p}'}{2} \left| T (\mathbf{p} + \mathbf{p}', \omega + \omega' + i\epsilon) \right| \frac{\bar{\mathbf{p}}' - \bar{\mathbf{p}}}{2} \right\rangle \right|^2$$

$$\times G^< (\mathbf{p}', \omega') G^> (\bar{\mathbf{p}}, \bar{\omega}) G^> (\bar{\mathbf{p}}', \bar{\omega}')$$

This is an exceedingly natural result. The lifetime is proportional to the cross section for a scattering process, $\mathbf{p}$, $\omega + \mathbf{p}'$, $\omega' \to \bar{\mathbf{p}}$, $\bar{\omega} + \bar{\mathbf{p}}'$, $\bar{\omega}'$. The differential cross section is composed of energy- and momentum-conserving delta functions times the squared magnitude of the direct scattering amplitude $\pm$ the exchange amplitude. This differential cross section is multiplied by the density of scatters $G^< (\mathbf{p}', \omega')$ and the available density of final states $G^> (\bar{\mathbf{p}}, \bar{\omega}) \, G^> (\bar{\mathbf{p}}', \bar{\omega}')$ and then integrated over all possible scatterers and final states.

$\Sigma^< (\mathbf{p}, \omega)$ has exactly the same structure except that $G^< (\mathbf{p}', \omega')$ is replaced by $G^> (\mathbf{p}', \omega')$ and $G^> (\bar{\mathbf{p}}, \bar{\omega}) \, G^> (\bar{\mathbf{p}}', \bar{\omega}')$ is replaced by $G^< (\bar{\mathbf{p}}, \bar{\omega}) \, G^< (\bar{\mathbf{p}}', \bar{\omega}')$.

The T matrix approximation is extremely useful when the potential has a hard core, e.g., Eq. (14.1). With a finite potential, we found that there was a term in $\Sigma \, (1, 1')$ proportional to $\delta \, (t_1 - t_{1'})$, which was, in fact, the Hartree–Fock contribution,

$$\Sigma_{\mathrm{HF}} (\mathbf{p}) = nv \, (k = 0) \pm \int \frac{d\mathbf{p}'}{(2\pi)^3} \, v \, (\mathbf{p} - \mathbf{p}') \, \langle n \, (p') \rangle$$

If, however, there is a hard core in the potential, the Hartree–Fock term diverges, since the $v(k)$ are infinite. There still is a finite term in Σ proportional to $\delta \, (t_1 - t_{1'})$, but instead of being the Hartree–Fock term, it is determined by T_0, the delta-function part of T in Eq. (14.6). Also, there is a term in T, and hence in Σ, proportional to $\frac{\partial}{\partial t_1} \delta \, (t_1 - t_{1'})$.

Brueckner and others have applied the T-matrix approximation to the calculation of the ground-state energy and density of nuclear matter. The results check nicely with the extrapolated properties of heavy nuclei.

The T approximation is conserving, i.e., it satisfies criteria A and B. Therefore, when stated in terms of $G(U)$, it may be used to describe nonequilibrium behavior. The Boltzmann equation for $g(U)$ derived from this approximation involves collision cross sections proportional to $|T|^2$. In the classical low-density limit, these reduce to the classical collision cross section.

14.2 Breakdown of the *T* Approximation in Metals

At very low temperatures, some metals exhibit the peculiar phenomenon of superconductivity. We now want to show how its appearance is signaled by the breakdown of the T approximation in a metal.

We can consider a metal to be a Fermi gas of electrons. The long-range part of the Coulomb interaction is effectively shielded out. For some metals, the residual interaction with the ions leads to a net effectively attractive interaction between the electrons. This effective interaction is highly velocity dependent. To a first approximation, it can be considered to act only between electrons whose energies lie in the range

$$|E(p) - \mu| < \hbar\omega_D \tag{14.24}$$

about the Fermi energy μ. The Debye energy, $\hbar\omega_D$, which is the maximum phonon energy in the metal, is comparatively small. It corresponds to a temperature of a few hundred degrees Kelvin, while μ is an energy of the order of 20, 000 degrees. The particles in this shell about the Fermi sea interact through a potential that may be taken to be

$$v(\mathbf{r}_1 - \mathbf{r}_2) = -v\delta(\mathbf{r}_1 - \mathbf{r}_2)$$

Such a potential can have no effect between electrons of the same spin. The exclusion principle prevents them from ever coming on top of one another. However, electrons of opposite spin can interact via this potential. There are, of course, no exchange process between particles of opposite spin. This is represented in our formalism by taking the total scattering matrix for all the particles in the process $\mathbf{p} + \mathbf{p}' \to \bar{\mathbf{p}} + \bar{\mathbf{p}}'$ having the same spin to be

$$\frac{1}{\sqrt{2}} \left[\left\langle \frac{\mathbf{p} - \mathbf{p}'}{2} \left| T(\mathbf{p} + \mathbf{p}', z) \right| \frac{\bar{\mathbf{p}} - \bar{\mathbf{p}}'}{2} \right\rangle - \left\langle \frac{\mathbf{p} - \mathbf{p}'}{2} \left| T(\mathbf{p} + \mathbf{p}', z) \right| \frac{\bar{\mathbf{p}}' - \bar{\mathbf{p}}}{2} \right\rangle \right]$$

$$\tag{14.25}$$

while the scattering matrix for the process in which $\mathbf{p}$ and $\bar{\mathbf{p}}$ have spin up while $\mathbf{p}'$ and $\bar{\mathbf{p}}'$ have spin down contains no exchange term and is simply

$$\left\langle \frac{\mathbf{p} - \mathbf{p}'}{2} \left| T(\mathbf{p} + \mathbf{p}', z) \right| \frac{\bar{\mathbf{p}} - \bar{\mathbf{p}}'}{2} \right\rangle \tag{14.26}$$

From Eq. (14.4), we can see that when $v(\mathbf{r}_1 - \mathbf{r}_2)$ is a delta function,

$$\langle 1, 2 | T | 1', 2' \rangle \sim \delta(1 - 2)\delta(1' - 2')$$

Therefore, in this case

$$\left\langle \frac{\mathbf{p} - \mathbf{p}'}{2} \left| T\left(\mathbf{p} + \mathbf{p}', z\right) \right| \frac{\bar{\mathbf{p}} - \bar{\mathbf{p}}'}{2} \right\rangle = T\left(\mathbf{p} + \mathbf{p}', z\right) \qquad (14.27)$$

so that the total scattering amplitude Eq. (14.25) for the same-spin particles vanishes. However, the scattering amplitude Eq. (14.26) for unlike spins is certainly nonzero.

To determine T in this case, we go back to Eq. (14.14). Since T is of the form (14.27) and, when nonzero $v\left(\mathbf{p} - \mathbf{p}'\right)$ is just $-v$, we see that

$$T\left(\mathbf{P}, z\right) = -v\left[1 + \int \frac{d\bar{\mathbf{p}}}{(2\pi)^3} \frac{d\bar{\mathbf{p}}'}{(2\pi)^3} \left\langle \bar{\mathbf{p}} \left| \mathcal{G}\left(\mathbf{P}, z\right) \right| \bar{\mathbf{p}}' \right\rangle T\left(\mathbf{P}, z\right) \right]$$

and consequently, where T is nonzero,

$$\left[T^{-1}\left(\mathbf{P}, z\right)\right]^{-1} + v^{-1} = \int \frac{d\omega'}{2\pi} \int \frac{d\omega}{2\pi} \int \frac{d\mathbf{p}'}{(2\pi)^3}$$

$$\times \frac{G^>\left(\mathbf{p} + \mathbf{P}/2, \omega\right) G^>\left(-\mathbf{p} + \mathbf{P}/2, \omega\right) - G^<\left(\mathbf{p} + \mathbf{P}/2, \omega\right) G^<\left(-\mathbf{p} + \mathbf{P}/2, \omega\right)}{z - \omega - \omega'}$$

$$(14.28)$$

For an attractive interaction, T has a very peculiar behavior at low temperatures. We shall see that when P, the total momentum of the particles taking place in the collision, is small, there appear complex poles in T for values of z near 2μ. To show this, we shall evaluate the integral in Eq. (14.28) at $\mathbf{P} = 0$, assuming that G can be replaced by G_0. Then

$$\left[T^{-1}\left(0, z\right)\right]^{-1} + v^{-1} = \int_{|E(p)-\mu|<\omega_D} \frac{d\mathbf{p}}{(2\pi)^3} \frac{1 - 2f\left(E(p)\right)}{z - 2E(p)}$$

where the limits of the integration are determined by the assumption that V only acts for energies in the range Eq. (14.24). Since the contributions to the integral all come from a narrow sheet about the surface of the Fermi sea, we can write

$$T^{-1}\left(0, z\right) = \frac{-v}{1 + v\rho_E \int_{-\omega_D}^{\omega_D} d\epsilon \, \frac{\tanh\left(\frac{\beta\epsilon}{2}\right)}{(z - 2\mu) - 2\epsilon}} \qquad (14.29)$$

where ϵ is single-particle energy measured relative to μ, i.e.,

$$\epsilon = \left(\frac{p^2}{2m}\right) - \mu$$

and $\rho_E = \frac{mp_F}{2\pi^2}$. Let us evaluate this integral for imaginary values of $z - 2\mu$, $z - 2\mu = iy$. Then Eq. (14.29) becomes

$$T^{-1}(0, z) = \frac{-v}{1 - v\rho_E \int_0^{\omega_D} d\epsilon \left(\tanh\left(\frac{\beta\epsilon}{2}\right)\right) \frac{4\epsilon}{(2\epsilon)^2 + y^2}} \qquad (14.30)$$

If the temperature is sufficiently high so that

$$v\rho_E \int_0^{\omega_D} d\epsilon \frac{\tanh\left(\frac{\beta\epsilon}{2}\right)}{\epsilon} < 1 \qquad (14.31)$$

then Eq. (14.30) will have no poles for real values of y, i.e., complex values of z. However, when the temperature is low enough so that

$$v\rho_E \int_0^{\omega_D} d\epsilon \frac{\tanh\left(\frac{\beta\epsilon}{2}\right)}{\epsilon} \geq 1$$

there will be poles for real values of y. For sufficiently low temperatures, this integral may be made arbitrarily large. For example, at zero temperature, $\beta = \infty$, $\tanh\frac{\beta|\epsilon|}{2} = 1$, and

$$\int_0^{\omega_D} d\epsilon \frac{4\epsilon}{y^2 + (2\epsilon)^2} = \frac{1}{2}\log\left(\frac{y^2 + 4\omega_D^2}{y^2}\right)$$

which we can make as large as we please by picking y sufficiently small.

Therefore, for high temperatures, the T approximation contains no complex poles and is perfectly consistent. For low temperatures, complex poles appear. The T matrix measures the probability amplitude for adding a pair of particles in a certain configuration, and then removing a pair in some other configuration. A complex pole in the upper half-plane in Eq. (14.29) then indicates that if a pair of particles with equal and opposite momenta are added at a certain time, the probability amplitude for removing such a pair increases exponentially in time. Then the T approximation as stated in Eq. (14.29) is no longer capable of correctly describing the system, except for very short times. The appearance of these complex poles signals that something about the system has radically changed. This change is actually the onset of superconductivity.

To estimate the critical temperature at which this change first occurs, we have to estimate the temperature at which the equality in Eq. (14.31) occurs. This estimate is most easily made if we use the experimental fact that the parameter v is roughly $1/4$. Then, the integral Eq. (14.31) will only be sufficiently large if $\beta^{-1} = k_\mathrm{B} T$ is small compared with ω_D, so that the hyperbolic tangent will be close to unity over most of the domain of integration. To get a rough estimate of the integral, we write

$$\tanh \left(\frac{\beta \epsilon}{2} \right) \approx \begin{cases} 1 & \text{for } \left(\frac{\beta \epsilon}{2} \right) > 1 \\ 0 & \text{for } \left(\frac{\beta \epsilon}{2} \right) < 1 \end{cases}$$

Then Eq. (14.31) determines the critical temperature $T_c = [k_\mathrm{B} \beta_c]^{-1}$ to be

$$1 = v \rho_E \log \frac{\beta_c \omega_\mathrm{D}}{2}$$

or

$$\beta_c^{-1} = k_\mathrm{B} T_c = \frac{\hbar \omega_\mathrm{D}}{2} e^{-\frac{1}{v \rho_E}} \approx \frac{\hbar \omega_\mathrm{D}}{2} e^{-4}$$

The critical temperature determined in this way is indeed quite small. In fact, it is typically of the order of 5 degrees, while the Debye temperature, $\hbar \omega_\mathrm{D}/k_\mathrm{B}$, is typically 300 degrees. This tremendous difference comes about because the coherent effects that lead to the complex pole and hence the instability in the normal state are an exceedingly delicate summation of small perturbations to produce a net large effect.

If we investigated the structure of T $(\mathbf{P}, z)$ in detail, we would discover that the complex pole first appeared at $\mathbf{P} = 0$, as indeed we have assumed in the foregoing analysis. This indicates that the instability first appears in the scattering of particles with equal and opposite momentum. We have already indicated that the complex pole appears only in the scattering of particles of opposite spin at total energy equal to 2μ. This complex pole appears because particles with equal and opposite momentum, opposite spin, and total energy 2μ form an essentially bound state. This pair formation is responsible for all the peculiar properties of superconductors.

Appendix A

Finite-Temperature Perturbation Theory

In these chapters, we always determined G by making use of some kind of equation of motion. However, there exists an alternative scheme for determining G based on an expansion of G in a power series of V and G_0. We described the first few terms of this expansion in Chapter 6. However, for many purposes, it is useful to know the structure of the entire expansion. We shall, therefore, describe this expansion in detail.

The basic elements in the expansion of $G\left(1, 1'; U\right)$ are the free particle propagator

$$G_0\left(1, 1'; U\right) = \quad 1' \longrightarrow 1$$

and the interaction:

$$i V\left(1 - 1'\right) = i v\left(\mathbf{r}_1 - \mathbf{r}_{1'}\right) \delta\left(t_1 - t_{1'}\right) = \quad 1' \dashrightarrow 1$$

$G(U)$ can be expressed as the sum of the values of all topologically different connected diagrams for which (a) one propagator line enters and one line leaves, (b) each potential line contains at both of its ends one entering and one leaving propagator line; i.e., the potential line appears only in the combination

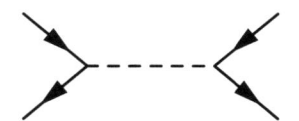

The point of connection between the two propagator lines and the potential line is called a vertex. Each vertex is labeled with a space–time point.

To calculate the value of a particular graph, for example,

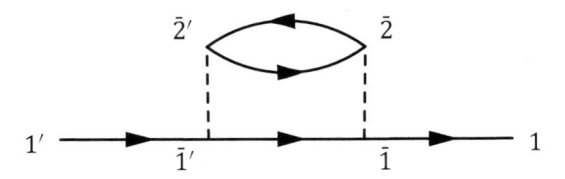

we do the following:

(1) Write down the product of all the propagators and interactions that appear in it, in this case

$$G_0(1, \bar{1})G_0(\bar{1}, \bar{1}')G_0(\bar{1}', 1')iV(\bar{1}' - \bar{2}')iV(\bar{1} - \bar{2})G_0(\bar{2}', \bar{2})G_0(\bar{2}, \bar{2}')$$

(2) Integrate the labels on all the vertices over all space and all times between 0 and $-i\beta$. In this case, we integrate the four barred variables.

(3) This gives the contribution of the diagram to $G(1, 1')$ for the case of bosons. For a fermion system, we multiply the result of the integration by a factor of $(-1)^\ell$, where ℓ is the number of closed loops composed of fermion lines in the diagram. In this example, there is one closed loop,

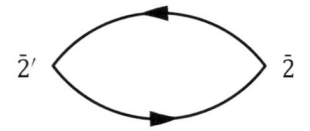

so we have to multiply by a factor of -1 fermions.

Therefore, this diagram contributes

$$\pm \int_0^{-i\beta} d\bar{1}d\bar{2}d\bar{1}'d\bar{2}'\, G_0\left(1, \bar{1}\right) G_0\left(\bar{1}, \bar{1}'\right) G_0\left(\bar{1}', 1'\right)$$
$$\times iV\left(\bar{1} - \bar{2}\right) iV\left(\bar{1}' - \bar{2}'\right) G_0\left(\bar{2}, \bar{2}'\right) G_0\left(\bar{2}', \bar{2}\right) \tag{A.1}$$

to $G(1, 1')$.

However, in equilibrium ($U = 0$), the physical information is most readily accessible not from $G\,(1 - 1')$ but from $A\,(p, \omega)$, which is easily determined from

$$G\,(p, z) = \int \frac{d\omega'}{2\pi} \frac{A\,(p, \omega')}{z - \omega'}$$

Therefore, what we really want is a diagrammatic expansion for $G\,(p, z)$. To get this expansion, we take the expansion for $G\,(1, 1'; U = 0)$, multiplied by $e^{-i\mathbf{p}\cdot(\mathbf{r}_1 - \mathbf{r}_{1'}) + iz_\nu(t_1 - t_{1'})}$, where

$$z_\nu = \frac{\pi \nu}{-i\beta} + \mu$$

$$\nu = \begin{array}{l} \text{even integer for bosons} \\ \text{odd integer for fermions} \end{array}$$

and integrate over all $\mathbf{r}_1$ and all t_1 in the interval $[0, -i\beta]$. In this way, we generate an expansion for $G\,(p, z_\nu)$.

The basic rules for calculating $G\,(p, z_\nu)$ are only slightly more complex than those for calculating $G\,(1, 1')$. In fact, we can derive these new rules by using the old rules and the fact that

$$G_0\,(1 - 1') = \frac{1}{-i\beta} \sum_\nu \int \frac{d\mathbf{p}}{(2\pi)^3} \frac{e^{i\mathbf{p}\cdot(\mathbf{r}_1 - \mathbf{r}_{1'}) - iz_\nu(t_1 - t_{1'})}}{z_\nu - \left(\frac{p^2}{2m}\right)} \tag{A.2}$$

We associate with every particle line in the diagram a momentum $\mathbf{p}$ and an "energy" z_ν. For example, the diagram we considered before is labeled

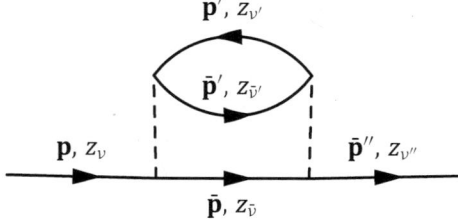

The "energies" and momenta of the lines are, respectively, summation and integration variables.

(1) For each particle line, we write a factor

$$G_0\,(p, z_\nu) = \frac{1}{z_\nu - \left(\frac{p^2}{2m}\right)} = \qquad \xrightarrow{\quad \mathbf{p},\, z_\nu \quad}$$

(2) For each potential line, with its associated particle lines, we write a factor

$$(2\pi)^3 \, \delta \left(\mathbf{p} + \mathbf{p}' - \bar{\mathbf{p}} - \bar{\mathbf{p}}'\right) (-i\beta) \, \delta_{\nu+\nu', \bar{\nu}+\bar{\nu}'}$$

which expresses the conservation of momentum and "energy" in the collision,

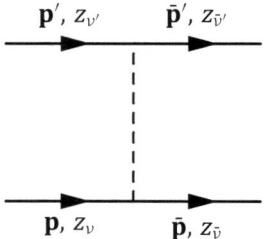

$$\mathbf{p}', z_{\nu'} \qquad \bar{\mathbf{p}}', z_{\bar{\nu}'}$$
$$\mathbf{p}, z_{\nu} \qquad \bar{\mathbf{p}}, z_{\bar{\nu}}$$

We also write a factor

$$i v \left(\mathbf{p} - \bar{\mathbf{p}}\right)$$

(3) To find the value of the diagram, we integrate over the momenta and sum over the possible "energies" of all lines, except one of the external lines, that is, one of the two lines that connect with only one vertex. Instead of summing over this external line, we set its "energy" and momentum equal to z_ν and $\mathbf{p}$. The energy sums are, of course, sums over ν. For each summation and integration, we also write a factor $\left(\frac{1}{-i\beta}\right) \left[\frac{1}{(2\pi)^3}\right]$

(4) Finally, for fermions we again multiply the resulting expression by $(-1)^\ell$, where ℓ is the number of closed loops.

In this way, we determine the contribution of the diagram to $G(p, z_\nu)$.

For the particular diagram we are considering, the particle lines give a factor

$$\frac{1}{z_\nu - \left(\frac{p^2}{2m}\right)} \, \frac{1}{z_{\nu'} - \left(\frac{p'^2}{2m}\right)} \, \frac{1}{z_{\bar{\nu}} - \left(\frac{\bar{p}^2}{2m}\right)} \, \frac{1}{z_{\bar{\nu}'} - \left(\frac{(\bar{p}')^2}{2m}\right)} \, \frac{1}{z_{\bar{\nu}'} - \left(\frac{(\bar{p}'')^2}{2m}\right)}$$

and[d]

$$i v \left(\mathbf{p} - \bar{\mathbf{p}}\right) i v \left(\bar{\mathbf{p}} - \mathbf{p}''\right)$$

---

[d]The typographic error $\bar{\mathbf{p}}'$ of the argument of the first potential factor $i v \left(\mathbf{p} - \bar{\mathbf{p}}'\right)$ in the original text is fixed by following the rule 2.

Since there is one closed loop, there is again a factor of ± 1. Therefore, the value of this diagram is

$$(\pm 1)\left(\frac{1}{-i\beta}\right)^4 \sum_{\substack{\bar{v},v' \\ \bar{v}',v''}} \int \frac{d\bar{\mathbf{p}}}{(2\pi)^3}\frac{d\mathbf{p}'}{(2\pi)^3}\frac{d\bar{\mathbf{p}}'}{(2\pi)^3}\frac{d\mathbf{p}''}{(2\pi)^3}$$

$$\times \frac{1}{z_v - \left(\frac{p^2}{2m}\right)}\frac{1}{z_{v'} - \left(\frac{p'^2}{2m}\right)}\frac{1}{z_{\bar{v}} - \left(\frac{\bar{p}^2}{2m}\right)}\frac{1}{z_{\bar{v}'} - \left(\frac{(\bar{p}')^2}{2m}\right)}\frac{1}{z_{\bar{v}'} - \left(\frac{(\bar{p}'')^2}{2m}\right)}$$

$$\times (2\pi)^3\,\delta\left(\mathbf{p} + \mathbf{p}' - \bar{\mathbf{p}} - \bar{\mathbf{p}}'\right)(-i\beta)\,\delta_{v+v',\bar{v}+\bar{v}'}$$

$$\times (2\pi)^3\,\delta\left(\mathbf{p}'' + \mathbf{p}' - \bar{\mathbf{p}} - \bar{\mathbf{p}}'\right)(-i\beta)\,\delta_{v''+v',\bar{v}+\bar{v}'}$$

$$\times i^2 v\left(\mathbf{p} - \bar{\mathbf{p}}\right)v\left(\bar{\mathbf{p}} - \mathbf{p}''\right) \tag{A.3}$$

We can see that $z_{v''}$ and $\mathbf{p}''$ are limited to be just equal to z_v and $\mathbf{p}$. Therefore, Eq. (A.3) is

$$\left(\frac{1}{z_v - \left(\frac{p^2}{2m}\right)}\right)^2 \Sigma_c^0\left(p, z_v\right) \tag{A.4}$$

where

$$\Sigma_c^0\left(p, z_v\right) = (\pm 1)\left(\frac{1}{-i\beta}\right)^3 \sum_{\bar{v},v',\bar{v}'} \int \frac{d\bar{\mathbf{p}}}{(2\pi)^3}\frac{d\mathbf{p}'}{(2\pi)^3}\frac{d\bar{\mathbf{p}}'}{(2\pi)^3}$$

$$\times \frac{1}{z_{v'} - \left(\frac{p'^2}{2m}\right)}\frac{1}{z_{\bar{v}} - \left(\frac{\bar{p}^2}{2m}\right)}\frac{1}{z_{\bar{v}'} - \left(\frac{(\bar{p}')^2}{2m}\right)}$$

$$\times (2\pi)^3\,\delta\left(\mathbf{p} + \mathbf{p}' - \bar{\mathbf{p}} - \bar{\mathbf{p}}'\right)(-i\beta)\,\delta_{v+v',\bar{v}+\bar{v}'}$$

$$\times i^2\left[v\left(\mathbf{p} - \bar{\mathbf{p}}\right)\right]^2 \tag{A.5}$$

The sums extend over v', $\bar{v}$, $\bar{v}'$ = even integers for bosons, odd for fermions.

If we now compute the frequency sums in Eq. (A.5), we find, after a considerable amount of algebra, that $\Sigma_c^0\left(p, z_v\right)$ is just the collisional self-energy in the lowest order. This lowest order is obtained by replacing the G's in the Born collision approximation of Chapter 5 by G_0's.

A useful method for doing these Fourier sums is to represent them as contour integrals in the complex plane. Consider the contour integral

$$I = \pm \oint_C \frac{dz}{2\pi} f(z)h(z) \tag{A.6}$$

where

$$f(z) = \frac{1}{e^{\beta(z-\mu)} \mp 1} \tag{A.7}$$

and $h(z)$ is an arbitrary function of z except for possible poles. Assume that the poles of $h(z)$ do not coincide with the poles of $f(z)$, which are at $z = z_\nu = \frac{\pi\nu}{-i\beta} + \mu$, and take the contour C in Eq. (A.6) to encircle all poles of f in the negative sense, but none of the poles of h. Since the residue of $f(z)$ at $z = z_\nu$ is $\pm\frac{1}{\beta}$, we have, on the one hand,

$$I = \frac{1}{-i\beta} \sum_\nu h(z_\nu) \tag{A.8}$$

Now on the other hand, if $zf(z)h(z) \to 0$ as $|z| \to \infty$, we can replace the contour C by the contour C' that encircles all the poles of $h(z)$ in the positive sense. Comparing these two evaluations of I, we find

$$\frac{1}{-i\beta} \sum_\nu h(z_\nu) = \mp \oint_{C'} \frac{dz}{2\pi} f(z)h(z) \tag{A.9}$$

To illustrate such a frequency summation, let us consider a simple diagram, the "bubble,"

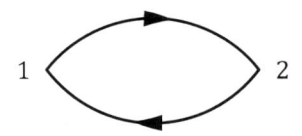

which, in space–time language, is

$$L^0(1, 2) = \pm i G_0(1, 2) G_0(2, 1) \tag{A.10}$$

This is a piece of the diagram we have been considering so far, and, it will be recalled, the bubble enters into the discussion of the random phase approximation.

Introducing the Fourier sum and integral representation of G_0, we find that

$$i L_0(1, 2) = \pm i \left(\frac{1}{-i\beta}\right)^2 \sum_{\nu'', \nu'} \int \frac{d\mathbf{p}}{(2\pi)^3} \frac{d\mathbf{p}'}{(2\pi)^3} \frac{1}{z_{\nu''} - \left(\frac{p^2}{2m}\right)} \frac{1}{z_{\nu'} - \left(\frac{(p')^2}{2m}\right)}$$

$$\times e^{-i(z_{\nu''} - z_{\nu'})(t_1 - t_2) + i(\mathbf{p} - \mathbf{p}')\cdot(\mathbf{r}_1 - \mathbf{r}_2)}$$

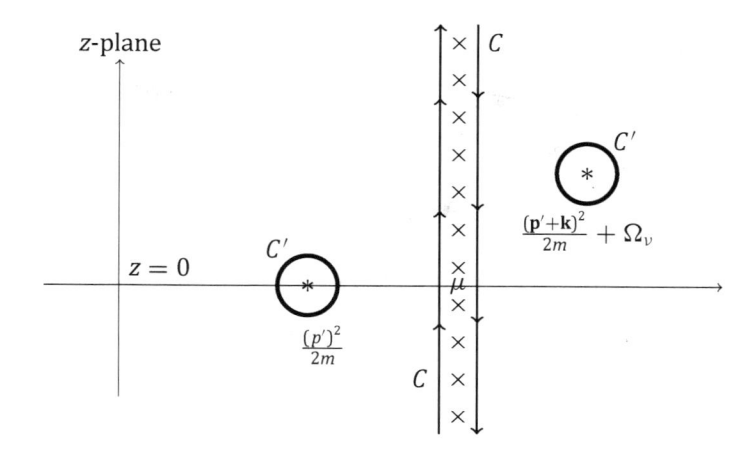

Figure A.1 The contour deformation from C to C' in the z-plane. The thin lines represent the contour C and their directions of integration are indicated by the arrows. The thick lines represent the contour C' with the direction of integration to be counter clockwise. The poles of the distribution function $f(z)$ (for fermions) are indicated by $\times$, while the poles from the particle lines are represented by $*$.

We multiply this expression by $e^{i\Omega_\nu(t_1-t_2)-i\mathbf{k}\cdot(\mathbf{r}_1-\mathbf{r}_2)}$ and integrate over all t_1 between 0 and $-i\beta$ and all $\mathbf{r}_1$. In this way, we pick out the Fourier coefficient:

$$L_0\,(k,\Omega_\nu) = \pm i\,\frac{1}{-i\beta}\sum_{\nu'}\int\frac{d\mathbf{p'}}{(2\pi)^3}\,\frac{1}{z_{\nu'}+\Omega_\nu-\frac{(\mathbf{p'}+\mathbf{k})^2}{2m}}\,\frac{1}{z_{\nu'}-\frac{(\mathbf{p'})^2}{2m}}$$

$$(A.11)$$

where

$$\Omega_\nu = \frac{\pi\nu}{-i\beta} = \text{even integer}$$

This we recognize as a portion of the expression (A.5).

We now apply Eq. (A.9) to the calculation of the sum in Eq. (A.11). In this case, the contours C and C' are as shown in Fig. A.1, since

$$h(z) = \frac{1}{z+\Omega_\nu-\frac{(\mathbf{p'}+\mathbf{k})^2}{2m}}\,\frac{1}{z-\frac{(p')^2}{2m}}$$

Then Eq. (A.5) becomes

$$
L_0\left(k, \Omega_\nu\right) = \frac{1}{i} \int \frac{d\mathbf{p}'}{(2\pi)^3} \oint_{C'} \frac{dz}{2\pi} f(z) \frac{1}{z + \Omega_\nu - \frac{(\mathbf{p}'+\mathbf{k})^2}{2m}} \frac{1}{z - \frac{(\mathbf{p}')^2}{2m}}
$$

$$
= \int \frac{d\mathbf{p}'}{(2\pi)^3} \frac{f\left(\frac{(\mathbf{p}')^2}{2m}\right) - f\left(\frac{(\mathbf{p}'+\mathbf{k})^2}{2m} - \Omega_\nu\right)}{\Omega_\nu + \frac{(\mathbf{p}')^2}{2m} - \frac{(\mathbf{p}'+\mathbf{k})^2}{2m}}
\tag{A.12}
$$

This equation tells us the values of the analytic function $L_0\left(k, \Omega\right)$ at the points

$$
\Omega = \Omega_\nu = \frac{\pi \nu}{-i\beta} \quad (\nu = \text{even integer})
$$

To discover $L_0\left(k, \Omega\right)$ from Eq. (A.12), we must analytically continue the right side of Eq. (A.12) to a function that is analytic for Ω not real and approaches zero as $|\Omega| \to \infty$. Just replacing Ω_ν by Ω in Eq. (A.12) is not satisfactory analytic continuation because it leads to an $L_0\left(k, \Omega\right)$ that does not approach zero as $\Omega \to \infty$ in all directions. The origin of this difficulty is that as $\Omega \to \infty$, $f\left(\left(\frac{p^2}{2m}\right) - \Omega\right)$ approaches ∓ 1 or 0, depending on whether $\Re\Omega$ is greater than or less than $\left(\frac{p^2}{2m}\right) - \mu$. The correct continuation is found by first replacing $f\left(\frac{(\mathbf{p}'+\mathbf{k})^2}{2m} - \Omega_\nu\right)$ by $f\left(\frac{(\mathbf{p}'+\mathbf{k})^2}{2m}\right)$ in Eq. (A.12). This does not change the value of $L_0\left(k, \Omega_\nu\right)$ since $e^{\beta\Omega_\nu} = 1$. Therefore, we can write Eq. (A.12) as

$$
L_0\left(k, \Omega_\nu\right) = \int \frac{d\mathbf{p}'}{(2\pi)^3} \frac{f\left(\frac{(\mathbf{p}')^2}{2m}\right) - f\left(\frac{(\mathbf{p}'+\mathbf{k})^2}{2m}\right)}{\Omega_\nu + \frac{(\mathbf{p}')^2}{2m} - \frac{(\mathbf{p}'+\mathbf{k})^2}{2m}}
\tag{A.13}
$$

We can now continue $L_0\left(k, \Omega_\nu\right)$ to $L_0\left(k, \Omega\right)$ by replacing Ω_ν by Ω in Eq. (A.13), since this continuation now leads to a function that approaches zero as $\Omega \to \infty$ in the upper or lower half-plane. Thus

$$
L_0\left(k, \Omega\right) = \int \frac{d\mathbf{p}'}{(2\pi)^3} \frac{f\left(\frac{\left(\mathbf{p}'-\frac{\mathbf{k}}{2}\right)^2}{2m}\right) - f\left(\frac{\left(\mathbf{p}'+\frac{\mathbf{k}}{2}\right)^2}{2m}\right)}{\Omega_\nu + \frac{(\mathbf{p}')^2}{2m} - \frac{(\mathbf{p}'+\mathbf{k})^2}{2m}}
\tag{A.14}
$$

This agrees with our earlier evaluation of $L_0\left(k, \Omega\right)$.

There is one remaining ambiguity in this graphical formalism, namely in the graphs that contain

$$\bar{2} \qquad = \pm i \int_0^{-i\beta} d\bar{2} \, V \left(1 - \bar{2}\right) G_0 \left(\bar{2}, \bar{2}\right) \qquad (A.15)$$

or

$$1' \longrightarrow 1 = i V \left(1 - 1'\right) G_0 \left(1, 1'\right) \qquad (A.16)$$

In both these cases, there appears $G_0 \left(1, 1'\right)_{t_{1'}=t_1}$, which is ambiguous since $G_0 \left(\mathbf{r}_1 t_1, \mathbf{r}_{1'} t_1^+\right) \neq G_0 \left(\mathbf{r}_1 t_1, \mathbf{r}_{1'} t_1^-\right)$. But in both cases, we should evaluate t_1 as $t_1^+ = t_1 + \epsilon$. Then Eq. (A.15) becomes

$$= \pm i \int dr_1 \, v \left(r_1 - r_2\right) \int \frac{d\mathbf{p}}{(2\pi)^3} \frac{1}{-i\beta} \sum_\nu \frac{e^{i z_\nu \epsilon}}{z_\nu - \frac{p^2}{2m}}$$

$$= -i v \int \frac{d\mathbf{p}}{(2\pi)^3} \oint_C \frac{dz}{2\pi} \frac{e^{i z \epsilon}}{e^{\beta(z-\mu)} \mp 1} \frac{1}{z - \frac{p^2}{2m}}$$

where $\epsilon = 0^+$.

Now notice that the integrand goes to zero exponentially as $z \to \infty$ in either the right or the left half-plane. Therefore, we can deform the contour C to encircle $\frac{p^2}{2m}$ in the positive sense and pick up no contribution at ∞. In this way, we find

$$
= -iv \int \frac{d\mathbf{p}}{(2\pi)^3} \oint_{C'} \frac{dz}{2\pi} f(z) \frac{1}{z - \frac{p^2}{2m}}
$$

$$
= v \int \frac{d\mathbf{p}}{(2\pi)^3} f\left(\frac{p^2}{2m}\right) = vn_0
$$

This is, of course, just the single-particle Hartree self-energy in the lowest order of approximation. Similarly, the diagram (A.16) is just the lowest-order single-particle exchange energy.

References and Supplementary Reading

Chapter 1

The annotator discussed quantum mechanics based heavily on P. A. M. Dirac, *The Principles of Quantum Mechanics*, 4th Edition (Clarendon Press, Oxford, 1998); Leonard I. Schiff, *Quantum Mechanics* (McGraw-Hill, New York, 1968); and J. J. Sakurai and San Fu Tuan, *Modern Quantum Mechanics*, Revised Edition (Addison-Wesley, Reading, Massachusetts, 1994). The required classical mechanics are presumably assumed that the readers have studied L. D. Landau and E. M. Lifshitz, *Classical Mechanics*, 3rd Edition (Elsevier, Amsterdam, 2005) and H. Goldstein, *Classical Mechanics*, 2nd Edition (Addison-Wesley, Reading, Massachusetts, 1980) along with the enough mathematical training by George B. Arfken, Hans J. Weber, and Frank E. Harris, *Mathematical Methods for Physicists: A Comprehensive Guide*, 7th Edition (Elsevier, Amsterdam, 2013). Copenhagen interpretation of quantum mechanics was first given by M. Born, *Z. Physik*, **37**, 863 (1926); *Nature* **119**, 354 (1927). The correspondence between the quantum mechanical expectation values and the classical mechanics was provided by P. Ehrenfest, *Z. Physik*, **45**, 455 (1927). The Principles of Uncertainty is given by W. Heisenberg, *Z. Physik*, **43**, 172 (1927). The idea of second quantization was given by P. A. M. Dirac, *Proc. R. Soc. London*, **114A**, 243 (1927). The non-relativistic quantum many-body theory was formulated by P. Jordan and O. Klein, *Z. Physik*, **45**, 751 (1927); P. Jordan and E. P. Wigner, *Z. Physik*, **47**, 631 (1928); V. Fock, *Z. Physik*, **75**, 622 (1932); while a rigorous treatment of relativistic quantum field theory can be found in J. M. Jauch and F. Rohrlich, *The Theory of Photons and Electrons*, Second Expanded Edition (Springer-Verlag, New York, 1976). The annotator discussed the second quantization based on Alexander L. Fetter and John Dirk Walecka, *Quantum Theory of Many-Particle Systems* (McGraw-Hill, New York, 1971) and John W. Negele and Henri Orland, *Quantum Many-Particle Systems* (Addison-Wesley, Redwood City, California, 1988). A fermionic many-particle wavefunction is expanded in terms of Slater's determinant formulated by J. C. Slater, *Phys. Rev.*, **34**, 1293 (1929).

Recent developments of nonequilibrium statistical mechanics are well guided by Jørgen Rammer, *Quantum Field Theory of Non-equilibrium States*

(Cambridge University Press, Cambridge, 2007); Alex Kamenev, *Field Theories of Non-Equilibrium Systems* (Cambridge University Press, Cambridge, 2011); and Gianluca Stefanucci and Robert van Leeuwen, *Nonequilibrium Many-Body Theory of Quantum Systems: A Modern Introduction* (Cambridge University Press, Cambridge, 2013).

Chapter 2

The discussion in the first four chapters is based to a large extent on the work of P. C. Martin and J. Schwinger, *Phys. Rev.*, **115**, 1342 (1959), where many earlier references are cited. The Green's functions were first introduced by T. Matsubara, *Progr. Theoret. Phys. (Kyoto)*, **14**, 351 (1955). The boundary conditions was derived by R. Kubo, *J. Phys. Soc. Japan*, **12**, 570 (1957). There is much work done along similar lines in Russia. See the review articles D. N. Zubarev, *Uspehki Fiz. Sauk*, **71**, 71 (1960) [translation *Soviet Phys. Uspekhi*, **3**, 320 (1960)] and A. I. Alekseev, *Uspekhi Fiz. Nauk*, **73**, 41 (1961) [translation *Soviet Phys. Uspekhi*, **4**, 23 (1961)] where extensive lists of references are given.

Chapter 3

For the basic notions of statistical mechanics, we refer the reader to Schrödinger's excellent little book, *Statistical Thermodynamics* (Cambridge University Press, London, 1946).

Chapter 4

For a discussion of the mathematical justification of the continuation of the Fourier coefficient function to all z, see G. Baym and N. D. Mermin, *J. Math. Phys.*, **2**, 232 (1961). The original Hartree and Hartree–Fock approximations are reviewed by D. R. Hartree, *Repts. Prog. Phys.*, **11**, 113 (1948).

Chapter 6

The variational derivative techniques were introduced by J. Schwinger, *Proc. Natl. Acad. Sci. U.S.A.*, **37**, 452 (1951). Perturbative expansions in v of G, Σ, and also some of the thermodynamic functions, e.g., the pressure, are very commonly used in many-particle physics. See, for example, E. W. Montroll and J. C. Ward, *Phys. Fluids*, **1**, 55 (1958); C. Bloch and C. DeDominicis, *Nuclear Phys.*, **7**, 459 (1958); J. M. Luttinger and J. C. Ward, *Phys. Rev.*, **118**, 1417 (1960). A very original approach to the problem of expanding Σ in terms of G is given by R. Kraichnan, Rep. HT-9, Devision of Electromagnetic Research, Institute of Mathematical Sciences, New York University, 1961.

Chapter 7

For discussions of the Boltzmann equation, see A. Sommerfeld, *Thermodynamics and Statistical Mechanics* (Academic Press, New York, 1956); J. Jeans, *Introduction to the Kinetic Theory of Gases* (Cambridge University Press, London, 1948); S. Chapman and T. G. Cowling, *Mathematical Theory of Non-Uniform Gases* (Cambridge University Press, London, 1939). These books also describe how dissipative phenomena, e.g., sound-wave damping and heat conduction, can be derived from the Boltzmann equation. The Landau–Vlasov equation is discussed by A. Vlasov, *J. Phys. (U.S.S.R.)*, **9**, 25 (1945).

The energy conservation law for $G(U)$ is demonstrated in the appendix to G. Baym and L. P. Kadanoff, *Phys. Rev.*, **124**, 287 (1961).

Chapter 8

The random phase approximation was developed by D. Bohm and D. Pines, *Phys. Rev.*, **92**, 609 (1953). An extensive list of references is given by D. Pines in *The Many-Body Problem* (W. A. Benjamin, New York, 1961). For work on zero sound, see L. D. Landau, *J. Exptl. Theoret. Phys. (U.S.S.R.)*, **32**, 59 (1957) [translation *Soviet Phys. JETP*, **5**, 101 (1957)]; K. Gottfried and L. Pičman, *Kgl. Danske Videnskab. Selskab, Mat.-fys. Medd.*, **32**, 13 (1960); J. Goldstone and K. Gottfried, *Nuovo cimento*, [X]**13**, 849 (1958).

Chapters 9, 10, and 11

We describe approximate equations for the linear responses of G to U in G. Baym and L. P. Kadanoff, *Phys. Rev.*, **124**, 287 (1961). At present we feel that for most applications, it is better to work with the real-time function, $g(U)$. The work in these three chapters is the result of the research of one of us (LPK) and is first being reported here.

Chapter 12

The Landau theory is best described in Landau's original papers, *J. Exptl. Theoret. Phys. (U.S.S.R.)*, **30**, 1058 (1956); **32**, 59 (1957); and **35**, 97 (1958). These are translated in *Soviet Phys. JETP* **3**, 920 (1957); **5**, 101 (1957); **8**, 70 (1959). The translations are reprinted in D. Pines, *The Many-Body Problem* (W. A. Benjamin, New York, 1961, pp. 260–278). Landau's arguments are extremely clear and convincing, but they are largely based upon physical intuition. J. M. Luttinger and P. Nozières, *Phys. Rev.*, in press)[a] have recently

---

[a]The publications appeared in P. Nozières and J. M. Luttinger, *Phys. Rev.*, **127**, 1423 (1962) and J. M. Luttinger and P. Nozières, *Phys. Rev.* **127**, 1431 (1962).

described how Landau's results may be justified within the framework of perturbation theory.

Chapter 13

Pines (*op. cit.*) discusses the consequences of the shielded potential in metals. He also gives a large bibliography on the subject. For a Green's function description of a metal, see G. Baym, *Ann. Phys.*, **14**, 1 (1961).

The calculation of P in the classical limit follows a method devised by S. Ichimaru and D. Pines. The method was described to us by Pines. A more general discussion of the relation of the partition function to the dielectric function is given by F. Englert and R. Brout, *Phys. Rev.*, **120**, 1085 (1960).

Chapter 14

The Bruckner theory of nuclear matter is presented in K. A. Brueckner and J. H. Gammel, *Phys. Rev.*, **109**, 1023 (1958). See also H. A. Bethe and J. Goldstone, *Proc. Roy. Soc. London*, **A238**, 551 (1957). The same problem was attacked with Green's function techniques by R. D. Puff, *Ann. Phys.*, **13**, 317 (1961).

The approach to the superconductor sketched here is worked out fully by L. P. Kadanoff and P. C. Martin, *Phys. Rev.*, **124**, 670 (1961).

Index